A Color Atlas of

# Diseases of Lettuce and Related Salad Crops
## Observation, Biology and Control

Dominique Blancard
Hervé Lot
Brigitte Maisonneuve
Institut de la Recherche Agronomique (INRA)
UMR Santé Végétale
Centre de Recherches de Bordeaux-Aquitaine, France

Foreword and Consultant Editor for the English edition
Edward J. Ryder
USDA Agricultural Research Station, Salinas, California

Translated from the French by Maggie Rosengarten

INRA EDITIONS

ELSEVIER
BOSTON • SAN DIEGO
Academic Press is an imprint of Elsevier
AP

First published in the United States of America in 2006 by
Academic Press, an imprint of Elsevier
30 Corporate Drive, Suite 400, Burlington, MA 01803, USA
525 B Street, Suite 1900, San Diego, CA 92101-4495, USA



First published in French as Maladies des Salades

ISBN 2–7380–1057–1

Ouvrage publié avec le concours du Ministère français chargé de la Culture –
Centre national du livre
This edition is published with the help of the French Ministry of Culture



ISBN–13: 978-0-12-372557-8
ISBN–10: 0-12-372557-7

**Library of Congress Cataloging-in-Publication Data**
A catalog record for this book is available from the Library of Congress

For information on all Academic Press publications
visit our web site at books.elsevier.com

Printed and bound in France

06 07 08 09 10 10 9 8 7 6 5 4 3 2 1

**Dedication**

*To Paul Rieuf,*

*general plant pathologist, recently passed away.*

# Acknowledgements

What pleasure may be derived from wandering through crops in search of diseased plants, as an 'academic' exercise, doubtless to the great annoyance of the growers and their technical assistants. I hope they will excuse me for doing so. Their patience and understanding have greatly contributed to the value of this book, and I am grateful to them.

The book is thus a product of these wanderings, but equally of the collaboration established with my two colleagues B. Maisonneuve and particularly H. Lot.

We would like to express our deep gratitude to E. Ryder, a world-renowned lettuce specialist, for his fine Foreword, and for his help in advising and providing additional information for the English edition.

We would also like to thank all those who have contributed, in their different ways, to enriching the book's quality, for:

- their valuable help in the course of various aetiological studies (S. Chamont, R. Sanson, C. Martin);
- the provision of photographs of viruses by electron microsocopy (B. Delecolle, **488–494**), of phytoplasma (M. Garnier, **487**), and of diseases in the field (C. Martin, **31, 318**; D. Izard, **313**, **314**; M. Villevieille, **446a, b**; O. Le Gall, **45–47**, **507**, **508** of Lactuca serriola);
- their critical and constructive review of the text ( F. Jailloux, P. Larignon, P. Lecomte);
- their financing a journey to the tropics to complete our studies and our photographic collection (P. Ricci and T. Candresse);
- their kindness, availability and efficiency in producing and publishing the book ( Editions INRA and more particularly D. Bollot, R. Boulidard, C. Colon, C. Parpinelli, S. Picard and J. Veltz; and Manson Publishing for the English edition).

# Contents

# Foreword – Edward J. Ryder

The word salad conjures up an image of freshness. And indeed, the salad is fresh. It is served uncooked and may be composed of combinations of leaves of various forms of lettuce, chicory, endive and other vegetables, mostly in shades of green, the cool, fresh colour.

Lettuce is the royalty of salad vegetables, and has in fact been called King Lettuce in the U.S. produce industry, and 'the queen of the salad vegetables' by Martin and Ruberte in their book on edible leaves of the tropics. It is grown commercially in many countries worldwide and in home gardens nearly everywhere. The production and use of chicory and endive is on a smaller scale, in fewer countries, principally in Europe and North America.

This book is the fourth on diseases of specific crops by Dominique Blancard, following similar exhaustive discussions of the diseases of tomato, cucurbits, and tobacco. His co-authors, H. Lot and B. Maisonneuve, contribute important information on the pathology, genetics, and breeding of salad crops.

## The origins and histories of salad crops

The history of lettuce is better known and documented than those of chicory or endive. There are two information trails that lead us back in time. Lettuce occurs in several forms, each a variation of a rosette of leaves on a shortened stem. Most bear resemblance to a type of lettuce depicted in tomb paintings, the earliest of which date back to the Fourth Dynasty of Ancient Egypt, about 2500 BC. The earlier paintings are less stylized, and more recognizable as lettuce, according to most scholars. They most resemble stem lettuce, a type that has thick, elongated, edible stems and narrow leaves. This type of lettuce is still grown in modern Egypt and may be the earliest one used for human consumption.

The other trail may lead back to the actual domestication of lettuce, but its direction is less clear. A type of lettuce, known as oilseed lettuce, is a similar wild lettuce but has certain characteristics that may be associated with domestication. One of these is seed size: the seeds may be 50% larger than the standard cultivated forms. They are pressed to express an edible oil that is used in Egypt for cooking. This practice is said to be several hundred years old. However, it is possible that it is considerably older and that the plant represents the transition form from wild to cultivated lettuce. Thus the origin of domesticated lettuce may have taken place in the Nile Valley, or alternatively, in the Tigris-Euphrates region, which is the area of greatest variation of weedy species of *Lactuca* and related forms.

As lettuce moved around the Mediterranean Basin, it appears likely that romaine lettuce in various forms appeared, including erect forms and types similar to leaf lettuce. The more specialized heading types (butterhead, crisphead, Latin) probably appeared later, north of the Mediterranean area. The last type to be differentiated was the iceberg lettuce from the Batavia type of crisphead. This took place in California in the 1940s.

Lettuce was known in Persia about 550 BC, and later in Greece and Rome, where several forms were described. The names given referred to colour differences and geographical origin. Lettuce was also taken to China, probably quite early, because the stem type is the predominant form found there now.

Chicory is widely distributed as a weed. It probably originated in the Mediterranean Basin and was used in Egypt, Greece, and Rome both as a salad vegetable and for medicinal purposes. It was probably grown in a semi-cultivated state for centuries. The first mention of the cultivation of chicory was in Germany in 1616.

In 1775, two French physicians discovered a use for the roots; they could be dried, roasted and ground, and used to make a coffee-like beverage.

Witloof chicory was apparently accidentally discovered by M. Breziers, of the Belgian State Botanical Garden, when he came across roots that had been left in the dark and sprouted new leaves, which were white and considerably milder flavoured than the green leaves.

Additional uses were developed. Chicory roots contain inulin, which can be converted in an industrial process to sugar, principally fructose, for use in making sweeteners. Finally, in New Zealand, it was found that chicory is a useful fodder for cattle. The versatility of chicory is quite remarkable.

The history of endive is more obscure. According to Sturtevant, it originated in India. Other authors name Sicily as its source. It was used as a salad vegetable in Egypt, Greece, and Rome supporting the idea of a Mediterranean origin. There are two types: broad-leaved (escarole) and narrow-leaved (frisée). The broad-leaved type is probably the older of the two.

## Classic research with the salad vegetables

Various plant species have served as models in the development of information useful in the understanding of general principles of plant growth and development. For lettuce, the elucidation of the principles of photodormancy stands out in the annals of classic plant research. Other original research in physiology, genetics, breeding, and disease resistance has been important primarily for the species itself.

## Research in physiology

The red-far red promotion-inhibition cycle of seed germination was first identified in lettuce by Flint and McAlister who showed that white or red light stimulated germination, while dark or far-red light inhibited germination. Borthwick and colleagues identified the action spectra for these phenomena, and proposed an interactive effect between light and a plant pigment they named phytochrome. Much work was done over the years in which the green leaf cultivar Grand Rapids became the standard experimental plant. The principles established were shown to be applicable to other plant species.

Further work on lettuce seed germination disclosed the inhibitive effects of high temperature, interactive effects of light and temperature, as well as the effects of various chemicals (alone or in combination with light and temperature) in promoting or inhibiting germination.

The phenomenon of seed stem elongation is well studied because, if premature, it has a deleterious effect on quality and harvestability, but is necessary for flowering and reproduction. Early work in Germany first showed the effects of long and short days on reproductive growth thus enabling the classification of cultivars into those requiring long photoperiods for reproductive growth, and those that were day length neutral. Several chemicals can be used to promote flowering and others to delay it. Most prominent of the enhancing chemicals are the gibberellins, as first demonstrated by Wittwer.

## Research in genetics and breeding

The most notable documented achievements in lettuce genetics research have been:

- the application of the gene-for-gene concept to the host-organism relationship for the downy mildew disease,
- the elucidation of anthocyanin genetics,
- the genetics of reproductive processes, and
- the development of a genetic map. In lettuce breeding, two major achievements are the development of the true iceberg lettuce, and the first use of the wild species, *Lactuca virosa*, in cultivar development. The classic genetic study in *Cichorium* was the first cross between chicory and endive. Breeding efforts were important in the development of procedures for the forcing of witloof chicory without soil cover and the conversion of inulin in the root to sugar.

Downy mildew has been a consistently serious problem in lettuce production, especially in Europe, for many decades. One of the difficulties in breeding for resistance to downy mildew of lettuce (incited by *Bremia lactucae*) has been the existence of physiologic races and the capacity of the fungus to replace old races with new virulent forms to challenge the resistance of newly introduced cultivars. This phenomenon had been observed in the wheat-stem rust relationship. It also occurred in flax and in 1953, Flor introduced the concept of the gene-for-gene relationship between the host and the flax rust organism. Crute and Johnson applied these principles to lettuce and downy mildew. They established the specific presence of resistance alleles in lettuce cultivars corresponding to avirulence alleles in *B. lactucae* and the replacement of the latter with virulence alleles in response to the introduction of new cultivars with new resistance alleles. Breeding for resistance to downy mildew is now guided by these principles in most programmes.

Anthocyanin pigmentation in lettuce was studied in the earliest work on lettuce genetics and was shown, in the specific crosses made, to be inherited as a single gene trait with red dominant to green. It remained for Ross Thompson, in a seminal paper, to unravel the complex genetic basis for anthocyanin occurrence and distribution in the leaves of lettuce plants. He showed that two complementary dominant alleles were required for colour, and that a third gene functioned only in the presence of both dominants to influence the pattern of red colour.

The transition from vegetative to reproductive growth is critical in rosette forming species like lettuce, chicory, and endive. In lettuce, much work has been done on the physiology of reproduction but less on the genetics. The earliest work, distinguishing long-day from day-neutral cultivars, by Bremer and Grana, identified the gene *tagneutral*, in which the dominant allele allowed bolting only under long days. Later studies identified genes for achene colour, flowering time, and male sterility. An important study that should be done is on the relationship between stem elongation and flowering. It is not clear whether the same genes play a role in both phenomena. Studies of gibberellin influenced dwarf lettuces show that flowering can occur with little or no stem elongation. Under certain environmental conditions or in certain genotypes, stem elongation may occur, but there may be a delay in subsequent expansion of the inflorescence. Normally, inflorescence expansion immediately follows stem elongation.

We can expect a surge in progress in all aspects of lettuce genetics as molecular studies continue to expand. Several notable achievements occurred or began in the laboratory of R.W. Michelmore (University of California). Genetic mapping was minimal until the identification of molecular markers, beginning with isozymes, followed by those based upon DNA hybridization and amplification. There is now a map combining morphological and other phenotypic traits and molecular markers in ten genetic groups, one more than the number of chromosomes.

The map positions of major genes for resistance to downy mildew, lettuce mosaic, corky root, and of quantitative trait loci (QTL) for root traits and water uptake have been established. The feasibility of transformation has been shown. Resistance genes, most notably *Dm-3*, have been isolated using map-based cloning techniques. Identification of resistance gene candidates through comparison with identified sequences from other species is in progress. It will eventually be possible to locate, characterize, and transfer many genes of interest for breeding, genetics, cultivar identification, and physiological studies in lettuce and related species.

The various lettuce types were developed over many centuries and represent unquestionably important breeding accomplishments that unfortunately, with one exception, were not recorded. The exception, and the last distinctive type to be developed, was the large firm crisphead type, commonly designated as iceberg. The first true iceberg lettuce was Great Lakes and was developed and released by Thomas W. Whitaker (USDA) in 1941. Iceberg lettuce quickly became the dominant type in the U.S. lettuce industry, and later also became important elsewhere.

Crosses between *Lactuca sativa* (cultivated lettuce) and *L. serriola* can be easily made. Crosses with *L. saligna* are more difficult, and it is very difficult to cross *L. sativa* and *L. virosa*. The first successful recorded cross with *L. virosa* was made by Ross C. Thompson in 1938, leading to the release of the cultivar Vanguard 20 years later. Vanguard, in turn, was the first cultivar of the Salinas-Vanguard group, presently the dominant form of iceberg lettuce, which is grown worldwide. Additional crosses were made to *L. virosa* by Brigitte Maisonneuve to transfer beet western yellows resistance, and by Albert Eenink, to transfer resistance to the lettuce aphid, *Nasonovia ribisnigri*.

## Classic research with *Cichorium* species

The species distinction between *Cichorium endivia* and *C. intybus* was established in a study by Charles M. Rick, in which he studied the frequency of crossing when *C. endivia* is the female parent and when *C. intybus* is the female. He noted the inheritance of traits in the segregating generation and also cytological characteristics. The seed produced on the endive plant was nearly all from self-pollination, while that on the chicory was nearly all hybrid due to the self-incompatibility of chicory. He concluded that they were two species.

For many years after the discovery that chicory leaves undergo regrowth in darkness, forcing was done in sand or soil to insure tight chicons. This process required much hand labour. J. E. Huyskes showed that covering soil was not required when specific forcing conditions were used and when suitable cultivars were developed. The hybrid cultivar Zoom was developed by Hubert Bannerot, and was for many years the standard cultivar of the industry.

There are two types of industrial chicory, one used for roasting of the roots to be used as a coffee substitute or additive, and the other for conversion of inulin to sugar. For the former use, selection and breeding work probably started quite early, perhaps in the latter part of the 17th Century when roasting became industrialized first in Holland, and later in Germany and France. Great impetus for the industry occurred in the Napoleonic years when coffee became scarce and expensive. Breeding goals included resistance to diseases as well as those with a bearing on product quality, especially dry matter content and bolting resistance.

Interest in use of chicory as a sugar source began because of a serious nematode problem in sugarbeet. At first, cultivars developed for roasting were used. However, demand increased for cultivars more suitable specifically for sugar production, and in 1972, Maison Florimond Desprez began a selection and breeding program, emphasizing not only inulin content and sugar production potential, but also bolting resistance and disease resistance.

## Research in disease resistance

In addition to the landmark research on downy mildew resistance of lettuce, the investigations of the nature of resistance to two other diseases are worth noting. The research on lettuce mosaic is notable for overall success in identifying resistance, incorporation into breeding programs, and substantial use of resistant cultivars. The studies on lettuce big vein have been equally notable in difficulties encountered in understanding the disease, identification of good sources of resistance, and in the slow adoption of resistant cultivars.

Lettuce mosaic was first identified in 1921. It became a serious worldwide economic problem, especially in areas where lettuce was grown in repeated plantings over long periods during the year. Two means were employed to reduce the effect of the disease. One was a seed indexing procedure, first adopted in California, to take advantage of the fact that the virus was transmitted through a small percentage of the seeds planted, providing primary inoculum sources for subsequent spread by aphids. The other was the use of genetic resistance, which was first adopted in Europe and later in the USA. Both methods have been successful in minimizing the problem. Work continues in searching for additional sources of resistance.

Big vein disease was first described in 1934 and for many years was considered to be an American problem. Later, it was found in lettuce grown in many countries. Early work focused on the nature of the agent and on its transmission. A root-feeding fungus, *Olpidium brassicae*, was shown to be the vector. The agent was at first considered to be a soil-borne virus, then a toxin, finally a virus-like agent. Later, a virus that was designated as lettuce big vein virus (LBVV) was identified. Most recently, a new virus called Mirafiori lettuce virus (MiLV) was described, and appears to be the one most closely associated with the disease. This aspect is still not settled. The development of resistant cultivars is the only method that has been seriously considered for control of big vein. Several crisphead and cos cultivars have been released with moderate resistance as compared to susceptible cultivars. This resistance can be partially overcome under severe conditions of low temperature and high soil moisture. The wild species *L. virosa* is highly resistant and is the best hope for future improvement.

## Conclusion

This foreword was largely written without benefit of seeing the book in its published form. However, Dominique Blancard has set high standards for himself with his previous reviews of the diseases of tomato, tobacco, and cucurbits. After reviewing the French edition, it is now apparent that he and his colleagues, together with the publishing branch of INRA, have presented us with an elegant and exemplary book on the diseases of lettuce and related salad crops.

## Bibliography


BARRONS K.C. and WHITAKER T.W. (1943) Great Lakes, a new summer head lettuce adapted to summer conditions. *Michigan Agri. Res. Sta. Quarterly Bul.* **25**: 1-3.

BORTHWICK H.A., HENDRICKS S.B., PARKER M.W., TOOLE E.H. and Toole V.K. (1952) A reversible photoreaction controlling seed germination. *Proc. Natl. Acad. Sci.* **38**: 662-666.

BREMER A.H. and GRANA J. (1935) Genetische untersuchungen mit salat. II. *Gartenbauwissenshaft* **9**: 231-245.

CRUTE I.R. and JOHNSON A.G. (1976) The genetic relationship between races of *Bremia lactucae* and cultivars of *Lactuca sativa*. *Ann. Appl. Biol.* **83**: 125-137.

DESPREZ B.F., DELESALLE L., DHELLEMES C. and DESPREZ M.F. (1994) Génétique et amélioration de la chicorée industrielle. *Comp. Rend. Acad. Agri. Fran.* **80**: 47-62.

EENINK A.H., DIELEMAN F.L. and GROENWOLD R. (1982) Resistance of lettuce (*Lactuca*) to the leaf aphid *Nasonovia ribisnigri*. 1. Transfer of resistance from *L. virosa* to *L. sativa* by interspecific crosses and selection of resistant breeding lines. *Euphytica* **31**: 291-300.

FLINT L.H. and MCALISTER E.D. (1937) Wavelengths of radiation in the visible spectrum promoting the germination of light-sensitive lettuce seed. *Smithsonian Miscellaneous Collections* **96**: 1-8.

FLOR H.H. (1955) Host-parasite interaction in flax rust – its genetics and other implications. *Phytopathology* **45**: 680-685.

GROGAN R.G. (1980) Control of lettuce mosaic with virus-free seed. *Plant Disease* **64**: 446-449.

HEDRICK U.P.(ed.) (1972) *Sturtevant's Edible Plants of the World*. Dover, NY.

HUYSKES J.A. (1963) Veredeling an witloof voor het trekken zonder dekgrond. *Inst. Veredeling Tuinbouwgewassen* **202**: 1-70.

JAGGER I.C. (1921) A transmissible mosaic disease of lettuce. *J. Agri. Res.* **20**: 737-741.

JAGGER I.C. and Chandler N. (1934) Big vein of lettuce. *Phytopathology* **24**: 1253-1256.

MAISONNEUVE B., CHOVELON V. and LOT H. (1991) Inheritance of resistance to beet western yellows in *Lactuca virosa* L. *Hort. Science* **26**: 1543-1545.

MARTIN F.W. and RUBERTE R.M. (1975) *Edible Leaves of the Tropics*. AID, Dept.State. and ARS-USDA, Mayaguez, Puerto Rico.

MICHELMORE R.W. (1995) Isolation of disease resistance genes from crop plants. *Current Opinion in Biotech.* **6**: 145-152.

RICK C.M. (1953) Hybridization between chicory and endive. *Proc. Amer. Soc. Hort. Sci.* **62**: 459-466.

RYDER E.J. (1999) *Lettuce, Endive and Chicory*. CABI Publishing, Wallingford, UK.

THOMPSON R.C. (1938) *Genetic Relations of Some Color Factors in Lettuce*. Tech. Bul. 620. USDA, Washington, DC.

THOMPSON R.C. and Ryder E.J. (1961) *Descriptions and Pedigrees of Nine Varieties of Lettuce*. Tech. Bul. 1241, USDA, Washington, DC.

VAN HEE L. and Bockstaele L. (1983) Chicorée á café. La culture de chicorée. *Rev. l'Agriculture* No. 3, 1015-1024.

WITTWER S.H. and BUKOVAC M.J. (1957) Gibberellin effects on temperature and photoperiodic requirements for flowering of some plants. *Science* **126**: 30-31.

# Introduction

The expansion of modern tools for detecting and characterizing pathogenic agents (serological, and especially molecular) has tended to overlook the true nature of plant disease diagnosis: a science which is based mostly on knowledge of the terrain and symptomatology.

This book, which complements the *Compendium of lettuce Diseases* published by R.M. Davis, K.V. Subbarao, R.N. Raid, and E.A. Kurtz in 1997, maintains the approach used in the previous three books on diagnosing diseases of tomato, the cucurbits, and tobacco. It incorporates the knowledge and approach a plant expert would use when attempting to identify a disease affecting a plant such as lettuce. So its main purpose is to allow the reader to do the following:

- **to identify parasitic and non-parasitic diseases** which attack leafy salad vegetables (l.s.v.), avoiding the numerous possible ways of confusing the diagnosis;
- **to consult recent data** on the biology of the phytopathogenic micro-organisms involved;
- and to be able to **select**, after becoming aware of the cause, the **protective methods** which can be used in order to control them.

The **first part** of this book has been designed as a real 'diagnostic tool', illustrated with 448 colour photographs and numerous diagrams to facilitate plant observation. The book is easy to consult because symptoms have been grouped together under simple headings. It is also educational and should ultimately allow you to adopt the approach gradually and make the many instinctive decisions which are essential for arriving at a reliable diagnosis.

> Many references are made in the text to observation and treatment in the nursery. Some planting systems, primarily in several European countries, use nurseries to raise seedlings from sowed seeds. These may be outdoors or under some sort of protective cover. The seed may be sowed in soil blocks, which are separated and lifted to be planted in the field, usually spaced in rows on flat ground. However, in many other countries, including the USA, Spain, and Australia, seed is sowed directly in the ground, usually on raised beds. When the seedlings have three or four leaves, the field is thinned with a hoe to the desired spacing within the row. In this context, reference to a nursery does not apply.

Future readers ought to be aware that this book should be used by following the '**utilization procedure**' described in the next few pages. Undoubtedly some readers, both experienced and inexperienced, will tend to want to 'short circuit' the proposed approach. If so they may be confused by the way the book has been organised, as it has not been structured according to type of disease but rather in accordance with the characteristic symptoms of the disorder.

Once you have identified the disease, you can go to the **second part** of the book to find a detailed list of most of the pathogenic agents, describing the following: their distribution and effect on growing l.s.v., the principal symptoms these agents cause, and the specific characteristics of their morphology and biology. The methods to be used effectively to protect l.s.v. immediately after diagnosis, or when planting the next crop, are also mentioned.

Almost all the **parasitic and non-parasitic diseases attacking l.s.v. all over the world** can be identified and controlled.

# How to use this book

## Warning

When faced with so many photos, you might be tempted to flick through this book to make your diagnosis, comparing the symptoms observed on the diseased lettuce with those which seem to resemble them in the book. However you should be aware that even if the diagnostic procedure proposed here is sometimes rather tedious, it is the best way to guarantee reliable identification, while teaching you to diagnose parasitic and non-parasitic diseases of l.s.v.

## Prepare your diagnosis carefully

Identifying diseases of l.s.v. is not easy, as you will find out at a later stage. In fact, it is very easy to confuse your diagnosis. In order to improve your results, we recommend that you adopt the following procedure:

1. **Select plants** whose symptoms are not too extensively developed, but are representative of the disease. It is essential to gather the whole plant, including its root system. Lift this very carefully and wash the roots.
2. **Collect as much information as possible:**
   - on the disease and its symptoms (distribution in the plot and over the plants - see pages 20–25, rate of development, climatic conditions which preceded its appearance or which appear to encourage its spread, and so on);
   - on the plant (characteristics of the variety, quality of seed, and so on);
   - on the plot (location, soil characteristics, previous crops, addition of humus or manure, and so on);
   - on the agricultural activities carried out (fertilization, method and frequency of irrigation, quantity of water applied each time, application of pesticides to the crops or in their vicinity, and so on).

## Carrying out the diagnosis

3. **Roughly evaluate the nature of the visible symptom or symptoms located on the l.s.v. (choose one or several of the following headings)**
   Although l.s.v. are relatively compact plants, we have arbitrarily separated the symptoms which can be observed on l.s.v. into to 2 preferred locations:
   - leaves;
   - leaves in contact with the ground and underground organs.

   When there has been damage to leaves, the numerous symptoms have been split into four sub-headings: 'Abnormalities in l.s.v. growth and/or in the shape of their leaves', 'Abnormalities in leaf colouration', 'Spots and damage on leaves' and 'Wilting, drying out, necrosis of leaves'.

   In the same desire for simplicity and clarity, we have subdivided the mainly soil-based symptoms into three sub-chapters: 'Damage on leaves, in contact with the soil, and the crown', 'Damage and abnormalities of roots', 'Internal and/or external damage and abnormalities of the taproot and stem'.
4. **Refer to the heading or headings you have chosen**
   At the start of each chapter or sub-chapter the following are mentioned:
   - **'symptoms studied';**
   - **'possible causes';**

   (*several hypotheses may cover several symptoms*).
5. **Select a symptom and turn directly to the pages concerned or consult all the symptoms listed in the sub-chapter or chapter.**
   At this point in the diagnosis you will need to be more precise in your definition of the symptom or symptoms observed. In addition, one or several 'possible causes' are associated with each symptom and these must be distinguished (*several hypotheses may correspond to one symptom*).

| Headings | Reference colours in the book | Where to find the observation guide and corresponding sections or sub-sections |
|---|---|---|
| **Leaves and head** | | |
| Abnormalities in the **growth** of leafy salad vegetables and/or in the **shape** of leaves | | page 29 |
| Abnormalities in leaf **colouration** | | page 47 |
| **Spots and damage on leaves** | | page 91 |
| **Wilting**, drying out, **necrosis** of leaves (may or may not be preceded or accompanied by **yellowing**) | | page 147 |
| **Leaves in contact with the soil and/or underground organs** | | |
| Damage to **leaves, in contact with the soil,** and the crown | | page 163 |
| **Root** damage and abnormalities | | page 185 |
| Internal and/or external damage and abnormalities of the **taproot** and **stem** | | page 207 |

6. **Determine the cause of the symptom.**
   In order to select from the different hypotheses, we suggest you:
   - compare the symptom or symptoms observed on the plants with those shown on the many photographs;
   - use the **'additional information for diagnosis'**;
   - make sure that you examine the symptoms described on relating pages, or even in other sections or sub-sections when recommended.

> Texts referring to diseases which are rarely or never found in France have been placed in boxes

## Understanding diseases and choosing appropriate methods of combating them

Once you have made your diagnosis, we recommend that you refer to the pages in the second part of the book which focus on the following aspects:

- **symptoms** (brief description, indicating the photograph numbers which illustrate these symptoms);
- **principal characteristics of the pathogenic agent** (frequency and extent of damage, survival, penetration, spread, conditions favourable to development, and so on);
- **treatment methods** (to be applied during growth and on the next crop).

As well as consulting the pages of the book, if you would like to have a complete overview of l.s.v. treatments, we recommend that you consult 'Summary of control methods for l.s.v. bio-aggressors in nurseries and during cultivation' page 347.

In the case of non-parasitic diseases, the methods you should use in order to limit their development often depend on the cause or causes at their origin (these will be defined in the first part of the book). In many cases, poor climatic management and/or inappropriate agricultural conditions, and so on, are responsible. This is the case, for example, with damage due to cold, numerous types of chemical injury, root suffocation, various deficiencies, and so on. To rectify this, any errors should be corrected and/or the plants made more 'comfortable'.

*The book concludes with an appendix summarizing some of the botanical characteristics of lettuce and in particular its resistance to bio-aggressors.*

Drawings, photos, and observations appear throughout the book and can normally facilitate diagnosis. They are accompanied by symbols indicating their purpose:

Diagnosis: defines the level of difficulty of the diagnosis

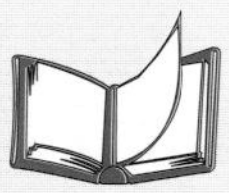

Diagnostic guide: specifies the nature and/or distribution of a symptom, supplies other diagnostic criteria

Shows and/or explains a symptom and its development

Suggests the use of an observation instrument (magnifying glass, stereoscopic microscope, microscope, and so on)

Part

# Diagnosis of parasitic and non-parasitic diseases of leafy salad vegetables

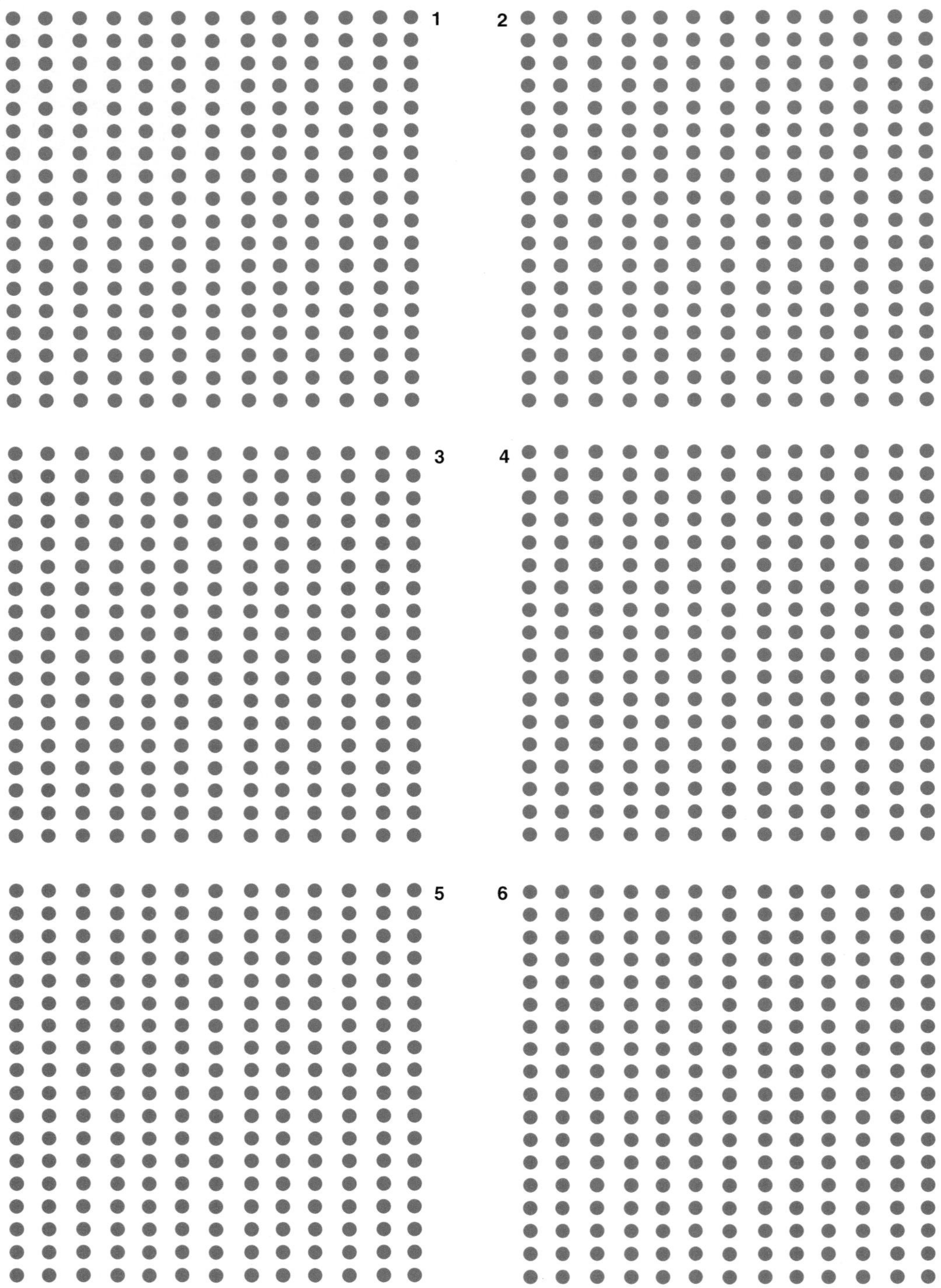


**Figure 1: Distribution of diseased lettuces within the crop**

**(healthy plants in green, diseased plants in red)**

1 Healthy crop
2 Diseased plants spread at random
3 One to several groups that may be widespread
4 One to several rows of diseased plants, varying in length
5 One section of the crop
6 Disease affecting most of the plot

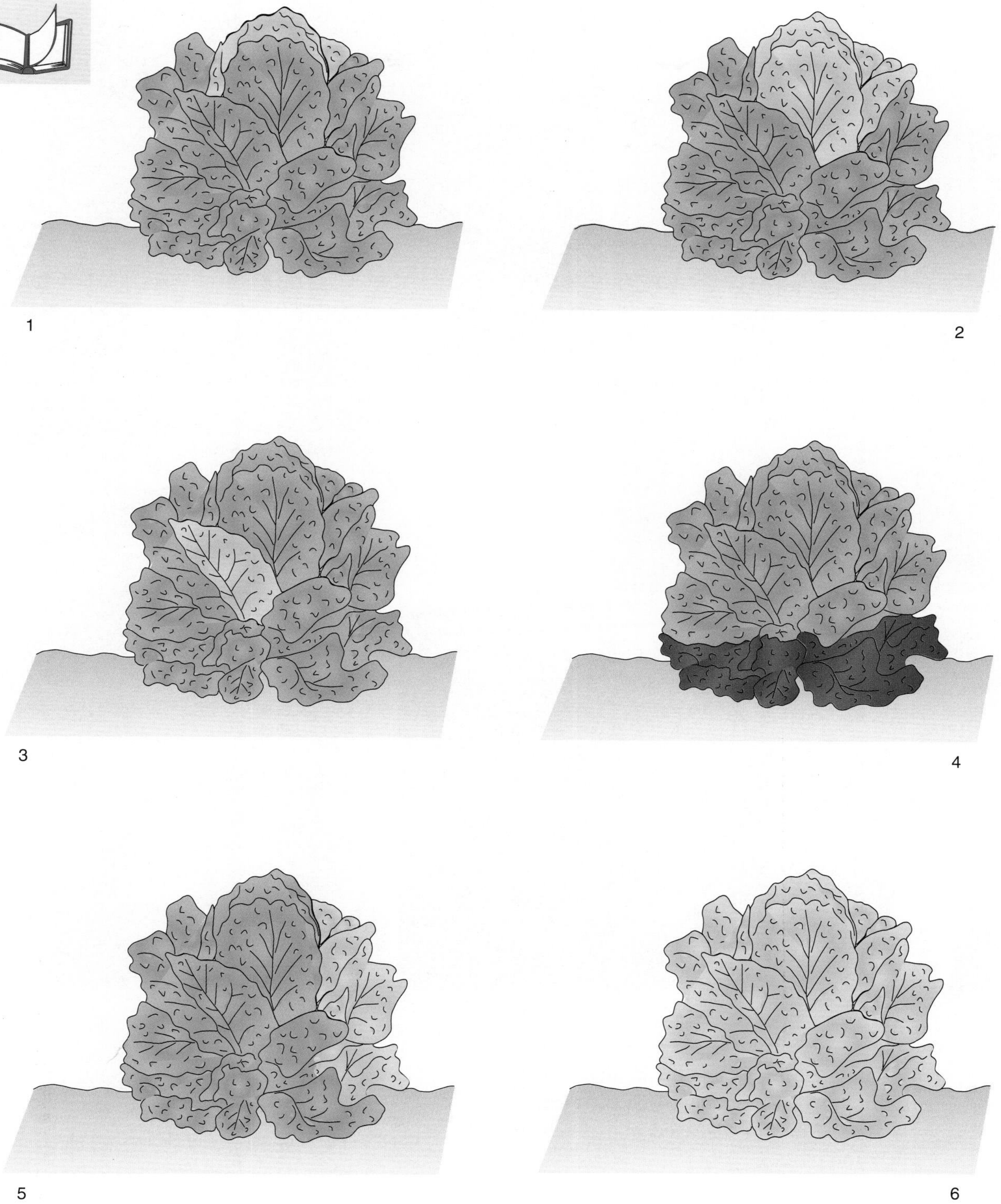


**Figure 2: Location of the foliar symptom(s) on examined lettuce(s)**

1 The apex, the terminal bud
2 Young leaves (top of plant)
3 Patchy and random
4 Old leaves (base of plant)
5 Leaves on one side of the plant(s) only (unilateral)
6 All leaves (widespread)

1

2

3

4

## Some examples of plant health contexts which are at the origins of different distributions of diseased plants within lettuce plots

**1** In this crop of lettuces, one single chlorotic plant can be clearly distinguished (in the foreground), suffering from a non-parasitic disease.
**Genetic abnormality**
**Random distribution, one single isolated plant**

**2** A few young isolated and dispersed plants are chlorotic and are wilting.
***Botrytis cinerea***
**Random distribution, a few dispersed plants**

**3** Several contiguous lettuces are smaller in size than the surrounding plants.
**Root suffocation**
**Distribution in groups**

**4** Numerous plants spread over 1 to 2 lines showing reduced growth compared with the surrounding plants.
***Rhizomonas suberifaciens***
**Distribution in rows**

5

6

7

8

**5** This plot of lettuces (oak leaf) seems to be subdivided into two distinct blocks. The bottom leaves of the lettuces located in the background are completely necrotic. They contrast with those in the foreground, from a different variety, which do not seem to have been sensitive to the mixture of pesticides sprayed onto the plants.
**Chemical injury**
**Distribution according to sector**

**6** In this plot we can see several small blocks of more chlorotic lettuces, with spots on their leaves. In fact, a grower received a mixture of plants of different varieties; the affected plants did not have a spectrum of resistance to *Bremia lactucae* capable of controlling the strain present in this greenhouse.
**Sectorial distribution**

**7** All the lettuces, slightly chlorotic, are more upright in appearance with several rolled leaves.
**Chemical injury**
**Widespread distribution**

**8** On all the lettuces in this plot we can see fairly marked yellowing of the leaves in the rosette.
**Beet western yellows virus (BWYV)**
**Widespread distribution**

**Table 1: Distribution of the principal parasitic diseases in lettuce plots**

| Pathogenic agents | Types of distribution of diseased lettuces in the field | | | | |
|---|---|---|---|---|---|
| | random | in group(s) + or − extensive | in row(s) | in sector(s) | widespread |
| **Airborne bacteria** | | | | | |
| *(Pseudomonas cichorii, Xanthomonas campestris* pv. *vitians,* and so on*)* | + | + | | * (if different types of lettuce in the same plot) | * |
| **Bacteria affecting the root and stem** *(Erwinia carotovora, Rhizomonas suberifaciens)* | + | + | | | * |
| **Leaf fungi** | | | | * (if mixture of genotypes with different resistances to mildew) | * |
| *Bremia lactucae* | + | + | | | |
| *Erysiphe cichoracearum* | + | + | | | |
| *Microdochium panattonianum* | | + | | | |
| **Fungi affecting lower leaves and crown** | | | | | |
| *Botrytis cinerea* | + | + | | | * |
| *Sclerotinia sclerotiorum* | + | + | | | * |
| *Sclerotinia minor* | + | + | | | * |
| *Rhizoctonia solani* | + | + | | | * |
| **Root fungi** | | | | | |
| *Pythium* spp. | + | + | | | |
| *Thielaviopsis basicola* | | + | | | * |
| **Viruses transmitted by seeds** (LMV) | + | | | | * |
| **Viruses transmitted by aphids** (LMV, CMV, and so on) | + | + | | | * |
| **Viruses transmitted by whiteflies** (BPYV) | + | + | | | * |
| **Viruses transmitted by thrips** (TSWV, and so on) | + | + | | | |
| **Viruses transmitted by fungus** (MiLV, LRNV) | + | + | | | |
| **Root nematodes** *(Meloidogyne* spp. *Pratylenchus* spp.*)* | | + | | | |

\+ infestation commonly observed
* very severe infestation

**Table 2: Distribution of the principal non-parasitic diseases in lettuce plots within the French plant health context**

| Non-parasitic diseases | Types of distribution of diseased lettuces in the field | | | | |
|---|---|---|---|---|---|
| | random | in group(s) + or − extensive | in row(s) | in sector(s) | widespread |
| Various types of phytotoxicity | + | | + | + | + |
| 'Tip burn' | + | + | | | + |
| Nutritional disorders (deficiencies, toxicities) | +/− | | + | | + |
| Root suffocation | | + | | | |
| Genetic abnormality | + | | | | |
| Sunstroke | + | | | | |
| Cold injury | | | | | + |
| Lightning injury | | + | | | |
| Hail injury | | | | | + |
| 'Hollow heart' | + | | | | |
| Vitrescence of the taproot | + | | | | |

\+ frequently observed occurrence
+/− sometimes observed

# Abnormalities and damage on the leaves and head

In order to simplify the diagnosis of diseases that cause symptoms on leaves, these have been divided into four sub-sections.

- Abnormalities in leafy salad vegetable growth and/or in the shape of their leaves (page 29)
- Abnormalities in leaf colouration (page 47)
- Spots and damage on leaves (page 91)
- Wilting, drying out, necrosis of leaves (may or may not be preceded or accompanied by yellowing) (page 147)

## Examples of abnormalities in leafy salad vegetable growth and in the shape of their leaves

**9** This lettuce, which is suffering from a **genetic abnormality**, has never developed properly. Its particularly small size is in sharp contrast with the surrounding lettuces.

**10** Lettuce growth is not affected in all cases. Leaf malformation may be sufficient to change the appearance of the plants.
**Lettuce big vein virus (MiLV)**

**11** Sometimes the leaves have undergone various changes affecting their appearance. In this case the surface of the blade has been eaten away to varying extents by a **pest** whose identity should be established. In this case the pest is a **slug**.

# Abnormalities in leafy salad vegetable growth and/or in the shape of their leaves

## Symptoms studied

- Abnormal plant growth (stunted, irregular, rank vegetation, and so on) (see page 31)
- Partially or totally deformed leaves (blistered, smaller, dentate, rolled, and so on) (see page 35)
- Ragged, serrated, and jagged leaves (see page 41)

## Possible causes

- Soil-based fungi and pests in the case of stunted plants (see section entitled 'Damage and abnormalities on leaves in contact with the soil and/or underground organs', page 159)
- *Pythium tracheiphilum*
- Bacteriosis
- Phytoplasma from the aster yellows group (fact file 20)
- Various types of virus
  - Lettuce ring necrosis agent (LRNA) (fact file 35)
  - Lettuce big vein virus (MiLV) (fact file 34)
  - Broad bean wilt virus (BBWV) (fact file 25)
  - Beet western yellows virus (BWYV) (fact file 23)
  - Cucumber mosaic virus (CMV) (fact file 22)
  - Turnip mosaic virus (TuMV) (fact file 27)
  - Dandelion yellow mosaic virus (DaYMP) (fact file 26)
  - Lettuce mosaic virus (LMV) (fact file 21)
  - Alfalfa mosaic virus (AMV) (fact file 24)
  - Endive necrotic mosaic virus (ENMV) (fact file 28)
  - Beet pseudo-yellows virus (BPYV) (fact file 30)
  - Other viruses (BiMoV, SYVV, TRV, TRSV, LCV, LIYV, SYNV, TBSV, INSV) (fact files 29, 31 and 33)
  - Lettuce dieback (TBSV, LNSV) (see pages 153 and 365)
- Nematodes (*Meloidogyne* spp., *Pratylenchus* spp., and so on) (fact files 36 and 37)
- Defoliation pests (slugs, insects, rabbits, birds, and so on)
- *Cuscuta* spp.
- Allelopathy (see page 198)
- Genetic abnormalities
- 'Multiple hearts'
- Cold injury
- Nutritional disorders
- Bolting
- Various types of chemical injury
- 'Acid soil'
- 'Saline soil'

## Arguments in support of the diagnosis

Diseases which cause abnormalities in leaf shape often have symptoms in common, and this makes them difficult to identify. This is especially relevant to leafy salad vegetables, because the distinctive structure of these plants, especially the very confined position of numerous leaves, makes observation difficult. We therefore suggest that you look at all the descriptions and photos in this section. These diseases also cause discolouration of the leaves. It is worth looking at this section too (page 47).

Symptoms of several viruses have been included in this section. As there is a high risk of confusing virus diagnosis, we advise you to have a serological or molecular test carried out by a specialist laboratory, in order to identify with certainty the virus(es) responsible.

*In many cases, in the presence of these types of symptoms, one or more hypotheses may be possible in relation to the disease(s) responsible, with the exception of a few rare problems with very characteristic symptoms.*

12 ***Pythium tracheiphilum*** takes hold at a very early stage in the vascular tissue of lettuce plantlets. Subsequently their growth slows down significantly. This lettuce affected by the fungus is dwarfed and is in striking contrast with the surrounding healthy, well-developed plants.

13 Various soil-based parasitic micro-organisms disrupt lettuce growth by damaging their root system. This is the case with the curly endive, located on the left, whose root system has been badly attacked by ***Thielaviopsis basicola***.

14 The viruses present in leafy salad vegetables not only cause a variety of symptoms (mosaic, blistering, and so on) but also adversely affect plant growth. This is the case with this lettuce, located on the left, which is affected by **lettuce mosaic virus (LMV)**.

15 The growth of this lettuce has been adversely affected at a fairly early stage. It is suffering from a **genetic abnormality** that is also responsible for rolling and malformation of the leaves.

# Abnormal plant growth (stunted, irregular, rank vegetation, and so on)

Generally speaking, parasitic micro-organisms, pests and non-parasitic diseases have various effects on leafy salad vegetables (l.s.v.), as you will realize after reading the various sections of this book. They affect growth in particular, but subsequently their appearance and growth habit are often changed and therefore become atypical.

## Slowed or blocked growth

L.s.v. growth can be affected to varying degrees depending on the nature of the disease and how early it attacks.

Plant development is sometimes adversely affected **at a very early stage**. This occurs in particular in plots that are seriously contaminated by a **soil-based parasitic fungus or by nematodes**. Plantlets, which are very rapidly confronted with the inoculum present in the soil, experience root damage and/or vascular invasion which adversely affects their development (see pages 185 and 207). As an example, ***Pythium tracheiphilum*** is frequently responsible for dwarfism in the lettuce because of its early colonization of young l.s.v. and its vascular location (**12**). Reduced growth can also be seen following early attacks of numerous types of fungi such as: ***Botrytis cinerea***, ***Sclerotinia* spp.**, ***Thielaviopsis basicola*** (**13**), and nematodes (***Pratylenchus* spp.**, ***Meloidogyne* spp.**, and so on).

**16** We find the same situation in the presence of chemical injury which has also caused reduced growth and malformation of the leaves in the heart of this young lettuce.

Among the other micro-organisms, viruses (LMV, CMV, TuMV, and so on) interfere in particular with plant development, whatever their means of transmission and the nature of their symptoms. **14** shows this characteristic very clearly. Reduced growth of plants will commence even earlier if infections occur at a very early stage in the course of the plant production cycle. It may be particularly marked if the virus (LMV) is transmitted by the seeds and multiplies prematurely (see pages 53, 59, 69, 78, and 152).

Phytoplasma from the Aster yellows group have been observed on l.s.v. in a limited number of countries (USA, Italy, and so on). If they attack at an early stage, the leaves in the head grow poorly. The plants remain dwarfed and the old leaves go yellow. Other symptoms may occur and you can find a list of these on pages 79 and 131.

Non-parasitic diseases produce the same effects on l.s.v. Some **genetic problems** lie at the origin of 'crazy plant' lettuces, as they are known. Their leaves are small and deformed, and grow very little. Ultimately the plants are stunted to a greater or lesser degree in comparison with the normal surrounding plants (**15**) (see page 37).

In certain situations, lettuces may initially grow completely normally. Then, suddenly, **in the course of their growth**, the shoot may stop growing or may only develop to a moderate extent. In this case, the size of the old leaves contrasts with that of the more recent growth. Attacks by a **virus** while the crop is growing, the accidental application of a herbicide (**16**), and so on, may cause this type of damage (for additional information about chemical injuries, see the heading 'Partially or totally deformed leaves', page 39)

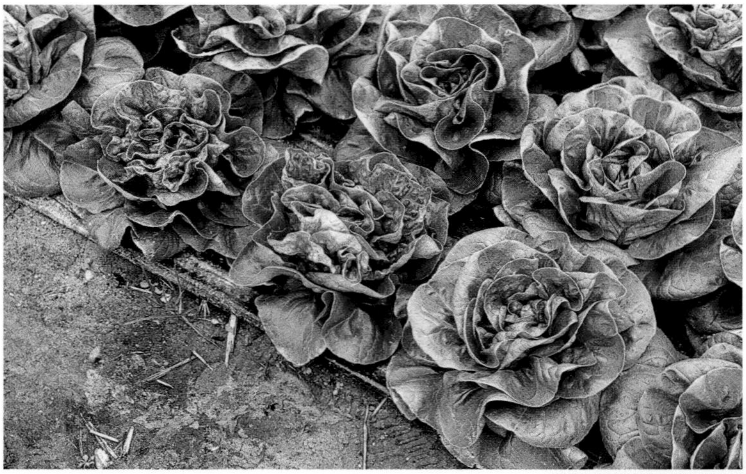

**17** Lettuce big vein virus **(MiLV)** has caused malformation of leaves in the head and noticeably changes the appearance and behaviour of affected lettuces.

**18** The effects of **MiLV** on 'oak leaf' lettuce are absolutely astonishing. On plants affected by the virus, numerous leaves are more filiform and their leaf blades are rolled up at the edge. This gives the plants a characteristic appearance which contrasts with that of healthy neighbouring plants.

**19** The growth of this escarole seems more rank and distorted than that of the surrounding plants.
**'Multiple hearts'**

**20** A longitudinal cross section of this escarole reveals the formation of 2 to 3 shoots.
**'Multiple hearts'**

As we have pointed out, **nutritional deficiencies**, although now rare in the field, can sometimes cause reduced growth of l.s.v. with the severity of the deficiency dictating the extent of the effect. For example some cause symptoms located on the shoot and young leaves, sometimes prematurely slowing the development of the lettuces. This is the case with deficiencies in **calcium**, **boron**, and so on. We definitely recommend consulting the section 'Partially or totally chlorotic or yellow leaves' (page 71) in order to obtain additional information about these nutritional disorders.

If the reader is interested in other edaphic factors, it has been observed that, in certain **acid soils** with a pH of less than 4, plants also develop fairly slowly (in addition to suffering from leaf chlorosis). Comparable symptoms are sometimes noted in groups of plants in certain areas of plots where water retention has occurred locally.

We ought to point out that several species of ***Cuscuta***, by developing on l.s.v., can affect their growth (see page 45).

Whatever the origins of the problem, affected plants often contrast sharply with healthy plants in their vicinity.

## Altered growth habit

21 The very hot temperatures recor- ded in the summer caused the prema- ture **bolting** of these lettuces.

In certain situations, plant growth may be more or less normal, but it is the particular shape of a few leaves that will change their growth habit. **Lettuce big vein virus (MiLV)** causes these symptoms as we can see on **17** and **18**.

Sometimes, lettuces can display more rank growth than normal and a rather atypical appearance. This may, for example, result in the development of two to three shoots at the same time (**19** and **20**). This physiological abnormality, known as **'Multiple hearts'**, is caused in particular by excessive temperatures occurring during seed germination or when leaf growth begins.

**Bolting**, which is a natural phenomenon, also changes the appearance of l.s.v. It corresponds to the lengthening of the internode of the stem, followed by the formation of flowers, and then the seeds. Unfortunately, it sometimes begins prematurely in certain crops (**21** and **22**), causing varying degrees of damage to the quality of the future crop.

**22** Lack of water may also trigger **bolting**; this has happened with this 'Latin' lettuce.

This is expressed as:

- absence of formation of head or imperfect head development;
- presence of a disagreeable taste associated with a higher latex content.

Several factors are likely to influence bolting:

- the variety used; if it is not perfectly suited to the production schedule, risks of bolting are increased. In this situation we may also note the formation of lateral suckers;
- temperature has an important effect. Tempe- ratures which are abnormally hot for the season, especially in the spring and summer, result in bolting. A period of drought can have the same effect.

L.s.v. can be rather sensitive to **soils that are too acidic or too saline**. Under these conditions, they may display slow growth, poor head development, or be slow to establish. Excessive salt also causes thickening of the blade in old leaves, which also have a more bitter taste.

These few examples will have shown you that the diseases described in this section often give l.s.v. (or their shoots temporarily or definitively) a distinctive appearance which contrasts with that of healthy surrounding plants. This appearance is sometimes very characteristic of one specific disease or, on the contrary, may be common to several diseases, and result in plant dysfunction. Consequently great caution must be exercised when attempting to identify their cause.

If, after reading the following pages, you are unable to identify the plant health problem affecting the development of your l.s.v., don't forget that several parasitic and non-parasitic diseases covered in the other sections of the book often cause slow and stunted growth of plants. In addition they cause other, much more characteristic symptoms, which need to be investigated and analysed.

23 Numerous leaves from the heart of this lettuce are extremely blistered, crinkled, and so on.
**Lettuce mosaic virus (LMV)**

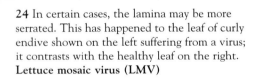

24 In certain cases, the lamina may be more serrated. This has happened to the leaf of curly endive shown on the left suffering from a virus; it contrasts with the healthy leaf on the right.
**Lettuce mosaic virus (LMV)**

25 The broader appearance and thickening of veins caused by **lettuce big vein virus (MiLV)** are a partial cause of the extremely blistered appearance of the lettuce leaf on the left.

## Various abnormalities of leaf shape associated with the presence of a virus

26 The growth of the young leaves of this lettuce has been slowed down; this gives the plant a characteristic appearance.
**Turnip mosaic virus (TuMV)**

# Partially or totally deformed leaves (blistered, reduced, dentate, rolled, and so on)

## Possible causes

- Phytoplasma from the Aster yellows group (see pages 29 and 31) (fact file 20)
- Various types of virus
  - Lettuce ring necrosis agent (LRNA) (fact file 35)
  - Lettuce big vein virus (MiLV) (fact file 34)
  - Broad bean wilt virus (BBWV) (fact file 25)
  - Beet western yellows virus (BWYV) (fact file 23)
  - Cucumber mosaic virus (CMV) (fact file 22)
  - Turnip mosaic virus (TuMV) (fact file 27)
  - Dandelion yellow mosaic virus (DaYMV) (fact file 26)
  - Lettuce mosaic virus (LMV) (fact file 21)
  - Alfalfa mosaic virus (AMV) (fact file 24)
  - Endive necrotic mosaic virus (ENMV) (fact file 28)
  - Beet pseudo yellows virus (BPYV) (fact file 30)
  - Other viruses (BiMoV, SYVV, TRV, TRSV, LCV, LIYV, SYNV, TBSV, INSV) (see pages 69, 78 and 152) (fact files 29, 31 and 33)
- Genetic abnormalities
- Nutritional disorders
- Various types of chemical injury

## Arguments in support of the diagnosis

It is fairly difficult to organize and separate symptoms that are considered to be abnormalities in the shape of l.s.v. leaves. In fact, a leaf may often be both blistered and rolled, which gives it a particular appearance. It is therefore very difficult to identify their cause solely on the basis of this type of malformation. Consequently we have grouped together under this sub-heading some of the principal afflictions causing leaf malformation. It is very often worth referring to other sub-sections in order to confirm your diagnosis.

- **Various types of virus**

As we explained previously, viruses, in addition to causing multiple abnormalities in colouration (mosaic, yellowing), produce changes in the shape of the lamina that are sometimes fairly spectacular. The nature and intensity of this damage fluctuates depending on the virus and strains concerned. For example, in the case of epidemics of lettuce mosaic virus (LMV), the new leaves formed on the plants are frequently blistered and swollen to varying degrees (**23**). Sometimes the blade is even more serrated (**24**). With severe attacks of lettuce big vein virus (MiLV), infected leaves also give the impression of being blistered (**25**). In fact this particular aspect of the blade is due to thickening of leaf veins. Turnip mosaic virus (TuMV), in addition to causing yellow mosaic patterns on the leaves, sometimes causes stunting of lettuces which also present narrower and smaller leaves (**26**).

These few examples, as well as those mentioned in the previous heading, should make you aware of the potential of certain viruses to cause leaf malformation. In order to make a more accurate diagnosis it is also worth referring to the chapter entitled 'Abnormalities in leaf colouration', page 47 and table 5 (also see pages 69 and 78).

**27** The lettuce on the right is slightly smaller than the adjacent healthy plant. Its leaves are smaller, deformed and slightly crinkled.
**Genetic abnormality**

**28** All the leaves of this chimeric lettuce are rolled.
**Genetic abnormality**

**29** At times the lamina is significantly rolled; consequently the size and appearance of the plant are significantly damaged.
**Genetic abnormality**

**30** Certain genetic abnormalities result in a reduction in the size of the lamina which is also more serrated.
**Genetic abnormality**

**• Phytoplasm from the Aster yellows group**

Although frequently associated with yellowing of leaves (see page 79), this phytoplasm causes a certain number of symptoms which change the appearance of plants and the shape of their leaves:

- leaves of the heart or crown are small, deformed (twisted, rolled, and so on);
- reduced growth of leaves in the heart;
- absence of head development;
- stunted plants, and so on.

Deposits of latex, pinkish to brown in colour, are visible under the veins of affected leaves (see page 131).

Intense necrosis of the phloem can be observed.

This micro-organism, transmitted by several species of leafhopper, has been reported on lettuce in a few countries: several US states, Canada, and Europe (Italy).

**• Nutritional disorders**

Lettuces are likely to be affected by several nutritional deficiencies. These generally take the form of abnormal colouring, which is, however, characteristic of various problems (it is essential to consult the section 'Partially or totally chlorotic or yellow leaves' page 71). Some of these also cause changes in the shape of the lamina which may take various forms:

- edge is slightly curved in towards the lower face (potassium);
- narrower (phosphorus);
- edge is slightly curved in towards the upper face (molybdenum, copper);
- more folded (calcium);
- smaller (sulphur).

*If you encounter this type of symptom, do not hastily assume that this is the result of a deficiency without having consulted a specialist and carried out physical and chemical soil analyses and/or growth analyses.*

**• Genetic abnormalities**

Certain mutations are responsible for phenotypical characteristics often considered undesirable by the breeder. They usually take the form of abnormalities in leaf colouration and shape. More specifically, as far as leaf blade malformation is concerned, leaves with the following characteristics may be noted (**27–30**):

- rather small in size and with a rather 'crinkled' appearance;
- rolled in relation to their axis or plane;
- small and more serrated;
- extremely small and filiform (narrow).

In addition, the tissues of affected plants can be thicker.

These 'chimera' must be viewed as curiosities that do not seriously affect the crop. In the majority of cases, a very low proportion of plants is affected (see **1**). They are classified under physiological diseases and must not be confused with viruses or phytotoxicity.

Other genetic abnormalities affecting the production of chlorophyll in lettuces are described on page 89, where you will also find additional information on this subject.

## Examples of leaf malformation associated with the effects of pesticides ('Pesticide injuries')

**31** Several lettuce plantlets with some leaves whose blades are more dentate and serrated.
**Chemical injury**

**32** The leaves from the head of this lettuce are smaller than normal and chlorotic.
**Chemical injury**

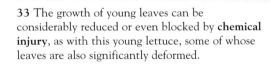

**33** The growth of young leaves can be considerably reduced or even blocked by **chemical injury**, as with this young lettuce, some of whose leaves are also significantly deformed.

**34** Foliar malformation is sometimes very marked. The cotyledons, and especially the young leaves of these few plantlets, have thicker tissues; the blade is very rolled up and twisted.
**Chemical injury**

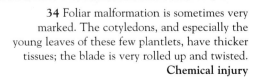

**35** On this lettuce, several leaves are rolled; the edge of the blade is also curved in towards the lower face. You can also see that the veins seem broader than usual.
**Chemical injury**

• **Various types of chemical injury**

Certain pesticides may cause chemical injury that may change the appearance of lettuces, cause foliar malformation, or a reduction or even total halt in plant growth. **31–36** will help you to make up your mind about the nature and intensity of the symptoms caused by certain compounds that are toxic to plants. More specifically you may observe:

- blades which are slightly dentate or more irregularly serrated (**31**);
- slowed development of the youngest leaves (**32**); in certain cases it has been completely blocked (**33**). Under these conditions, plants eventually take on a stunted appearance;
- varying degrees of distortion and/or rolling of young leaves, which are sometimes shorter (**34**). You can see that the tissues are significantly thickened;
- spoon-shaped leaves curved inwards and/or rolled in the shape of cups (**35**);
- rolling of the whole blade;
- twisted, crinkled appearance of all leaves (**36**).

In the presence of chemical injury we suggest that you ask yourself the following questions:

- were weeds eliminated from the previous crop using persistent herbicides?
- has a herbicide treatment been carried out close to your crop?
- did you rinse your application equipment properly?
- do you keep your sprinkler properly maintained (cleaning, calibration, and so on)?
- did you use the correct product, at the correct dose?
- did you follow the instructions for use indicated on the packaging?
- did you mix incompatible products with each other or did you use too many products together?
- were the applications carried out under unsatisfactory conditions (strong wind, temperatures which are too high or too low)?

N.B. Irrigation water may be polluted by a herbicide.

**Pesticide injuries**

36 A herbicide, with particularly deforming effects, has given this lettuce a crinkled appearance.

**Chemical injury**

## What should you do in the event of chemical injury?

Although there is no miracle solution in this situation, you can take the following steps:

- carefully define the origins of the chemical injury;
- prevent it from occurring again;
- do not remove the plants immediately. Look after them in the normal way and observe their development; the latter will depend above all on the nature and dose of the product(s) responsible and any residual amounts persisting in the soil, the stage of lettuce growth and, the type and variety being grown.

No other specific measure can be recommended.

***Chemical injuries not only cause the type of malformation described above, but also other symptoms (especially some degree of yellowing); you can read about these aspects in the sections entitled 'Abnormalities in leaf colouration' (page 47), 'Spots and damage on leaves' (page 91) and 'Wilting, drying out, necrosis of leaves' (page 147).***

**37** Several leaves from this lettuce have an abnormally serrated margin. Previously the tissues had gone brown, then black, and had liquefied.
**Bacteriosis**

**38** The outer leaves of these lettuces have scattered spots, the central tissues eventually decompose and fall.
***Microdochium panattonianum***

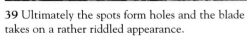

**39** Ultimately the spots form holes and the blade takes on a rather riddled appearance.
***M. panattonianum***

**40** The blade of this leaf has been damaged quite extensively in several places. The tissues exposed to the air are oxidizing and assuming an orange colour. You can also distinguish a few irregular holes.
**Slug damage**

**41** Careful observation of the plants or their immediate environment will reveal the slug(s) responsible.
**Slug damage**

# Ragged, serrated and jagged leaves

## Possible causes

- Various cryptogamic diseases and in particular ***Microdochium panattonianum*** (see the section 'Yellow, orange to light brown spots and damage on leaves', page 111)
- Bacteriosis
- Leaf eating predators (slugs, insects, rabbits, birds, and so on)
- *Cuscuta* spp.
- Hail injury
- Various types of chemical injury

## Arguments in support of the diagnosis

- **Bacteriosis**

Under particularly damp conditions, certain bacteria, covered in the chapter entitled 'Spots and damage on leaves', may develop around the margins of the blade and liquefy the tissue. The blade becomes abnormally serrated (**37**). For additional information about these bacteria, see page 97.

- ***Microdochium panattonianum*** **('Anthracnose')**

This fungus, responsible for anthracnose, causes the spots that develop, in particular, on the lower leaves of lettuces, both on the blade and on the veins. The affected portions of the blade go brown fairly quickly, become necrotic, and fall. The leaves then have numerous perforations that give them a very characteristic riddled appearance (**38** and **39**). This symptom is referred to as 'Shot hole'.

Other symptoms are described in the chapter 'Spots and damage on leaves' on pages 116 and 117.

- **Hail injuries**

Damage is frequently observed on l.s.v. crops after a heavy rainfall which was accompanied by hail. In fact the leaves, whose tissues are by nature tender, are particularly vulnerable to hailstones. When these batter the blade they produce holes, tearing, and localized necrotic damage, corresponding to the point of impact. Certain leaves may be completely jagged, particularly the most spread out external leaves. Plants are very vulnerable as harvest-time approaches.

As a general rule a large proportion, or even all the plants in the plot, are affected.

**42** The blade of several leaves of this lettuce is perforated and sometimes serrated around the edge.
**Moth damage**

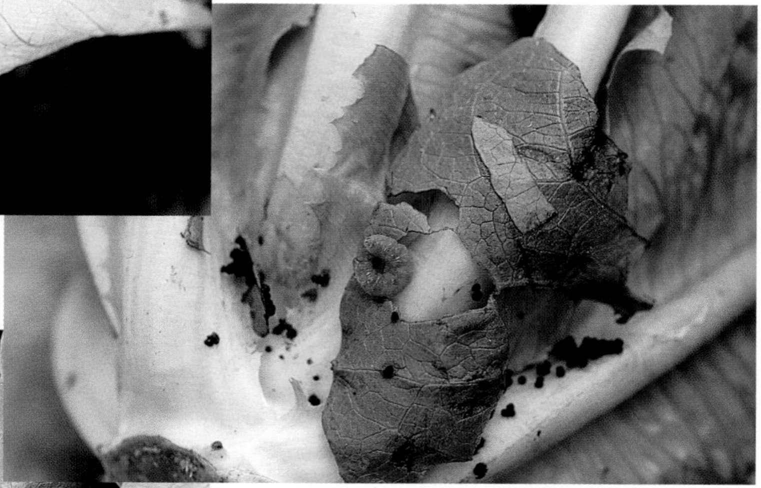

**43** The guilty party, in the case in question a green moth caterpillar, is not far away. You can also see the presence of black faeces.
**Moth damage**

**44** Numerous leaves of this lettuce have been eaten by a **rabbit**.

**Table 3: Principal insects causing damage in the nursery, after planting and while the crop is growing, which may be the cause of holes in leaves**

| Organs attacked and type of damage | Name | Description of pest |
|---|---|---|
| Crown eaten and leaves in proximity with the soil nibbled away. | ***Bourletiella hortensis*** (garden collembola) | Adults, black to dark green in colour, 1.5 mm in length. Large head with long antennae. Prominent black eyes, circled with yellow. |
| Crown eaten and leaves serrated. | ***Agrotis* spp., and so on** (grey worm, soil living moths) | Caterpillars 3.5 cm in length, fleshy, variable in colour, greyish to greenish, surmounted by dark patches or bands. |
| Leaves cut. | ***Autographa gamma*** (gamma moth) | Caterpillars 3.5–4.5 cm in length, green to blackish green. They have 3 pairs of false abdominal legs. |

### • Defoliating pests ('Leaf-cut pests')

Some predators are likely to eat into or devour lettuce leaves. Their activities frequently result in the partial or total disappearance of fairly significant, regular portions of the blade.

**Slugs** are particularly fond of lettuce leaves, whether cultivated under protection or in the open field, and whatever their stage of development. They sometimes cause serious damage. Look out for:

- destruction of plantlets if attacks are very early;
- fairly extensive consumption of the blade (**40**) making the lettuces unmarketable.

If you look carefully you will probably find the slugs responsible (**41**). Several species of slug may be rife in market gardens:

- ***Deroceras reticulatum***, the grey or reticulated slug. It is 3–5 cm in length and is greyish-brown in colour;
- ***D. laeve***, smaller and more mobile slug, only 2.5 cm in length;
- ***Arion hortensis***, the black garden slug. Between 2–4 cm in length, it is more difficult to identify because it frequently chooses to hide away in several centimetres of soil.

These slugs reproduce throughout the year, except during very cold and/or very dry periods.

**Other pests**, particularly various adults, caterpillars, and **insect** larvae, attack l.s.v. leaves (*Table 3*). They may attack very early, in the nursery, or during the days following planting out, or affect adult lettuces. Various types of damage may be involved, depending on the insects and the stage of development of the plants:

- wilting of plantlets, disappearance of plantlets;
- roots eaten away;
- severing of the crown;
- consumption of the lamina resulting in the appearance of fairly numerous rounded holes (**42**).

As is the case with slugs, the guilty party or parties are usually fairly close by (**43**). By carefully examining the affected plant(s) and their environment, the pest(s) responsible can be found. The principal pests implicated are indicated in *Table 3*.

Sometimes **rabbits** enjoy l.s.v. and they may cause significant damage. They partially consume numerous leaves (**44**).

In addition to attacking lettuce seeds and plantlets, various **bird** species may be responsible for irreversible leaf damage. Repeated pecking results in holes or a jagged appearance (**45** and **46**).

**45** A chicken making its way through a plot of lettuces.
**Chicken damage**

**46** It is as keen on this host as various other types of bird; it greedily consumes the leaves which have a jagged appearance.
**Chicken damage**

**47** This lettuce, invaded by a dodder, has not grown as much as surrounding plants.
***Cuscuta* sp. damage**

**48 *Cuscuta* sp.** is a parasitic plant with a leafless stem.

**49** Its flowers, the colour of which varies according to species, are gathered together in clusters.
***Cuscuta* sp.**

• ***Cuscuta* spp. ('Dodder')**

Numerous species of dodder have been identified throughout the world on a wide variety of hosts. This plant sometimes lives as a parasite on l.s.v. which are progressively colonized by its creeping stems. Plants affected in this way grow poorly (**47**). On the leafless stems of dodder (**48**), you can also see suckers which allow it to feed at the expense of its hosts without having to resort to photosynthetic activity. The plants also have small flowers, which vary in colour according to species (white, yellow, pink), gathered together in clusters (**49**). Countless seeds are produced and remain dormant in the soil for several years. They are easily spread, like the majority of pathogenic agents in the soil, by water, earth particles, and so on.

It is particularly difficult to control this parasite. Measures also commonly used for soil-based parasites e.g. crop rotation, crops of non-host plants, pre-emergence herbicides, and so on; can be used in a complementary way.

# Abnormalities in leaf colouration



## Symptoms studied

- Mosaic leaves (mosaics and similar symptoms)
- Partially or totally chlorotic or yellow leaves

## Possible causes

- Vascular diseases (Verticillium, Fusarium, and so on.)
- *Pythium tracheiphilum* (fact file 11)
- Phytoplasm from the Aster yellows group (fact file 20)
- Lettuce ring necrosis agent (LRNA) (fact file 35)
- Lettuce big vein virus (MiLV) (fact file 34)
- Broad bean wilt virus (BBWV) (fact file 25)
- Beet western yellows virus (BWYV) (fact file 23)
- Lettuce mottle virus (LMoV) (fact file 29)
- Cucumber mosaic virus (CMV) (fact file 22)
- Turnip mosaic virus (TuMV) (fact file 27)
- Dandelion yellow mosaic virus (DaYMV) (fact file 26)
- Lettuce mosaic virus (LMV) (fact file 21)
- Alfalfa mosaic virus (AMV) (fact file 24)
- Endive necrotic mosaic virus (ENMV) (fact file 28)
- Beet pseudo-yellows virus (BPYV) (fact file 30)
- Other types of virus (BiMoV, SYVV, TRV, TRSV, LCV, LIYV, SYNV, BYSV, LNSV, LNYV, TBSV, TSWV) (fact files 29, 31 and 33) (also see page 152)
- Lettuce dieback (TBSV, LNSV) (see pages 153 and 365)
- *Pemphigus bursarius* (see page 205)
- Genetic abnormalities
- Nutritional disorders
- Various types of chemical injury

## Delicate diagnosis

Numerous diseases, particularly several types of virus, may cause changes in the colouration of lettuce leaves. The abnormalities most frequently observed on leaves are mosaics, or similar symptoms, and yellowing. These abnormalities in colouration may be very distinctive in the case of certain diseases, but are sometimes fairly similar even though the problems are very different. This situation makes it very difficult to identify the diseases (parasitic or non-parasitic) responsible for these symptoms. We therefore suggest that you carefully consult all the illustrations in this sub-section, which is particularly well illustrated, and take the time to consider the various hypotheses formulated in the section entitled 'Arguments in support of the diagnosis'.

In addition, these problems frequently cause changes in leaf shape. It is therefore well worth reading the sub-section dealing with this type of abnormality as well, in order to complete your diagnosis.

Symptoms of several types of virus have been included in this section. There is a high risk of confusion between types of virus affecting lettuces, so we advise you to have serological or molecular tests carried out by a specialist laboratory, in order to identify with certainty the virus(es) responsible.

**Figure 3: Appearance and location of principal abnormalities in leaf colouration observed on l.s.v.**

1 Healthy leaf with uniform colouration
2 Abnormality occurring between the veins
3 Abnormalities starting from the veins
4 Abnormality located at the base of the leaf
5 Abnormality located around the margin of the blade
6 Abnormality located at the tip of the blade
7 Abnormalities occurring close to the veins (on the right 'vein banding', on the left 'vein mottling')
8 Abnormality covering the left part of the leaf in a random way and the right part in a uniform way

50

51

52

53

## Examples of abnormalities in the lettuce leaf colouration

**50** Mosaic escarole leaf.
**Turnip mosaic virus (TuMV)**

**51** Lettuce leaves partially chlorotic between the veins.
**Chemical injury**

**52** Yellowing of veins and adjacent tissues of a Batavia lettuce leaf.
**Lettuce big vein virus (MiLV)**

**53** Yellowing and whitening of sectors of several lettuce leaf blades.
**Genetic abnormality**

## Various types of mosaic on l.s.v. (true mosaics or similar symptoms) associated with the presence of a virus

**54** Marked mottling on butterhead lettuce.

**55** Mosaic on iceberg.

**56** Mosaic in very marked spots.

**57** Mosaic in small spots on endive ('spotted' mosaic).

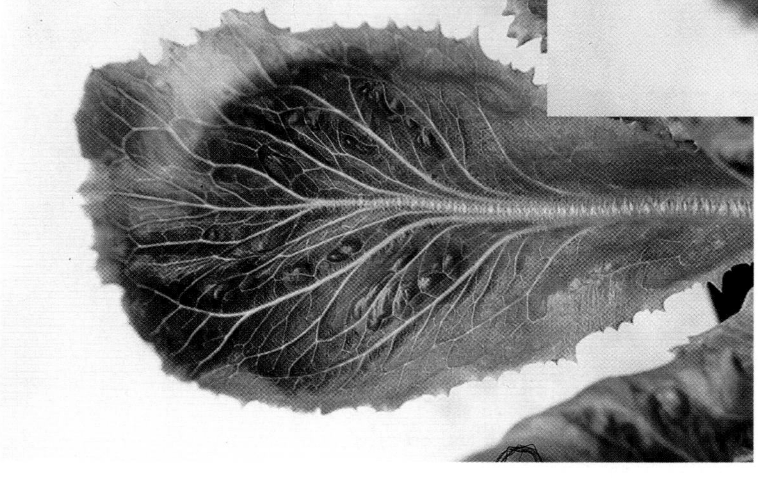

**58** Mosaic in broad light green to dark green areas.

# Mosaic leaves (mosaic and similar symptoms)

## Possible causes

- Lettuce big vein virus (MiLV) (fact file 34)
- Broad bean wilt virus (BBWV) (fact file 25)
- Lettuce mottle virus (LMoV) (fact file 29)
- Cucumber mosaic virus (CMV) (fact file 22)
- Turnip mosaic virus (TuMV) (fact file 27)
- Dandelion yellow mosaic virus (DaYMV) (fact file 26)
- Lettuce mosaic virus (LMV) (fact file 21)
- Alfalfa mosaic virus (AMV) (fact file 24)
- Endive necrotic mosaic virus (ENMV) (fact file 28)
- Other types of virus (BiMoV, SYVV, TRV, TRSV, LNSV) (fact files 29 and 33)
- Genetic abnormalities (consult the heading 'Partially or totally chlorotic or yellow leaves', page 89)
- Various types of chemical injury (consult the heading 'Partially or totally chlorotic or yellow leaves', page 83)

Non-specialists, as well as numerous plant experts, often gather together under the name 'mosaic', symptoms which may differ significantly in terms of intensity and appearance. Photos **54–64** will help you to evaluate some symptoms that are frequently associated with the notion of mosaic, and are considered as such. If you encounter these symptoms you will be justified in suspecting the presence of a virus. However, you should be very careful when diagnosing the specific virus(es) involved. Numerous types of virus actually produce a wide variety of fairly comparable symptoms on plants, thus leading to confusion.

In order to help you with your diagnosis, you will find on the following pages the principal symptoms that you are likely to observe on l.s.v. We have also indicated, each time, the virus with which they are particularly associated.

***It is quite difficult to assess the presence of mosaic on lettuce leaves, especially on sunny days. In this situation we advise you to look at the leaves when illuminated from behind, taking care to shade them.***

In addition we recommend that you consult the sub-section entitled 'Abnormalities in l.s.v. growth and/or in the shape of their leaves', because viruses not only change the colour of the lamina, but also cause varying degrees of malformation.

**59** This lettuce, which has been attacked by a virus, presents several abnormalities in colouration (bands of dark green tissue all along the veins or 'vein banding', 'vein clearing', chlorosis, blisters, and so on), is frequently classified as 'mosaic' lettuce.

60

61

62

63

64

## Other symptoms associated with the presence of a virus in lettuces and sometimes considered as 'mosaics'

**60** Localized clearing of several portions of veins and adjacent tissues.
**61** Widespread yellowing of secondary veins ('vein yellowing').
**62** Broad uniform bands of dark green tissue located on both sides of the veins ('vein banding').
**63** Patterns, arabesques, chlorotic rings ('line pattern', 'ring').
**64** Chlorotic and necrotic damage limited to the peri-veinal tissues ('etch', and so on).

## Arguments in support of the diagnosis

- **Lettuce mosaic virus (LMV)**

LMV, which is transmitted by aphids and, in the case of certain strains, by seeds, is considered to be one of the most damaging types of virus affecting lettuces. Its attacks can take place very early, in the seedling stage, particularly if the seed lot is contaminated or following early infection of the plants by viruliferous aphids. In this case, growth of lettuces is significantly disrupted and it is quite frequent for them to remain dwarfed or for the head not to form. LMV, as its name indicates, is above all responsible for varying degrees of mottling and mosaic (**65** and **66**). Vein clearing, sometimes intense, and vein banding are added to these symptoms, accentuating the mosaic appearance of the leaves (**67** and **68**). Under certain climatic conditions and/or in the presence of particularly aggressive strains, it is fairly frequent to observe necrotic damage on the leaves, which appears as minuscule spots and/or necrotic bands or broad areas of degenerated tissue (**69**).

These various abnormalities in leaf colouration are often accompanied by changes in the shape of the lamina. The latter is frequently blistered, crinkled, rolled, and more serrated; lettuces affected in this way have a very specific appearance. These symptoms are listed under the heading 'Abnormalities in l.s.v growth and/or in the shape of their leaves'. We recommend that you refer to this, pages 31 and 35.

As we have suggested previously, the nature and intensity of the symptoms caused by LMV fluctuate according to climatic contexts, the strains involved (**70–72**), and the types and genotypes of lettuces cultivated (**73–76**).

Numerous varieties of lettuce are tolerant to lettuce mosaic virus; this element can help you to make your diagnosis. However, you need to be vigilant and not hastily set aside the hypothesis of LMV in the presence of a tolerant variety. In fact, strains capable of combating this resistance have been identified in several European countries, particularly in France and Spain, but also Brazil, Chile and, more rarely, the USA.

A virus, which may be confused with LMV, has been reported in Brazil. This is **Lettuce Mottle Virus (LMoV)**. It has been observed on LMV-resistant lettuce varieties and seems to attack in association with the latter. Some characteristics of this virus are indicated in *Table 4*.



**Table 4: Some characteristics of the mottling virus of lettuce**

| Virus | Symptoms | Vector | Viral particle shape |
|---|---|---|---|
| **Mottling viruses of lettuce**<br>**Lettuce mottle virus (LMoV)* (LeMoV)**<br>**Virus still poorly characterized** | Absence of symptoms on certain cultivars Vein clearing, mottling, mosaic. Deformation and size reduction of young leaves. The plants develop very small heads. | Vector not known |  |

*: This virus has now been given the acronym LeMoV; it is probably identical or extremely similar to DaYMV.

## Lettuce mosaic virus (LMV)
## Variability of most common symptoms

**65** Classically, lettuces attacked by **lettuce mosaic virus (LMV)** have leaves with varying degrees of mosaic patterns and blistering.

**66** Careful observation of the lamina will allow you to note the presence of vein clearing and blisters of darker colour.
**LMV**

**67** In addition to vein clearing, the lamina, slightly mottled and chlorotic, shows localized 'vein banding'.
**LMV**

**68** LMV may also cause swelling of the lamina and intense vein yellowing, spreading to the surrounding tissues.
**LMV**

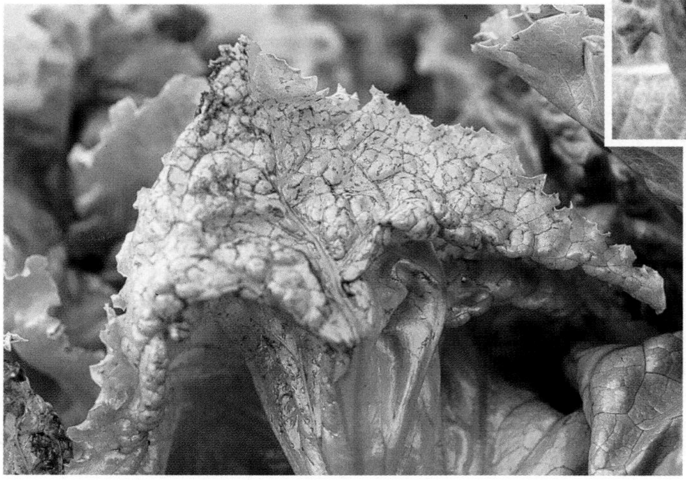

**69** Following changes in climate or attacks of certain strains, necrotic damage can be seen on the leaves. This may be minuscule, located on the veins, for example, or spread over broad areas.
**LMV**

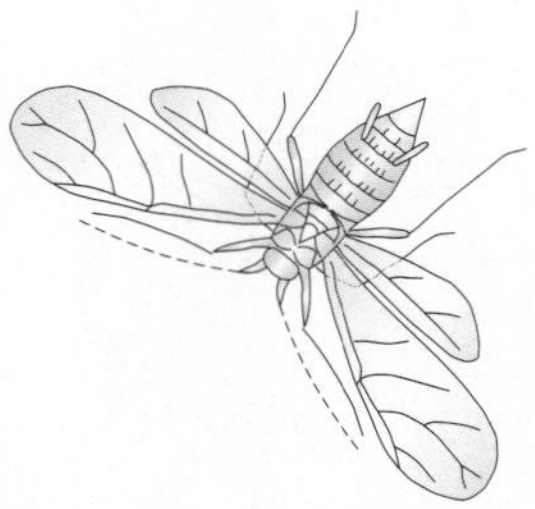

70 Slowing down of growth, slight mosaic pattern and crinkled leaves are characteristic of the effects of this strain of LMV (healthy plant on the left).
**LMV**

71 This other strain causes a yellow mosaic pattern and more marked vein clearing. We can see that the leaves are flat but the lamina is more serrated and sometimes even rolled.
**LMV**

72 This third strain has caused extensive deformity; numerous blisters and swellings are scattered on the lamina.
**LMV**

**Lettuce mosaic virus (LMV)**
**Variability of symptoms between strains**

*These lettuces are infected by three strains of the same pathotype; the effects of the latter on lettuce are very different.*

## Lettuce mosaic virus (LMV)
## Variability of symptoms on one genotype possessing the resistance gene *mo1*[1]

**73** This healthy lettuce, not contaminated by **LMV**, displays normal growth.

**74** The **LMV 0** strain does not cause any discernible symptoms on this variety of resistant lettuce.

**75** The **LMV 1** strain can overcome resistance and causes clearing of veins and yellow mosaic of the leaves in the head.

**76** The **LMV E** strain, which can overcome resistance, is one of the most virulent and damaging strains for lettuces. We often see numerous blistered, deformed leaves, displaying a very marked mosaic pattern and vein lightening.

**• Cucumber mosaic virus (CMV)**
This other virus transmitted by aphids, although not transmitted by the seed, is also very widespread on lettuces. Just like LMV, its attacks may occur very early, in summer nurseries or after planting out. Lettuce growth is then greatly reduced, and the plants remain stunted.

CMV more frequently causes mottling and fairly subtle mosaic patterns that may pass unobserved or diminish over time as the plants grow (**77** and **78**). More marked symptoms are displayed when specific climatic conditions apply. Symptoms are all the more marked if the temperature difference between day and night is very great. This is why this virus is seen especially in autumn. In this case, a yellow mosaic pattern and extensive leaf necrosis cover the leaves almost entirely (**79**). The heads of affected plants are sometimes reduced in size.

The nature and intensity of the symptoms caused by CMV are likely to vary according to growing procedures, strains of this virus, and the type of l.s.v. cultivated. This virus is generally very difficult to differentiate from LMV. Mixed infections of LMV and CMV are frequent in lettuces.

## Cucumber mosaic virus (CMV)

**77** The lettuces located on the right of the photo have been infected by **cucumber mosaic virus (CMV)**. The plants are less developed and have a more spread out appearance compared with the healthy control plants located on the left.

**78** Chlorotic mottling and crinkling of the lamina are very often signs of the presence of **cucumber mosaic virus (CMV)** in lettuces.

**79** Symptoms are often more marked following climatic periods when there is a big difference between night-time and daytime temperatures. Yellowing of the lamina and necrotic damage, localized in patches, appears on the leaves.
**Cucumber mosaic virus (CMV)**

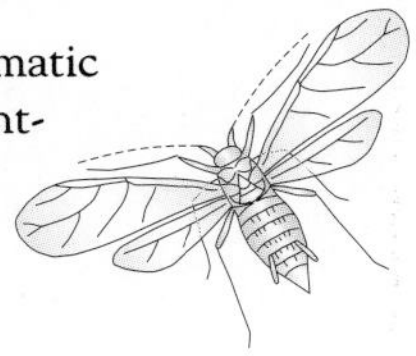

### • Broad bean wilt virus (BBWV)

The symptoms caused by this virus are relatively subtle; they can sometimes pass unobserved. Mottling of young leaves and slight discolouration of outer leaves indicate the presence of the virus in the plants. Plant growth is sometimes slowed down (**80** and **81**).

Leaf malformation, chlorotic mottling, and vein necrosis have been reported in the USA (New York). In this country, head development may not take place or the growth of the plants may remain blocked. Tolerant plants show slight mottling or no symptoms. Cold nights, alternating with mild days, aggravate the symptoms.

Some confusion in diagnosis has been reported between this virus and LMV and CMV.

**80** BBWV is not very damaging to lettuces in France. Slower growth and discolouration of the leaves characterize the presence of this virus on the plant located on the right of the photo.
**Broad bean wilt virus (BBWV)**

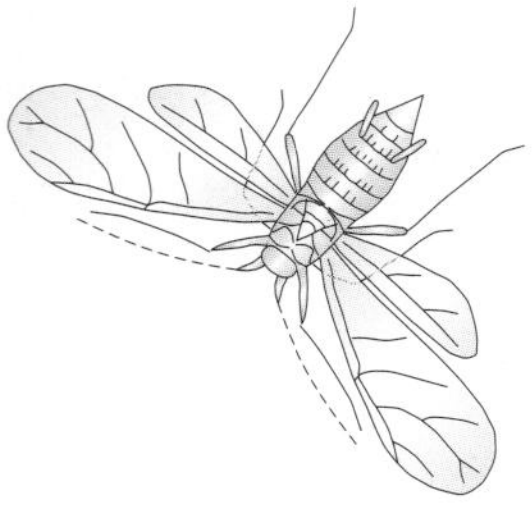

**Broad bean wilt virus = BBWV**

**81** The blade of the external leaves is slightly chlorotic; it also shows a few yellower patches. The young leaves are slightly mottled.
**Broad bean wilt virus (BBWV)**

**Table 5: Some characteristics of the principal types of viruses causing mosaic or similar symptoms on l.s.v.**

| Virus | Frequency and severity | Method of transmission (vectors and seeds) | Seasonal distribution | Spatial distribution at the start of the attack | Existence of resistant varieties |
|---|---|---|---|---|---|
| **LMV** | Very frequent and very serious. | Aphids, in a non-persistent way. (seed +)* | Summer, autumn, throughout the year if seeds are contaminated. Essentially in the open field. | Random plants or one to several groups. | Numerous resistant varieties. Take care, certain strains overcome resistance. |
| **CMV** | Frequent and sometimes serious. | Aphids, in a non-persistent way. (seed −)* | Essentially autumn open field crops. | One to several groups. | No. |
| **BBWV** | Rare and not very serious. | Aphids, in a non-persistent way. (seed −)* | Summer and autumn crops. | Isolated plants distributed at random. | Tolerant varieties. |
| **DaYMV** | Underestimated, frequency and seriousness to be defined. | Aphids, in a non-persistent way. (?) (seed ?) | All crops, more serious in the autumn and sometimes winter. Open field and under protection. | Isolated plants distributed at random. | No. |
| **MiLV** | Frequent. | *Olpidium brassicae* (seed −) | Particularly winter crops. | Isolated plants or groups of a few plants. | Tolerant cultivars. |
| **ENMV** | Virus detected very recently and therefore underestimated, frequency and seriousness to be defined. | Aphids, in a non-persistent way. (seed −) | Mainly summer crops, especially in the open field. | Plants distributed at random. | Most varieties of lettuce are resistant. Escarole and curly endive are sensitive. |
| **TuMV** | Fairly frequent, sometimes serious. | Aphids, in a non-persistent way. (seed −) | Essentially open field autumn crops. | Plants distributed at random or in groups. | Most varieties of lettuce are resistant. Escarole or curly endive are sensitive. |
| **AMV** | Rare and not serious. | Aphids, in a non-persistent way. (seed −) | Summer open field crops. | Isolated plants or in groups or grown close to fields of alfalfa. | No. |

* **seed +**: virus which can be transmitted by seed; **seed −**: virus which cannot be transmitted by seed

**82** This lettuce affected by **DaYMV** reveals poor growth; its appearance is abnormal because of numerous slightly deformed leaves.
**Dandelion yellow mosaic virus (DaYMV)**

**83** Affected leaves reveal various types of damage: chlorotic spots, necrotic damage, mild necrosis, sometimes giving the lamina a slightly 'bronzed' appearance.
**DaYMV**

**84** Endive is also affected; this escarole displays a fairly intense mosaic pattern.
**DaYMV**

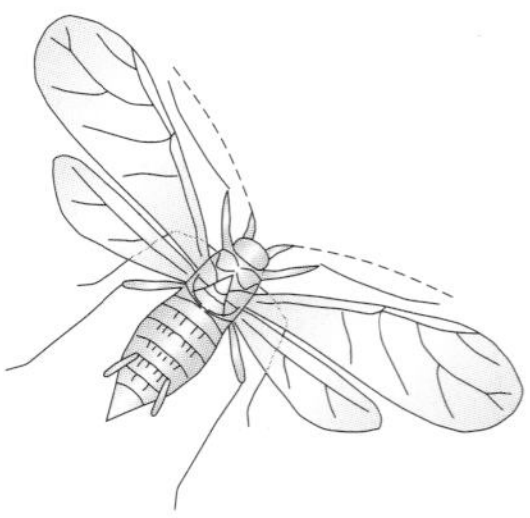

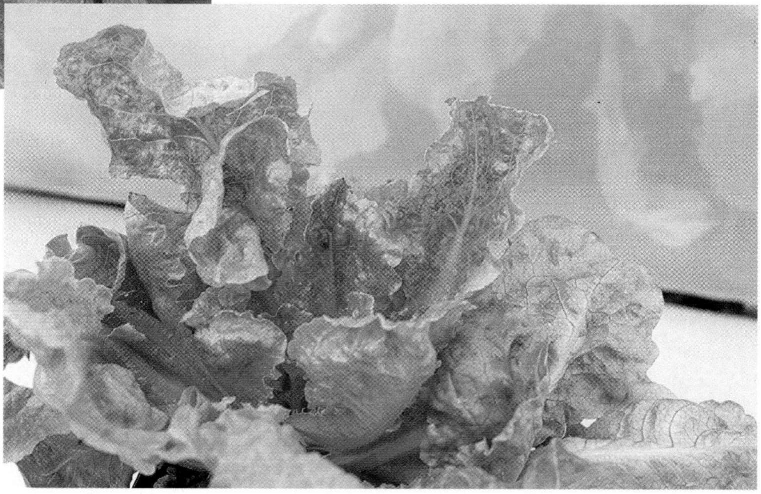

**85** In certain cases, chlorotic spots may be larger and very marked. Their location is frequently associated with the vein network.
**DaYMV**

• **Dandelion yellow mosaic virus (DaYMV)**

The symptoms and incidence of this virus on l.s.v. are still poorly understood. It began to be detected and its seriousness taken into consideration when the growing of varieties of lettuce resistant to LMV became widespread. Since then, it has become possible to observe symptoms of this virus, which up until then had been masked by symptoms of LMV.

In a fairly general way, growth of plants is reduced (**82**). Leaves may be deformed to varying degrees, particularly during autumn and winter (see the section entitled 'Abnormalities in l.s.v. growth and/or in the shape of their leaves'). On the blade we can see various abnormalities in colouration such as: chlorotic spots, vein clearing, and rather chlorotic mottling (**84** and **85**). Minuscule necrotic damage ('etches') appear on or between the veins; spots of the same type are also sometimes visible (**83**). Some young leaves have a bronzed appearance.

Lettuce, and also escarole or curly endive, can present symptoms of this kind. Damage due to **DaYMV** seems to be more serious in the open field than under protection.

Lettuce mottling has been reported in Brazil, particularly on varieties resistant to LMV. This mottling, attributed to lettuce mottle virus (LMoV), may in fact be caused by **DaYMV**.

**86** The earliness of the attack and climatic conditions influence the appearance of symptoms. These are sometimes moderate. This is the case on this lettuce, which displays fairly widespread mottling particularly visible on the old leaves.
**DaYMV**

**87** This lettuce, grown in winter, is badly affected by **DaYMV**. Its growth has been stopped and its young leaves are very deformed. The oldest leaves are rather chlorotic.
**DaYMV**

88 

89 

90 

91 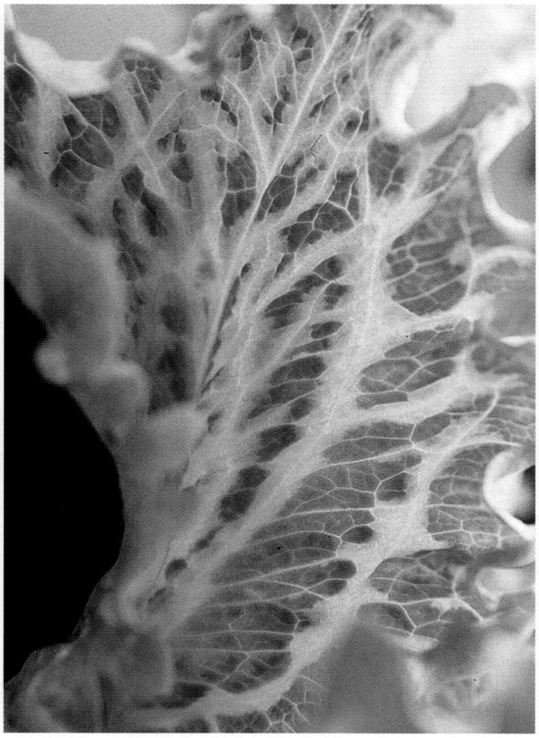

92 

**88** Several leaves of this lettuce have veins which appear broader than normal.
**Lettuce big vein virus (MiLV)**

**89** More careful observation reveals that the veins and adjacent tissues are progressively lightening.
**MiLV**

**90** The worst affected leaves ultimately become thickened; they are also blistered, crinkled and their margin is more serrated.
**MiLV**

**91** Ultimately, broad bands of tissue around the edge of the veins are affected, giving the blade its 'big vein' appearance.
**MiLV**

**92** Affected plants have a fairly twisted appearance that contrasts with that of healthy neighbouring plants.
**MiLV**

- **Lettuce big vein virus (Mirafiori lettuce virus (MiLV))**

This virus, transmitted by a non-pathogenic soil fungus (*Olpidium brassicae*), causes relatively characteristic symptoms in the lettuce. These are essentially located on the veins, which, as well as adjacent tissues, become progressively 'clear' (**89**). Ultimately, broad bands of tissue edging the veins are affected, giving a 'big vein' appearance to the leaves (**91**). This symptom gives the disease its name: **lettuce big vein disease or 'big vein'**. Once this symptom has spread to the blade, the leaves appear thicker, blistered and crinkled, with a dentate edge (**90**). The appearance of affected plants is in strong contrast with that of healthy surrounding plants (**88** and **92**).

The symptoms caused by this virus are essentially observed on lettuces grown under protection during the winter. Under these growing conditions, cold, damp soil encourages the development of the vector fungus *Olpidium brassicae*, especially if it has not been disinfected.

We should bear in mind that this fungus, visible in the cells of the root cortex (see page 119) also transmits another serious lettuce virus, **lettuce ring necrosis agent (LRNA)**. This virus is influenced by the same growing conditions as big vein disease. It is therefore hardly surprising that they are found at the same time in the same plantings and, sometimes, on the same plants. Consequently we suggest that you consult the headings 'Partially or totally chlorotic or yellow leaves', 'Yellow, orange to light brown spots and damage on leaves' and 'Abnormalities in l.s.v. growth and/or in the shape of their leaves'.

**We should point out that lettuces affected by a virus might be more sensitive to other diseases. We have noted this on batavias already invaded by MiLV; plants suffering from the virus being more systematically attacked by a bacteria: *Pseudomonas cichorii* (93 and 94).**

**93** Among these batavias, only the plant with 'big vein' associated with parasitism from **MiLV** has been attacked by ***Pseudomonas cichorii***.

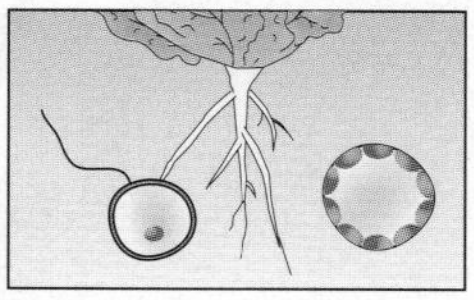

**94** On the leaves of this batavia, we can clearly see the presence of broad wet, dark brown to black lesions, characteristic of bacteriosis on lettuce, in addition to 'big vein' symptoms.
**Lettuce big vein virus (MiLV) + *P. cichorii*.**

**Lettuce big vein virus (Mirafiori lettuce virus (MiLV))**

95 A necrotic mosaic is starting to develop on this young lettuce attacked by ENMV.
**ENMV**

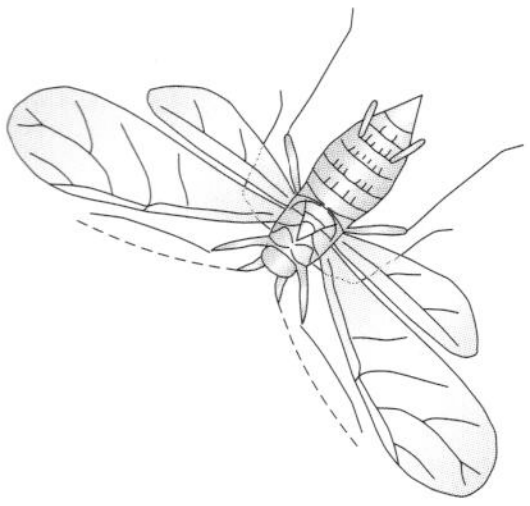

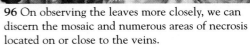

96 On observing the leaves more closely, we can discern the mosaic and numerous areas of necrosis located on or close to the veins.
**ENMV**

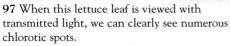

97 When this lettuce leaf is viewed with transmitted light, we can clearly see numerous chlorotic spots.
**ENMV**

98 This young infected lettuce also shows vein clearing.
**ENMV**

## Endive necrotic mosaic virus (ENMV)

- **Endive necrotic mosaic virus (ENMV)**

Another underestimated virus, ENMV was discovered in Europe in the middle of the 1990s, initially on escarole and curly endive. We detected it in France in 1999, in Provence on a crop of lettuce. Only a few varieties are sensitive.

When ENMV attacks young plants, they become stunted and their leaves are deformed, displaying broad areas of necrosis. On adult plants, we observe fairly systematically areas of mosaic that are necrotic to varying degrees (**95** and **96**). The latter are accompanied or preceded by chlorotic spots and vein clearing (**97** and **98**).

The symptoms caused by this virus are fairly comparable, on endive, to those of LMV or TuMV. Consequently, confusion in diagnosis is probable, so you need to be particularly vigilant.

- **Turnip mosaic virus (TuMV)**

This virus is encountered in summer and particularly autumn endive crops. This means it essentially affects escarole and curly endive, but also a few varieties of iceberg lettuce. In Provence we also isolated it on certain cultivars of batavia, but never on romaine. It does not seem to attack butterhead lettuce.

Young infected plants reveal deformed leaves that have chlorotic, circular to elongated, lesions located between or next to the veins (**99**). If attacks are very early, the growth of the plants is halted and they subsequently die.

Comparable damage is found on older plants. Initially we note mottling, then limited chlorotic lesions of variable size on the lower leaves. These lesions progressively expand and converge, with the blade becoming entirely yellow (**100–103**). Locally, reddening (**104**) and/or brown necrotic spots appear (**105**).

Ultimately, the plants are very chlorotic and confusion in diagnosis is possible, particularly with beet western yellows virus (BWYV). Unlike BWYV, the leaves of plants infected by TuMV have discoloured veins and their blades are normal to soft in consistency, and under no circumstances brittle. These few observations can prevent errors in diagnosis.

99

100

101

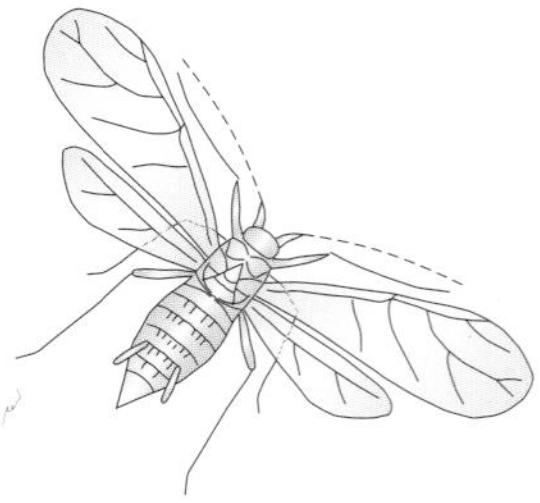

## Turnip mosaic virus (TuMV)

**99** This young lettuce has deformed leaves that curve inwards; chlorotic damage can also be seen when the leaves are viewed with transmitted light.
**TuMV**

**100** The growth of this older plant has also been disrupted; we can see a very marked yellow mosaic.
**TuMV**

**101** Yellow mosaic is essentially affecting the lower and intermediate leaves of this escarole.
**TuMV**

102   103

104  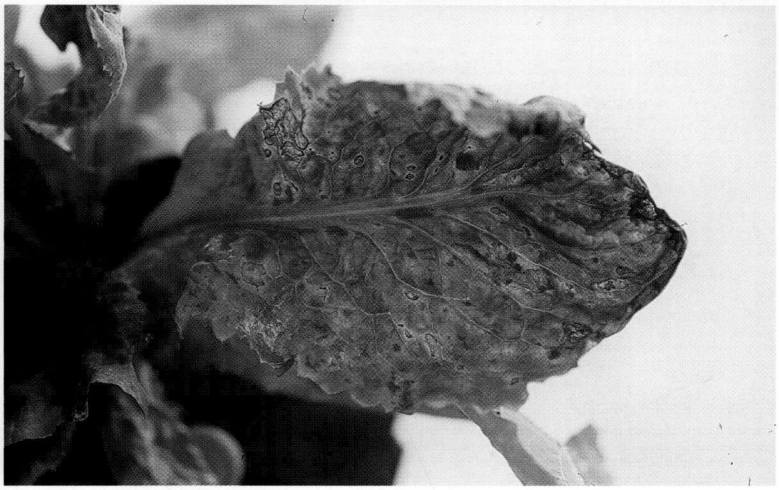 105

**102** The morphological variability of the chlorotic lesions is clearly emphasized on these lettuce leaves. They are small in size, and sometimes circular.
**TuMV**

**103** Chlorotic lesions extend and converge on the lamina. Entire sectors of the blade take on a bright yellow colouration, whereas the veins become discoloured.
**TuMV**

**104** On this variety of batavia, reddening of the lamina accompanies the appearance of chlorotic lesions.
**TuMV**

**105** Ultimately necrotic patches, brown to dark brown in colour, appear in the most chlorotic sectors of the lamina.
**TuMV**

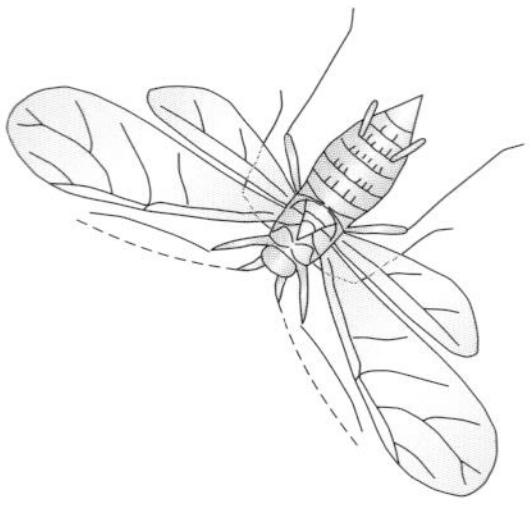

106 In numerous hosts, such as l.s.v., AMV causes symptoms which are always bright yellow to white in colour, no matter what their location and nature (spots, rings, arabesques, bands, and so on).
**AMV**

107 Yellowing may be inter-veinal in the form of spots that spread progressively.
**AMV**

108 In certain cases, rings and vein yellowing reveal the presence of this virus in l.s.v.
**AMV**

109 Symptoms are equally impressive on curly endive. Yellowing of the lower leaves is also apparent.
**AMV**

## Alfalfa mosaic virus (AMV)

**• Alfalfa mosaic virus (AMV)**

AMV is responsible for bright yellow mosaic patterning on lettuces, sometimes called 'calico' mosaics (**106**). Thus we can observe rings and/or spots, yellow to white in colour, on the lower leaves (**107** and **108**). Yellowing, then whitening sometimes begins close to the broadest veins and then spreads. These spots are often irregular and small in size. This discolouration does not under any circumstances affect the whole lamina. These symptoms are very different from those caused by TuMV and viral yellowing (BWYV, BPYV), so the diagnosis of this virus does not normally pose any problem.

In general, attacks of AMV likely to cause damage to lettuces are only noted in plots located close to infected alfalfa fields. Alfalfa is in fact a preferred host of this virus; inoculum pressure can therefore be very strong in areas where this species is grown.



**Table 6: Other principal types of virus responsible for leaf discolouration on lettuces**

| Virus | Symptoms | Vectors | Viral particle shape |
|---|---|---|---|
| **Bidens mottle virus (BiMoV)**<br>***Potyviridae Potyvirus*** | Clearing of leaf veins, mottling sometimes developing into vein necrosis. Plants attacked early remain dwarfed. | Several species of aphids, in a non-persistent way. *Myzuspersicae* is particularly effective. | 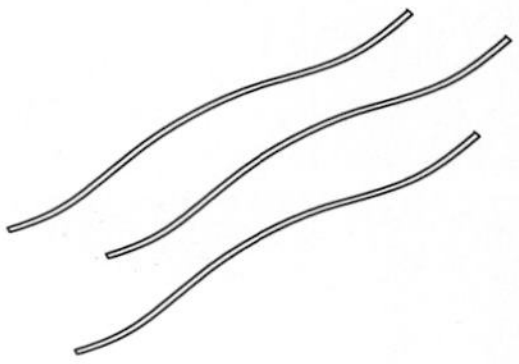 |
| **Sowthistle yellow vein virus (SYVV)**<br>***Rhabdoviridae Nucleorhabdovirus*** | Clearing of veins especially around the periphery of the lamina, which may lead to confusion with 'big vein' symptoms. Bands of dark green tissue all along the veins ('vein banding'). Plants infected early grow very poorly. | Transmitted by the annual sowthistle (aphid) *Hyperomyzus lactucae*, in a persistent way. | 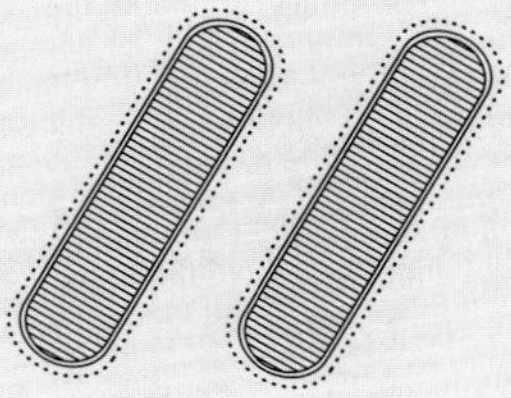 |
| **Tobacco rattle virus (TRV)**<br>***Tobravirus*** | Mottling, spots, rings, and irregular patterns, varying degrees of yellow. Growth of plants is slowed down and the leaves are flatter than normal. | Several species of nematodes of the *Trichodorus* spp. and *Paratrichodorus* spp. genus. These nematodes remain infectious for several months, or even several years. (Weak transmission by the seed in a few types of weeds.) | 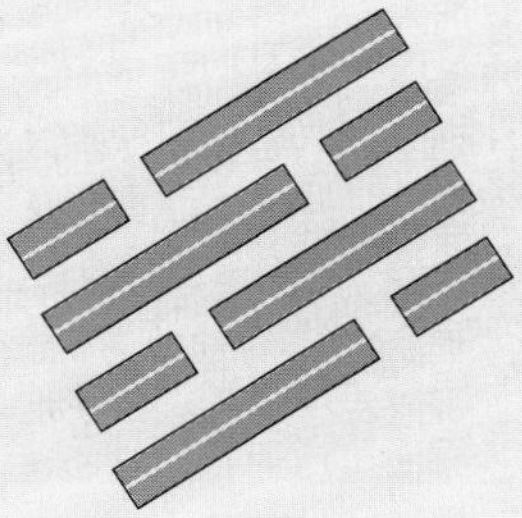 |
| **Tobacco ring spot virus (TRSV)**<br>***Comoviridae Nepovirus*** | Widespread yellow mottling of lower leaves. Irregular rings and patterns, yellow in colour, varying in intensity, comparable to that observed with TRV. Plants that have been attacked are frequently stunted and very squat. | Transmitted by nematodes, ectoparasites of the *Xiphinema* genus, particularly *Xiphinema americana*. In certain hosts, it can be transmitted after feeding by grasshoppers and thrips. (Transmission by seed is possible in lettuce.) |  |

110

111

112

113

114

## A few examples of yellowing on lettuce leaves

**110** This lettuce leaf is scattered with numerous small chlorotic spots.
**Turnip mosaic virus (TuMV)**

**111** Several dark brown necrotic spots are visible on this leaf. They are surrounded by a yellow halo varying in intensity.
***Septoria lactucae***

**112** On this lettuce leaf, we can see yellowing of the blade starting from the veins.
**Chemical injury**

**113** This leaf of curly endive has only gone partially yellow on one side due to the effect of a **vascular disease**.

**114** The blade of this lower leaf displays peripheral yellowing which is progressively becoming widespread. The veins remain green.
**Beet pseudo-yellows virus (BPYV)**

# Partially or totally chlorotic or yellow leaves

## Possible causes

- *Pythium tracheiphilum* (see page 211) (fact file 11)
- Phytoplasm from the Aster yellows group (fact file 20)
- Lettuce ring necrosis agent (LRNA) (fact file 35)
- Beet western yellows virus (BWYV) (fact file 23)
- Turnip mosaic virus (TuMV) (fact file 27)
- Dandelion yellow mosaic virus (DaYMV) (fact file 26)
- Alfalfa mosaic virus (AMV) (fact file 24)
- Endive necrotic mosaic virus (ENMV) (fact file 28)
- Beet pseudo-yellows virus (BPYV) (fact file 30)
- Tomato spotted wilt virus (TSWV) (see page 152) (fact file 32)
- Other types of virus (LCV, LIYV, BYSV, SYNV, LNYV) (fact files 29 and 31)
- *Pemphigus bursarius* (see page 205)
- Genetic abnormalities
- Nutritional disorders
- Various types of chemical injury



Partial or widespread yellowing of one or several leaves, also known as chlorosis, is a symptom frequently observed on lettuces and is not very specific to any given disease. It can in fact appear in a wide variety of ways:

- it can be limited to a small surface area in the form of a spot (**110**) (see page 111) or be combined with a necrotic area, surrounding it with a yellow halo which varies in intensity (**111**) (for example, see page 102);
- it can sometimes affect one side of one leaf, this unilateral yellowing may be characteristic of a vascular disease (**113**) (see page 207);
- it can develop from the veins (**112**) or between the veins (**114**) (inter-veinal chlorosis).

It may also start with young leaves of the shoot or with old leaves from the base of the plants.

Sometimes yellowing can be widespread over the whole plant. It sometimes takes on very different intensities, going from light green to bright yellow, occasionally developing until the leaves are white or red.

It is a symptom that reveals an abnormality in plant function, frequently as a result of the following:

- one or several parasitic attacks occurring either directly and locally on the leaf (airborne disease for example), or on other parts of the plants, particularly on the roots or the stem;
- viral attacks (yellowing in particular) or non-parasitic diseases, such as a deficiencies or chemical injury.

***It is often difficult to determine the cause of yellowing: consequently be very careful when carrying out your diagnosis.***

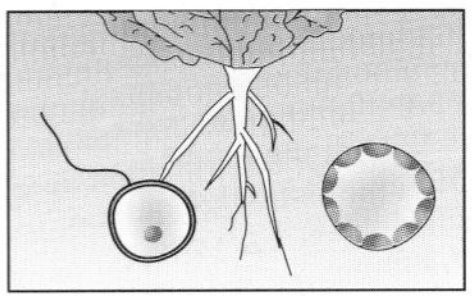

115 Small spots and a few chlorotic rings are visible on the lower or intermediate leaves of this lettuce.
**LRNA**

116 On the lower face of the blade this damage is oily to damp and orange in colour.
**LRNA**

117 Ultimately, whole leaves become affected; on butterhead lettuce they may then have an intense yellow colour.
**LRNA**

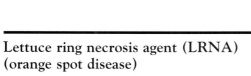

118 Diseased tissues quickly become necrotic and dry out.
**LRNA**

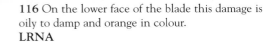

## Lettuce ring necrosis agent (LRNA) (orange spot disease)

## Arguments in support of the diagnosis

- **Lettuce ring necrosis agent (LRNA)**

This virus is responsible for **orange spot disease**. It causes chlorotic spots and rings on the upper face of the lamina of the lower leaves of lettuces (**115**). Under the leaves, this damage is oily and rather orange in colour; this symptom lies at the origins of the French name for the disease (**116**). If conditions are favourable, on butterhead lettuce, a large proportion of the lamina takes on a fairly marked yellow colour (**117**); affected tissues quickly become necrotic and dry out (**118**).

LRNA is transmitted by a soil fungus: *Olpidium brassicae*. It only affects lettuces grown in the winter under protection. Heavy, poorly drained, cold soils encourage the development of the vector fungus and the expression of the symptoms of this virus. It is fairly frequent to observe LRNA in plots suffering from **big vein disease (MiLV)**. As their respective symptoms are very different, confusion in diagnosis is not possible (also see pages 63 and 119). This is not the case with **dandelion yellow mosaic (DaYMV)**. In winter conditions this virus, transmitted by mites, causes stunting of plants accompanied by yellowing of the leaves (**119**). The incidence of this virus at this time of the year is not very common as it has only been detected in France fairly recently (see page 61). It is rare elsewhere.

**119** Within these healthy lettuces, one plant displays several yellow necrotic leaves; growth of the leaves in the heart has stopped and they are deformed.
**Dandelion yellow mosaic virus (DaYMV)**

## Beet western yellows virus (BWYV)

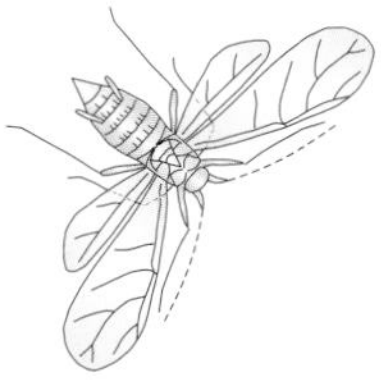

**120** If infection is very early, yellowing between the veins may be observed on young plants. On this lettuce plant two lower leaves reveal this type of symptom. The earlier the attack, the more effects yellowing will have on

## Beet pseudo-yellows virus (BPYV)

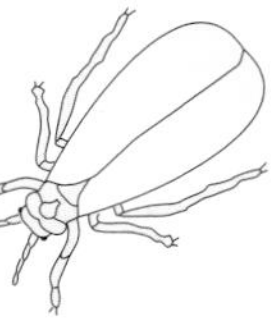

**121** Yellows may occur after planting; this has happened with this lettuce that has been attacked at the 15-leaf stage. Whatever the stage of plant development, yellowing always occurs on lower leaves.

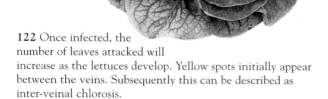

**122** Once infected, the number of leaves attacked will increase as the lettuces develop. Yellow spots initially appear between the veins. Subsequently this can be described as inter-veinal chlorosis.

**123** Consequently, a high proportion of leaves go yellow. Yellowing has spread to the blade and has taken over a large number of leaves.

> **Yellows of salad vegetables**
> Symptoms are similar whatever the stage of plant development

**124** If the symptoms of yellows are very advanced, the lower leaves may be completely discoloured, sometimes white, and may become necrotic at the edge of the lamina. The veins and one band of adjacent tissue remain green. Affected leaves break easily when folded.

• **Summer and winter yellows of l.s.v.**
**Beet western yellows virus (BWYV)**
**Beet pseudo-yellows virus (BPYV)**

Two different types of virus are often responsible for l.s.v. yellowing. As the symptoms are identical, for some time they were confused with those of certain nutritional disorders (**120–129**). They occur during the year in two different growth periods, which allows them to be distinguished fairly easily.

**BWYV**, transmitted by mites in a non-persistent way, is responsible for **summer yellows** of l.s.v. Initially we see the appearance of chlorotic spots between the veins of lower leaves. These spots progressively expand; ultimately they can be described as inter-veinal yellowing in order to characterize the effects of this virus on different types of l.s.v. The veins and a limited band of adjacent tissue remain green. The lower leaves affected in this way may become totally discoloured and go white. In addition they are thicker and break when folded.

**BPYV**, transmitted by aphids (*Trialeurodes vaporariorum*), essentially attacks under protection in winter. It causes symptoms that are completely comparable to those of BWYV. In the same way as for the latter, inter-veinal yellowing of the lower leaves indicates the effects of BPYV. This virus is associated with the presence of aphids in greenhouses. In fact, serious epidemics are often encountered when lettuce/cucumber, and, to a lesser degree, lettuce/melon rotation, is being practised. These two hosts are also very sensitive to BPYV, displaying the same yellowing of the lower leaves.

**125** The chlorotic parts of the blade sometimes take on a reddish colour. This is the case, for example, in periods of low temperatures.

**126** On batavia, as on other lettuces, it is the lower leaves that begin to go yellow first.

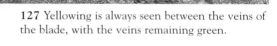

**127** Yellowing is always seen between the veins of the blade, with the veins remaining green.

**128** The same is true for the romaine ...

**129** ... and the escarole.

## Lettuce yellows
## The symptoms are similar whatever the type of lettuce

**Table 7: A few characteristics of the principal diseases responsible for yellowing on l.s.v. leaves**

| Diseases | Frequency of the disease | Number of plants affected | Distribution on the plot | Location of initial symptoms |
|---|---|---|---|---|
| **Beet western yellows (BWYV)** | Fairly frequent, essentially in summer in the open field. | From a few random plants to numerous plants. | Large groups or throughout the plot. | On lower leaves. |
| **Beet pseudo-yellows (BPYV)** | Fairly frequent, especially under protection in winter. | From a few random plants to numerous plants. | Groups limited to more or less scattered areas of the crop. | On lower leaves. |
| **Orange spots (LRNA)** | Rare except in France where it is fairly frequent, essentially protection in winter. | Variable depending on the pressure from *Olpidium brassicar*, a soil fungus, which is a virus vector. | Isolated plants dispersed over the whole plot; more or less scattered zones of the crop. | Mainly on lower leaves. |
| **Dandelion yellow mosaic (DaYMV)** | Frequency unknown, virus recently discovered. | Not very numerous under protection during the winter. | Isolated plants distributed at random. | Mainly on intermediate leaves. |
| **Nutritional disorders** | Not very frequent. | Fairly numerous to very numerous. | Fairly widespread. | On lower or upper leaves. |
| **Chemical injury** | Frequent. | Variable. | Variable. | On lower or upper leaves. |
| **Genetic abnormalities** | Rare. | Very few (often one isolated plant). | Random. | On lower or upper leaves. |

Table 8: Other types of virus responsible for leaf yellowing on l.s.v.

| Virus | Symptoms | Vector | Viral partical shape |
|---|---|---|---|
| **Lettuce chlorosis virus (LCV)**<br>***Closteroviridae Crinivirus*** | Clearing of veins; yellowing, reddening and rolling of leaves. Plants may remain dwarfed. | Transmitted by *Bemisia tabaci* and by *B. argentifolii* with the same efficacy in a semi-persistent way. | 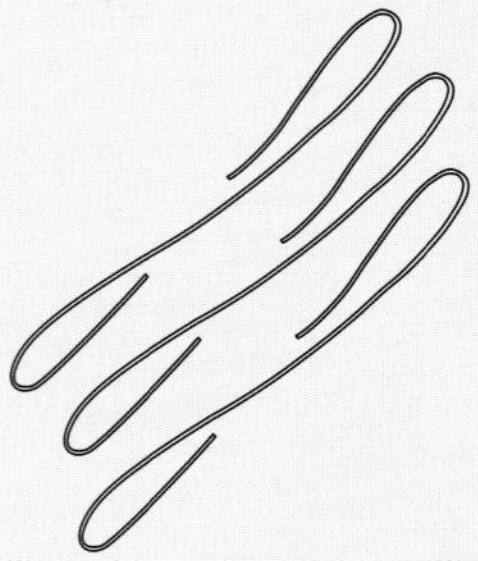 |
| **Lettuce infectious yellows virus (LIYV)**<br>***Closteroviridae Crinivirus*** | Clearing of veins; yellowing, reddening and rolling of leaves. Plants may remain dwarfed. | Transmitted by *B. tabaci* in a semi-persistent way. | 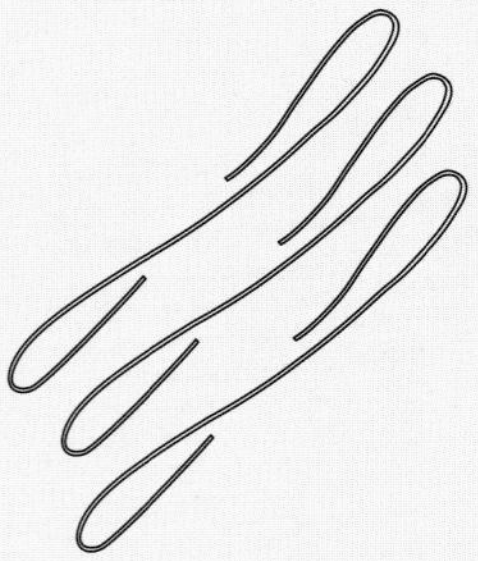 |
| **Beet yellow stunt virus (BYSV)**<br>***Closteroviridae Closterovirus*** | Progressive yellowing of old leaves and abrupt collapse of plants that ultimately die. A longitudinal cross section of the taproot and stem reveals the presence of internal necrotic lesions. | Effectively transmitted by *Hyperomyzus lactucae,* in a semi-persistent way. *Myzuspersicae, Macrosiphum euphorbiae,* and so on may be vectors in an unpredictable way. | 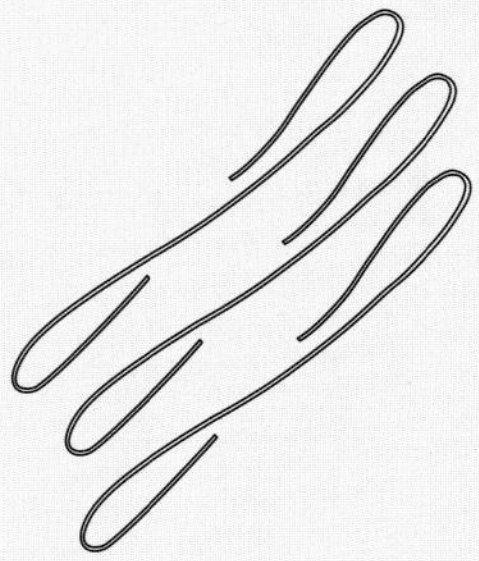 |
| **Sonchus yellow net virus (SYNV)**<br>***Rhabdoviridae Nucleorhabdovirus*** | Clearing of veins and yellowing of leaves. Broad yellow spots, interveinal, on lower leaves. Plants infected early may remain dwarfed. | Transmitted by the aphid *Aphis coreopsidis,* in a persistent way. | 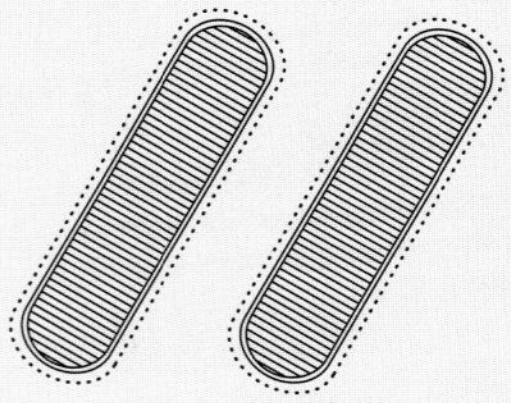 |
| **Lettuce necrotic yellows virus (LNYV)**<br>***Rhabdoviridae Cytohabdovirus*** | Pale green to chlorotic plants, with a fairly flattened appearance. Mottling sometimes present on lower leaves. Plants that have been attacked before formation of the head may present internal necrotic damage and | Essentially transmitted by the annual sowthistle aphid *H. lactucae,* in a persistent way | 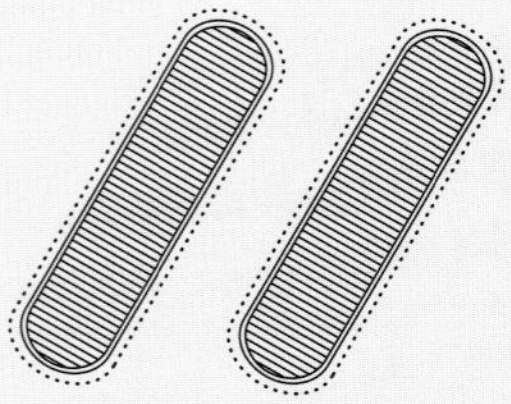 |

• **Phytoplasma from the Aster yellows group**

This micro-organism, transmitted by several species of leafhopper, causes various symptoms which can be classified as abnormalities in leaf shape or colouration. As far as changes in colouration are concerned, we note in particular chlorosis of the lamina affecting the leaves in the head and/or the crown, depending on how early contamination takes place. In addition to going yellow, leaves are small in size and deformed (twisted, rolled, and so on) (see page 37).

Deposits of latex, pinkish to brown in colour, are visible under the veins of affected leaves. Intense necrosis of the phloem can be seen after carrying out a transversal cross section of the stem.

Serious attacks on lettuce have been observed in a few countries: several US states, Canada, and Europe (although only in Italy).

**Table 9: Symptoms of deficiencies appearing firstly on lower leaves and subsequently becoming widespread to all leaves of l.s.v.**

| Deficiency | Symptoms | Principal functions disturbed |
|---|---|---|
| **Potassium (K)** | Chlorotic spots, more or less marked, located on the periphery of the blade. Locally, areas of tissue ultimately go brown and necrotic. Leaves are also flatter. Their edge may be slightly curved in towards the lower face of the blade. Plant growth is reduced to a greater or lesser degree. | Osmotic regulation of the plant (ionic balance of cells, water retention in tissues, water transport), regulation of cellular pH (element of plant quality: thicker walls). |
| **Magnesium (Mg)** | Inter-veinal chlorosis beginning with the periphery of the blade, which may ultimately result in whitening of the leaves, while the big veins tend to retain their green colouring. Lettuce growth reduced to a greater or lesser degree depending on the severity of the deficiency. | Synthesis of chlorophyll constituents, co-factor of numerous enzymes. |
| **Nitrogen (N)** | Leaves are pale green, and the oldest ones display a more marked yellowing, tending to become necrotic and dry out. Plant growth is restricted. | Protein synthesis (constituent of amino acids, proteins, nucleic acids, and so on), constituent of chlorophyll molecule, and so on. |
| **Phosphorous (P)** | Leaves are dark green, and sometimes have a bronzed to purplish appearance, particularly in red cultivars. The oldest leaves go yellow and dry out. Some of these may be more filiform and give the plant a rosette appearance. Plants generally grow poorly. | Constituent of enzymes, proteins, phospholipids, and nucleic acids; vital role in the life of the plant and its reproduction. |
| **Molybdenum (Mo)** | Yellowish brown necrotic damage developing around the blade periphery of old leaves, giving the leaves a raised edge. Growth of vegetation and root system may be significantly reduced. | Metal component of two enzymatic systems: nitrate reductase and nitrogenase. |
| **Copper (Cu)** | Inter-veinal chlorosis beginning around the blade periphery. Subsequently tissues may become necrotic and dry out and blade edge may become raised. Plants are sometimes less turgid and their growth is reduced. | Element involved in numerous enzyme systems, in the formation of the cell wall, transport of electrons and oxidation reactions. |

**Table 10: Symptoms of deficiency mainly appearing on the shoot and young leaves of l.s.v.**

| Deficiency | Symptoms | Main functions disturbed |
|---|---|---|
| **Calcium (Ca)** | Young leaves are sometimes darker in colour and are more folded than normal. Grey to brown lesions can appear around blade edge, leading to the death of certain leaves after spreading extensively. Plant growth is greatly reduced, particularly leaves in the head, which gives the lettuces a distinctive appearance. | Constituent of the cell wall in the form of pectate and calcium oxalate, involved in cell elongation and division, influences cellular pH, structural stability and permeability of cell membranes. |
| **Boron (Bo)** | Young leaves are thicker and rigid, slightly mottled around the lamina edge. The terminal bud may become necrotic or remain stunted. Lettuces are often darker than normal and their growth is greatly reduced. | Involved in the transport of sugars across cell membranes and in the synthesis of cell wall constituents. It influences transpiration, development and elongation of cells, synthesis of proteins in particular, and interacts with auxins. |
| **Manganese (Mn)** | Lettuces are slightly chlorotic and their growth is only slightly restricted. Young leaves may be scattered with small necrotic patches. In the case of severe deficiency, old leaves become chlorotic, but their veins remain green. | Associated with the development of oxygen during photosynthesis, electron transport, constituent of several enzyme systems. |
| **Sulphur (S)** | Young leaves are pale green, smaller, thick and rigid. Consequently, affected plants have a rosette appearance and vein colouration is virtually unchanged. | Involved in the composition of two amino acids indispensable to protein synthesis, involved in the formation of vitamins and hormones and in oxidation-reduction reactions, constituent of several co-enzymes and certain lipids. |
| **Iron (Fe)** | Inter-veinal chlorosis of young leaves, followed by older leaves whose principal veins remain fairly green and contrast with the inter-veinal tissue. Ultimately the whole blade may discolour. Lettuces are pale green in colour and their growth is more or less halted. | Essential in the synthesis of chlorophyll, is involved in fixing nitrogen, photosynthesis, electron transfer and in several enzyme systems mainly controlling respiration. |

* **Nutritional disorders**

As with numerous cultivated plants, l.s.v. require various mineral elements if they are to grow well into quality products. Their short cycle and rapid growth make their fertilization fairly complex (particularly feeding with nitrogen). This is particularly influenced by the following:

* the root system of l.s.v., which is fairly small and fairly superficial;
* the richness and balance of the soil in terms of fertilizing elements (soil analyses are therefore essential);
* the type of l.s.v. grown and the variety chosen;
* density of planting;
* the nature of the irrigation system and the way it is managed;
* and any other factor disturbing the growth of plants.

Faced with this situation, in the field we may sometimes encounter plants suffering from deficiencies or excesses of nutritional elements. These problems are classified as **'non-parasitic diseases'** and are grouped together under the terminology **'nutritional disorders'**. The latter are revealed fairly frequently by yellowing, the nature and distribution of which, varies depending on the type of lettuce. In order to make you aware of the symptoms they may cause and the physiological functions they are likely to disturb, we have described, in *Tables* 9 and 10, several types of deficiency sometimes encountered in l.s.v.

When the word deficiency is used, we often tend to associate true deficiencies with induced deficiencies.

**True deficiencies** (element in a too small a quantity in the soil) are increasingly rare; their visual diagnosis is very delicate because, without exception, the symptoms they cause are discolouration and leaf yellowing of varying degrees of intensity, aspects which are very difficult for a non-specialist to assess.

In the majority of cases this concerns **induced deficiencies** (elements which are present yet not available), whose diagnosis is difficult. In addition to determining the nature of the deficiency, the cause or causes must also be identified. The latter may be diverse, for example incorrect irrigation (too much or not enough water), the temperature or pH of the soil is too high or too low, root systems are in poor condition, and so on.

If you are confronted with this type of symptom, you must not reach too hasty a conclusion in favour of any particular deficiency without having spoken to your technician or without having **consulted a specialist** and **carried out the essential physical and chemical soil and plant analyses**.

***Deficiencies are especially likely to occur in crops which have been empirically fertilized, or in the absence of any soil analysis.***

### • Various types of chemical injuries

Among the pesticides used in agriculture, **herbicides** and, to a lesser degree, **insecticides** and **fungicides**, are likely to cause damage to lettuces, which is sometimes extensive.

We suggest that you consult a set of photographs showing abnormalities of colouration, essentially yellowing, caused by certain herbicides on lettuce leaves (**130–139**). This yellowing can vary in intensity and distribution over the leaves:

- yellowing in spots, in areas varying in extent, sometimes developing into tissue necrosis;
- yellowing of veins and adjacent tissues;
- yellowing of young leaves from the shoot;
- diffuse inter-veinal yellowing of the lamina;
- diffuse yellowing of the whole lamina;
- yellowing and drying out between the veins, developing rapidly;
- more or less uniform yellowing of the lamina between the veins, sometimes developing into whitening;
- whitening of the lamina.

The use of a herbicide on a crop, or close to one, is never a completely harmless operation. The risk of chemical injury can never be completely excluded (consult *Table 11*).

Other pesticides, for example mixtures of **insecticides** and **fungicides**, substances such as **fertilizer**, can also cause chemical injury to lettuce. They also cause yellowing, as well as other symptoms which you will find listed on pages 39, 109, and 155.

The origins of chemical injury are fairly difficult to determine. In fact, the grower very often rejects the possibility of having made an error or having caused the damage which has been detrimental to the plant. Studying the distribution of the symptoms caused by this chemical injury over time and space can, in the majority of cases, allow its cause to be confirmed.

## Distribution of symptoms over time

The period between supplying the product which initiates the chemical injury and the appearance of the initial symptoms may vary considerably:

- very short (the relation of cause and effect is rapid), immediately after application of a pesticide on the crop or close to it (in the form of browning);
- fairly long as is the case, for example, of poor choice of previous annual or perennial crop (from which weeds have been eliminated using a herbicide which persists in the soil, or unsatisfactorily leached following a dry winter; a perennial crop which has been subjected to weedkiller for several years, with this situation leading to product accumulation in the soil) or after supplying straw from a cereal crop which has undergone herbicide treatment or manure made from straw of this same kind.

130

131

132

133

134

## Examples of leaf yellowing associated with the effects of herbicides (herbicide injuries)

130 Small chlorotic spots starting on or close to the veins.
**Chemical injury**

131 Widespread yellowing and blocked growth of young leaves.
**Chemical injury**

132 Marked, irregular inter-veinal yellowing of a high proportion of the lamina.
**Chemical injury**

133 Uniform yellowing of the extremity of old leaves.
**Chemical injury**

134 Yellowing of the lamina starting from the veins.
**Chemical injury**

**Table 11: Some characteristics of major types of herbicides used in agriculture**

| Types of herbicide and examples | Principal symptoms observed on the plants | Main cell damage and functions disturbed |
|---|---|---|
| **Photosynthesis inhibitors (diquat, paraquat**<br><br>*(leaf herbicides, on contact migrate very little)* | • Wilting, drying out, leaf necrosis. | • Damage to cell membranes (plasma, tonoplast, and so on).<br>• $CO_2$ assimilation halted.<br>• Inactivation of electron transfer system.<br>• Increased membrane permeability.<br>• Reduction of chlorophyll and carotenoid content, and so on. |
| **Pigment synthesis inhibitors (carotenes) (aminotriazole, pyridazinones)**<br><br>*(leaf herbicides, migrate into the plants)* | • Slowed growth.<br>• Formation of light yellow to white (albino) stems and leaves. | • Selective action on chloroplasts which are deprived, to a greater or lesser degree, of their internal lamellae system.<br>• Inhibition of carotene synthesis.<br>• Subsequent disappearance of chlorophyll which is no longer protected.<br>• Reduction of nitrogenous metabolism.<br>• Reduction of certain enzyme activity (catalase, and so on). |
| **Cell division inhibitors (anti-mitotics) (carbamates, pendimethalene, butraline, carbetamide, propyzamide)**<br><br>*(herbicides incorporated in the soil, vapourizing, metabolic poisons similar to colchicine)* | • Halting of growth and swelling of meristems.<br>• Root extremities form club shape.<br>• Inhibition of lateral root formation. | • Inhibition of mitosis which remains blocked at the metaphase.<br>• Absence of microtubules. |
| **Fatty acid inhibitors (thiocarbamates, butam, difenamide, propyzamide)**<br><br>*(herbicides incorporated in the soil, absorbed by the leaves and roots)* | • Inhibition of plant germination and emergence.<br>• Reduced growth.<br>• Leaf malformation and appearance of necrosis around lamina margin, and so on. | • Reduction in the quantity of epicuticular waxes.<br>• Inhibition of fatty acid synthesis.<br>• Increase in cuticle permeability.<br>• Increased water absorption by roots.<br>• Inhibition of gibberellic acid synthesis.<br>• Inhibition of protein synthesis, and so on. |
| **Herbicides with auxinic action (hormonal) (pichlorane, MCPA, 2,4-D, 2, 4, 5-T, dicamba)**<br><br>*(herbicides absorbed by the leaves, migrating in the phloem towards the meristems, high concentrations of contact fungicides)* | • Slight wilting of plants.<br>• Numerous malformations and fasciation of the whole plant.<br>• Leaf malformation and appearance of necrosis around the edge of lamina, and so on. | • Halting of primary meristem activity and activation of secondary meristems.<br>• Hypertrophy of vascular tissues.<br>• Excretion of $H^+$ ions, intake of potassium (K) and water.<br>• Activation of parietal hydrolases.<br>• Increase in ADN synthesis and proteins (enzymes).<br>• Increase in ethylene production. |
| **Amino acid synthesis inhibitors (glyphosphate, sulfonylures-chlorsulfuron, imidazolines)**<br><br>*(leaf herbicides, typically symplastic, rapidly degraded in the soil, action is independent of photosynthesis)* | • Slowed growth.<br>• Chlorosis of lamina and leaf malformation.<br>• Appearance of leaf necrosis, and so on. | • Disorganization of chloroplasts in apical parts.<br>• Reduction of chlorophyll content.<br>• Reduction of auxin content of treated tissues.<br>• Disruption of potassium, calcium, and magnesium absorption.<br>• Inhibition of 5-enolpyruvylshikimate-3-phosphate synthetase and, therefore, the formation of phenylalanine, tyrosine, and tryptophan. |

135

136

138

137

139

## Pesticide injuries

**135** On some of the leaves of all these lettuces (as is often the case with chemical injury) we can note the presence of chlorotic damage around the edge of the lamina.
**Chemical injury (fungicide)**

**136** Pesticides often accumulate around the edge of the lamina. It is not surprising to find large portions of the blade chlorotic and necrotic, as a result of **chemical injury (fungicide)**.

**137** Out of the two varieties of oak leaf cultivated, only one has suffered from the last treatment carried out: a mixture of an insecticide and a fungicide.
**Chemical injury**

**138** When the leaves are turned over we can see that the damage is much more widespread than it appeared on the lamina. Numerous orange to brown lesions are scattered over the latter.
**Chemical injury (fungicide)**

**139** The outermost leaves have become progressively yellow and have dried out.
**Chemical injury**

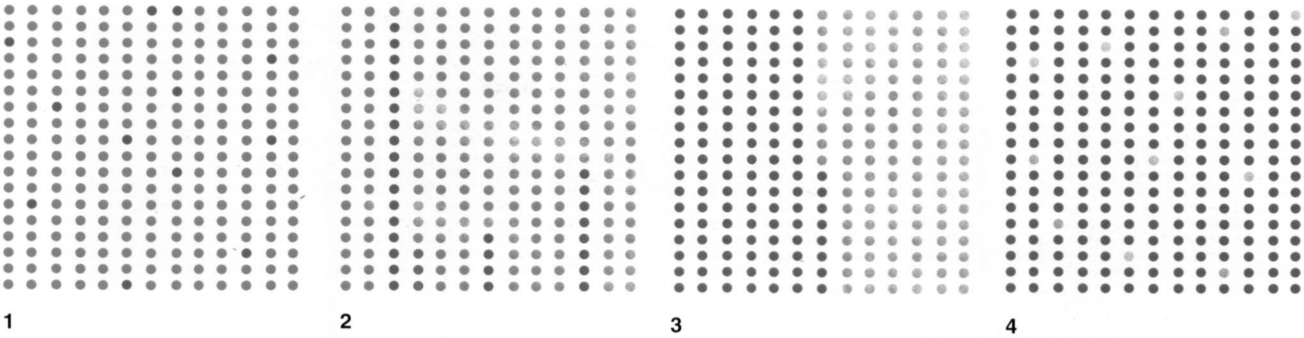


**Figure 4: Possible distributions of l.s.v. suffering from chemical injury within the crop** (healthy plant in green, diseased plant in red)



**Spatial distribution of symptoms**

This may vary depending on the composition of the phytotoxic compound, the way it is applied and its location on the plant (see fig. 4).

- If the phytotoxic compound has been applied to the leaves (leaf herbicide, insecticide or fungicide which has been overdosed or applied under incorrect conditions, and so on), the distribution of diseased plants may be:
  - widespread and uniform (4);
  - at the beginning of the row (2);
  - on one side of the plants.

If the compound is present in the soil in the form of residues (root herbicide, and so on), the distribution of affected plants may be:

- widespread and more or less uniform (4);
- distributed randomly over the whole plot (1).

**There are differences in sensitivity between varieties and types of l.s.v. (137).** If you are growing several varieties at the same time, you may then observe sectorial distribution (3) of sick plants and healthy plants corresponding to different varieties.

We recommend that you also investigate all the weeds still present in the crop or other plants being grown close by which may have suffered the same chemical injury, and may therefore be showing the same symptoms. If this is the case, this partly confirms the hypothesis of chemical injury.

In the presence of chemical injury, we suggest that you ask yourself the following questions:

- Were weeds eliminated from the previous crop using persistent herbicides?
- Have herbicide treatments been applied close to your crop?
- Did you rinse your application equipment properly?
- Do you keep your sprinkler properly maintained (cleaning, calibration, and so on)?
- Did you use the correct product at the correct dose?
- Did you follow the instructions for use shown on the packaging?
- Did you mix incompatible products with each other or too many products together?
- Were the applications carried out under unsatisfactory conditions (strong wind, temperatures which are too low or too high)?

N.B. Irrigation water may be polluted by a herbicide.

**What should you do following chemical injury?**

Although there is no miracle solution in this situation, you can adopt the following measures:

- clearly define the origin of the chemical injury;
- prevent this recurring;
- do not remove the plants immediately, look after them in the normal way and observe their development; the latter will depend above all on the nature, dose, and persistence of the product or products responsible, the stage of l.s.v. growth, the type and variety being grown.

No other specific measure can be recommended.

***Chemical injury causes other symptoms, in addition to yellowing, and you can refer to these in the sections entitled 'Abnormalities in the growth of l.s.v. and/or the shape of leaves' (page 39), 'Spots and damage on leaves' (page 109) and 'Wilting, drying out, necrosis of leaves' (page 155).***

140

141

142

143

144

## Genetic abnormalities

**140** In this lettuce crop, one single isolated plant is affected by a **genetic abnormality**.

**141** The leaves are more or less covered with irregular areas of a light green to creamy yellow colour. This symptom may lead to confusion with a mosaic.
**Genetic abnormality (Variegation)**

**142** The sick plants have a few yellow leaves locally deprived of chlorophyll pigments.
**Genetic abnormality (Variegation)**

**143** In certain cases, the tissues are completely deprived of pigment and take on a white colour.
**Genetic abnormality (Albinoism)**

**144** In other situations, almost all the leaves may be affected.
**Genetic abnormality (Variegation)**

## • Genetic abnormalities

A certain number of genetic abnormalities occur on lettuces. Very often, the number of plants affected is very low and, in many situations, the plant in question constitutes a real curiosity (**140**). Generally speaking, the colour and shape of the leaves are relatively characteristic and do not lead to any confusion.

**Leaf chimera** produce yellowing, or even whitening of sectors of the lamina, of varying degrees of intensity. Thus, the leaves have broad irregular yellow-green to creamy-white areas, giving them an attractive mottled and mosaic appearance (**variegation**) (**141–144**). Seriously affected plants may become completely white (**albinoism**) (**143**) and die. This discolouration is due to an abnormality in the development of chloroplasts in leaf tissues.

These genetic abnormalities only concern the plant or plants affected and are not, under any circumstances, transmissible to neighbouring plants. They may be transferred to descendants, if the plant or plants affected are used in cross-breeding, which you are advised to avoid. Generally speaking, affected plants must be removed as soon as they are detected.

Other genetic abnormalities produce morphological aberrations in l.s.v.; refer to these in the paragraph entitled 'Genetic anormalities' on page 37.

# Spots and damage on leaves

## Symptoms investigated

- Brown or black spots on leaves showing varying degrees of necrosis
- Yellow, orange to light brown spots and damage on leaves
- Spots and damage located mainly on the principal vein or on the periphery of leaves
- Spots with powdery areas, matting, mould on leaves, damp patches leading to rotting of the head

## Possible causes

- *Alternaria cichorii* (fact file 4)
- *Botrytis cinerea* (fact file 5)
- *Bremia lactucae* (fact file 1)
- *Cercospora longissima* (fact file 4)
- *Erysiphe cichoracearum* (fact file 2)
- *Microdochium panattonianum* (fact file 3)
- *Mycocentrospora acerina* (fact file 4)
- *Myrothecium roridum*
- *Puccinia opizii*
- *Puccinia hieracii* var. *hieracii* f. sp. *endiviae*
- *Puccinia dioicae (Puccinia extensicola* var. *hieraciata)*
- *Pythium* spp. (see page 180)
- *Sclerotinia sclerotiorum* (fact file 6)
- *Septoria lactucae* (fact file 4)
- *Stemphylium botrytosum* f. *lactucum* (fact file 4)
- *Erwinia carotovora* subsp. *carotovora* (fact file 18)
- *Pseudomonas cichorii* (fact file 15)
- *Pseudomonas marginalis* pv. *marginalis* (fact file 17)
- *Pseudomonas viridiflava*
- *Xanthomonas campestris* pv. *vitians* (fact file 16)
- Phytoplasma from the Aster yellows group (fact file 20)
- Lettuce ring necrosis agent (LRNA) (fact file 35)
- Tomato spotted wilt virus (TSWV) (fact file 32)
- Tobacco ring spot virus (TRSV) (fact file 33)
- Tobacco rattle virus (TRV) (fact file 33)
- Tobacco streak virus (TSV) (fact file 32)
- Tobacco necrosis virus (TNV) (fact file 35)
- Tomato black ring virus (TBRV) (fact file 33)
- Lettuce necrotic spot virus (LNSV) (fact file 33)
- Predator damage (slugs, leaf miners, thrips)
- 'Brown rib'
- Cold injury
- 'Russet spotting'
- 'Tip burn'
- 'Pink rib'
- Various types of chemical injury
- Atmospheric pollution (refer to page156)
- Excessive salinity
- 'Brown stain'
- Latex spots
- Vitrescence of the lamina

## Fairly difficult diagnosis

Numerous pathogenic agents are likely to cause spots and damage to leaves which may be rather similar to those caused by various non-parasitic diseases. It is therefore fairly easy to confuse the diagnosis. In order to facilitate your identification we suggest following the procedure described below:

- observe the spots on several leaves and on several plants in order to evaluate their development over time (size, colouration, presence of a halo, distribution over the plant and over the leaves, and so on) (refer to the observation guide);
- always look at the lower face of the leaves in order to note any presence of fruiting bodies from a fungus or other elements facilitating diagnosis;
- then consult all the sections dealing with spots.

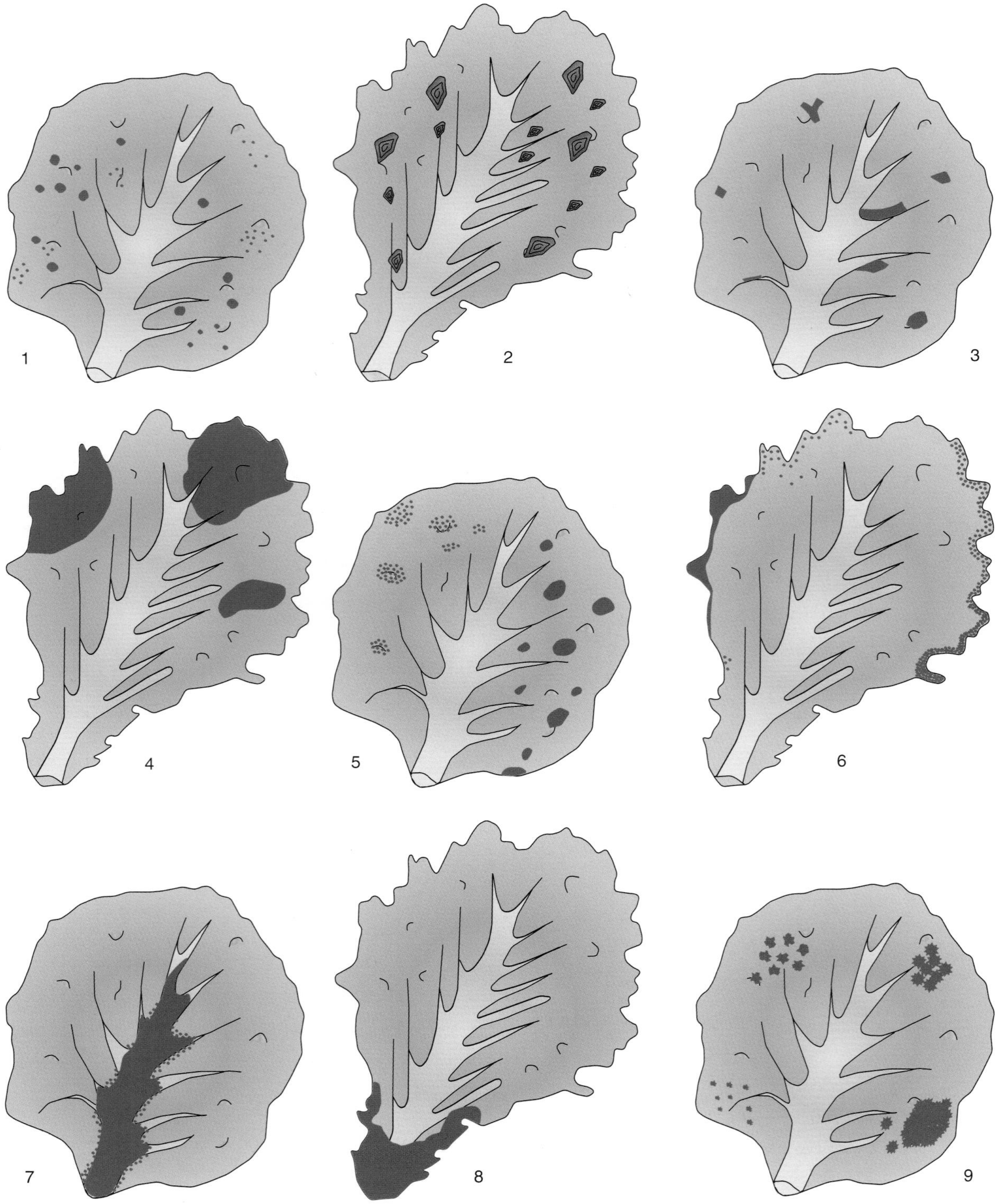


**Figure 5: Appearance, location, and possible development of spots and damage observed on lettuce leaves**

1 Random or small sized spots
2 Well defined spots, sometimes concentric
3 Angular spots, outlines are defined by the veins
4 Broad spots with damage to areas of tissue
5 Rather diffuse spots (on the left) or clearly defined (on the right)
6 Spots, damage located on the edge of lamina
7 Damage located all along the principal vein
8 Damage beginning at the base of leaves
9 Merging spotss

145

146

147

148

Spots

## Examples of spots and damage on lettuce leaves

**145** Small damp brown spots affecting the edge of several lettuces leaves.
***Pseudomonas* sp.**

**146** Damp, brown lesions affecting the principal vein and attacking the secondary veins and the blade of several leaves.
***Pseudomonas cichorii***

**147** Chlorotic spots, delimited by the veins, covered by white mould.
***Bremia lactucae***

**148** Damp rot invading the head of a lettuce; grey mould is partially covering it.
***Botrytis cinerea***

**149** Spots which are variable in size, damp, and brown affecting several old leaves on this lettuce.
***Pseudomonas* sp.**

**150** Several angular spots, defined by the veins, light brown in colour, have caused partial chlorosis of the blade.
***Septoria lactucae***

**151** On this leaf, fairly rounded dark brown spots, surrounded by a clearly defined yellow halo, reveal an attack by ***Cercospora longissima***.

**152** Some lower leaves are covered with small lesions which are both chlorotic and necrotic.
**Lettuce ring necrosis agent (LRNA)**

**153** Inter-veinal spots of a dark chestnut to brown colour are damaging the most exposed portions of the lamina.
**Chemical injury**

# Brown or black spots on leaves showing varying degrees of necrosis

## Possible causes

*Alternaria alternata*

- *Alternaria cichorii* (fact file 4)
- *Botrytis cinerea* (fact file 5)
- *Cercospora longissima* (fact file 4)
- *Mycocentrospora acerina* (fact file 4)
- *Myrothecium roridum*
- *Puccinia hieracii* var. *hieracii* f. sp. *endiviae*
- *Puccinia prenanthis*
- *Septoria lactucae* (fact file 4)
- *Stemphylium botryosum* f. *lactucum* (fact file 4)
- Foliar bacteria (also consult page 139)
  - *Pseudomonas cichorii* (fact file 15)
  - *Pseudomonas fluorescens*
  - *Pseudomonas marginalis* pv. *marginalis* (fact file 17)
  - *Pseudomonas viridiflava*
  - *Xanthomonas campestris* pv. *vitians* (fact file 16)
- Lettuce ring necrosis agent (LRNA) (see pages 73 and 127) (fact file 35)
- Tomato spotted wilt virus (TSWV) (see page 152 'Wilting, drying out, necrosis of leaves') (fact file 32)
- Tobacco streak virus (TSV) (see page 152 'Wilting, drying out, necrosis of leaves') (fact file 32)
- Lettuce necrotic spot virus (LNSV) (fact file 33)
- Marginal necrosis of leaves ('tip burn') (refer to pages 132 and 135)
- Various types of chemical injury

While they are growing, l.s.v., whose leaf tissues are often tender and fragile, display a number of types of spot, and damage of varying extent. These are first of all oily; subsequently they take on a brown to black colouring and become necrotic. Exact identification of their cause is often difficult. Under this section we have summarized the principal parasitic problems which may be implicated. As you will realize, the risk of confusion is significant. In certain cases observation under a microscope will be necessary in order to diagnose accurately the nature and morphology of the structures and fungal spores present on and in the damaged tissues. Should this be necessary we suggest that you contact a specialist laboratory. In addition, in order to facilitate your identification, we recommend that you consult other headings in the section entitled 'Spots and damage on leaves'.

154

155

156

157

158

## Examples of bacterial symptoms on l.s.v. leaves

**154** Some brown spots have appeared on the leaves in the head of this lettuce.
**Bacteriosis**

**155** When we examine the upper face of the lamina, the spots appear to be chlorotic, and damp; they rapidly become necrotic.
**Bacteriosis**

**156** On the lower face of the lamina, the same spots are more oily and are rust brown in colour.
**Bacteriosis**

**157** If conditions are particularly wet in places where bacteria rapidly colonize tissues, the latter become soft, wet, and black in colour.
**Bacteriosis**

**158** In some cases, the vein of the leaves may be affected. It goes brown and/or black progressively from the base of the leaf to the top or from nearby foliar necrosis, which is what has happened on the photograph.
**Bacteriosis**

## Arguments in support of the diagnosis

- **Leaf bacteria**

Spots which are at first transparent, and sometimes damp, rapidly becoming dark brown to black (**154–168**), can develop on the lamina of l.s.v., often following damp climatic periods or sprinkler irrigation. If humidity persists these spots quickly spread, sometimes causing rot limited to a few leaves. This may spread to the head and eventually reach the interior of the stem (**168**). These symptoms are often bacterial in origin, but it is difficult to associate them with any particular type of bacteria. In fact, isolating procedures carried out at the junction of healthy and diseased tissues often reveal one or several bacteria acting together: ***Pseudomonas cichorii*, *Xanthomonas campestris* pv. *vitians*, *P. marginalis* pv. *marginalis*, *P. viridiflava*, *P. fluorescens*,** and so on. Under these conditions it is very difficult to know which one is really responsible for the damage. The descriptions of the principal symptoms caused by these different bacteria on lettuces are given in *Tables 12* and *13*.

*P. cichorii* and *X. campestris* pv. *vitians* are generally considered to be bacteria which are pathogenic to lettuce. Other bacteria probably behave as secondary invaders, or opportunists (*P. marginalis pv. marginalis*) or simply as saprophytes (*P. viridiflava*). The last two bacteria are very pectinolytic; they help to increase rotting, especially if conditions are damp.

Differences in sensitivity to bacteriosis have been noted in the field between types of lettuce (**163**), and also between varieties. Sometimes we see bacterial attacks concentrated on plants which have already been affected by another disease (for example 'big vein'; refer to page 63).

In fact, whatever the nature of the symptoms and the bacterium or bacteria responsible, accuracy of diagnosis is only of relative importance since, in the majority of cases, the same corrective methods must be implemented.

159

160

161

162

163

## Examples of bacterial symptoms on lettuce leaves

**159** This lettuce leaf displays small brown spots, which are slightly angular.
***Pseudomonas* spp., *Xanthomonas campestris* pv. *vitians***

**160** Several damp black spots sometimes have irregular outlines which give them the appearance of 'stars'.
***P. cichorii***

**161** Spots located on the edge of the lamina have rapidly converged; black rot is now taking over the interior of the lamina.
***P. marginalis* pv. *marginalis***

**162** Numerous dark brown spots, more or less circular and converging in places, are scattered over this leaf. Some spots are starting to dry out and have a wrinkled appearance.
***X. campestris* pv. *vitians***

**163** The bottom leaves of the lettuces (on the left) are covered with bacterial spots; this is not the case with the batavias (on the right) in spite of strong selection pressure. This is a good example of differences in sensitivity to types of **bacteriosis**.

Table 12: Some characteristics of the principal types of bacteria attacking l.s.v. leaves

| | *Pseudomonas cichorii* | *Xanthomonas campestris* pv. *vitians* | *Pseudomonas marginalis* pv. *marginalis* |
|---|---|---|---|
| **Symptoms** | • Small chlorotic spots, which rapidly become necrotic, dark brown to black in colour. They are shiny, in the shape of a circle or polygon, and sometimes star-shaped. They can be delimited by the secondary veins which also go brown.<br>• Change in the rather firm consistency of the principal vein which also takes on a brown to black colour. The tissues near and at the base of the vein sometimes show the same type of damage. The taproot normally remains healthy. Other secondary bacteria sometimes colonize the tissues and destroy them. If the tissues are able to dry out their colour lightens and they become paper-like.<br>• Spots, extensive damage and necrosis, brown and shiny, occasionally appear on the leaves of the heart, without any change in consistency. **('Varnish spot')** | • Minuscule damp, angular spots form on the well-developed leaves surrounding the head. As they grow they become more or less circular, taking on a dark brown to black colouration. They converge and cover significant portions of the blade, leading to total destruction of certain leaves. Once they are dry, the tissues also become paper-like and tear. | • Faded wilting of the lamina periphery. Quite quickly affected zones show some some necrotic spots which go brown and black. Ultimately the tissues can dry out, become lighter in colour and paper-like. If conditions are very damp, a black oily rot invades the leaves. Some secondary veins sometimes appear darker.<br>• the bacterium is capable of attacking the crown and stem if conditions are particularly damp. It causes an olive-coloured soft rotting of the pith, comparable to that caused by *Erwinia carotovora* **('Butt rot')**. |
| **Location of symptoms and pre-disposing conditions** | All the leaves may be attacked. This bacterium affects the leaves in the head, and the leaves on the crown to a lesser extent. Lettuces are particularly receptive during rainy periods or after sprinkler irrigation at the rosette stage. Symptoms of the 'varnish spot' type are generally only observed as harvesting approaches. | *Xanthomonas campestris* pv. *vitians* prefers to attack the leaves in the crown. It mainly attacks during hot damp periods. Rainfall or sprinkler irrigation mark the start of its attacks. | Not all the leaves are necessarily affected. Initially lesions are mainly distributed around the periphery of the lamina. This opportunistic bacterium easily establishes itself on tissues suffering from 'tip burn' (see page 133). It only causes rotting of the pith on certain varieties of lettuce during overcast and damp periods. |

* *Pseudomonas viridiflava* (Burkholder) Dowson has also been recorded on lettuce in the saprophytic state. It is more likely to develop on old leaves and tends to progress to the periphery of the lamina and all along the veins. It can cause spots which are fairly similar to those caused by *Pseudomonas cichorii* and *P. marginalis* pv. *marginalis*, as well as damp rot due to the pectinolytic activities of certain strains.

164

165

166

167

168

## Possible locations of bacterial symptoms on lettuces

164 ***Xanthomonas campestris* pv. *vitians*** is more likely to develop on the lower leaves, causing spots which are more less circular and brown in colour.

165 Numerous converging spots cover the intermediate leaves and those of the head of this plant. ***Pseudomonas* spp., *X. campestris* pv. *vitians***

166 The principal vein of numerous leaves of this curly chicory have gone black over practically the whole of their length. ***Pseudomonas cichorii***

167 Damp black rotting progressively spreads to several intermediate leaves. ***Pseudomonas* spp., *X. campestris* pv. *vitians***

168 Dark brown rotting, soft in consistency, has invaded the pith of this lettuce. ***Pseudomonas* pv. *marginalis*, *Erwinia carotovora*** can cause this type of damage.

Table 13: **Some characteristics for determining fairly common pathogenic agents and non-parasitic diseases that cause black to brown spots and leaf damage and may lead to confusion in diagnosis**

| **Certain characteristics** | **Various types of bacteria** | **Lettuce ring necrosis agent (LRNA)** | **Marginal necrosis** | **Various types of chemical injury** |
|---|---|---|---|---|
| **Relative occurrence** | ++ | + | +++ | +/– |
| **Frequency on l.s.v.** | Common. | Not common, mainly under protection. | Very common. | Occasional. |
| **Plant sensitivity stage** | Any time after planting. | Any time after planting. | Close to harvest-time. | At any point during growing season, both in the nursery and in the open field. |
| **Location of symptoms on the plant** | Random, also close to and all along the principal vein (see page 127). | Mainly on the leaves of the crown, also close to and along the principal vein (see page 127). | Leaves surrounding the head (damp necrosis); old leaves from the crown and periphery of the lamina (dry necrosis or 'tip burn'). | On both young this depends on the nature and properties of the product responsible. |
| **Location of diseased plants within the plot** | In groups, rapidly becoming widespread following rainfall or sprinkler irrigation. | Mainly in groups or scattered plants. | Scattered plants or widespread phenomenon affecting numerous plants. | Variable distribution (see page 87). |

## *Septoria lactucae*

**('Septoria leaf spot')**

**169** The bottom leaves of this Latin lettuce, a particularly sensitive lettuce, have a scattering of angular, necrotic spots.
***S. lactucae***

**170** The spots, delimited by the veins and brown in colour, are encircled by a chlorotic halo.
***S. lactucae***

**171** On the lower face, these spots are lighter in colour.
***S. lactucae***

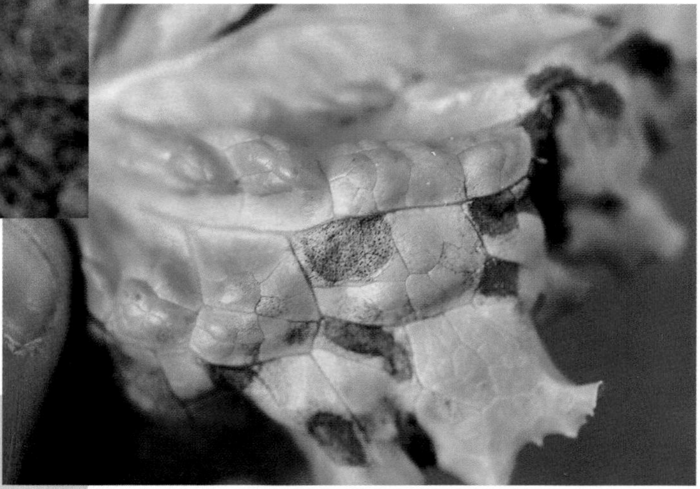

**172** Careful observation of the lamina can reveal tissues encrusted with minuscule black structures: these are the pycnidia of the parasitic fungus.
***S. lactucae***

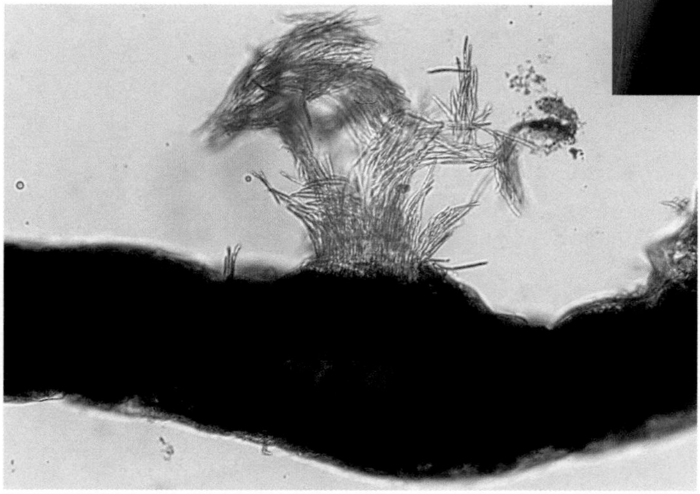

**173** On this cross section of a leaf, a cluster of filiform spores progressively leaving a pycnidium can be clearly distinguished.
***S. lactucae***

- ***Septoria lactucae* (Septoriosis, 'Septoria leaf spot')**

This worldwide fungus attacks only very sporadically where lettuce is grown intensively. Initially it causes small irregular, chlorotic spots, mainly on old leaves (**169**). These progressively increase and take on an olive to light brown colour. A yellow halo, varying in intensity, surrounds them (**170** and **171**). Inside the spots the naked eye can distinguish minuscule black structures, pycnidia (asexual reproductive organ of the fungus) (**172**). Once dead, affected tissues dry out, crumple and/or split. Subsequently the leaves have a fairly large number of perforations.

The pycnidia of *Septoria lactucae* are globular and measure between 100 to 200 μm in diameter. During wet periods they produce numerous filiform, compartmentalized spores (25–40 × 1.5–2 μm) coming out agglomerated by an ostiole in the shape of a long cyrrhus (**173**). Several other species of *Septoria* have probably been identified on lettuce (see fact file 4).

- ***Mycocentrospora acerina* ('Mycocentrospora leaf spot')**

This fungus, more often found on carrots and celery, is rarely seen.

It causes spots which frequently begin on the edge of the lamina or along the principal vein of leaves located near the ground. The spots, which are initially small, transparent and rather rounded, rapidly increase and become brown to dark brown in colour (**174** and **175**). The necrotic tissues dry out, become paper-thin and fall off. Ultimately, numerous holes perforate the lamina. If conditions are particularly favourable, the fungus colonizes a large proportion of the lamina, affecting entire leaves and plants. Rotting begins, transforming the tissues into a black, deliquescent mass which progressively takes over the head. Sometimes the fungus attacks the taproot via the point of insertion of one or several rotted leaves. If this happens the plant suddenly wilts; this syndrome is somewhat reminiscent of an attack of *Botrytis cinerea* or *Sclerotinia* spp. in the crown. Make sure you don't confuse your diagnosis!

## *Mycocentrospora acerina* ('Mycocentrospora leaf spot')

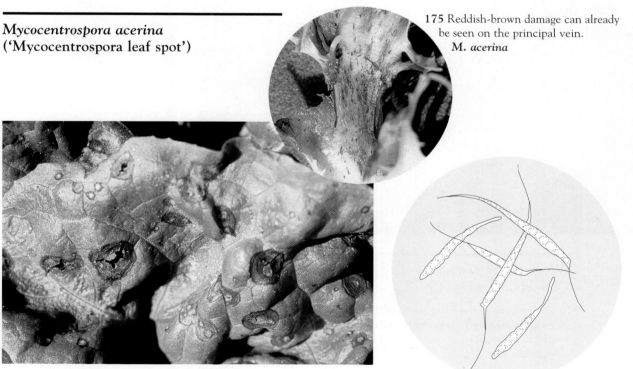

**175** Reddish-brown damage can already be seen on the principal vein. ***M. acerina***

**174** Several transparent to brown spots can be seen on this lettuce leaf. Some of these are starting to split. Eventually they will form perforations. ***M. acerina***

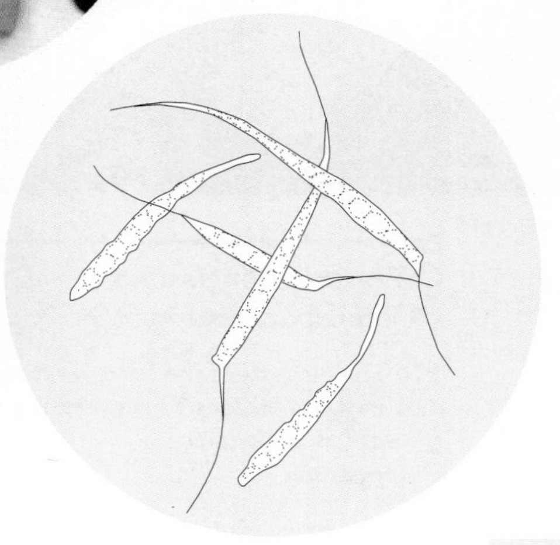

**Figure 6: Appearance of the conidia of *M. acerina***

The conidia, characteristic of this fungus, can be collected by scratching the damaged tissues. They are pluriseptate, hyaline, very elongated and have a thin, tapered lateral appendage..

176

177

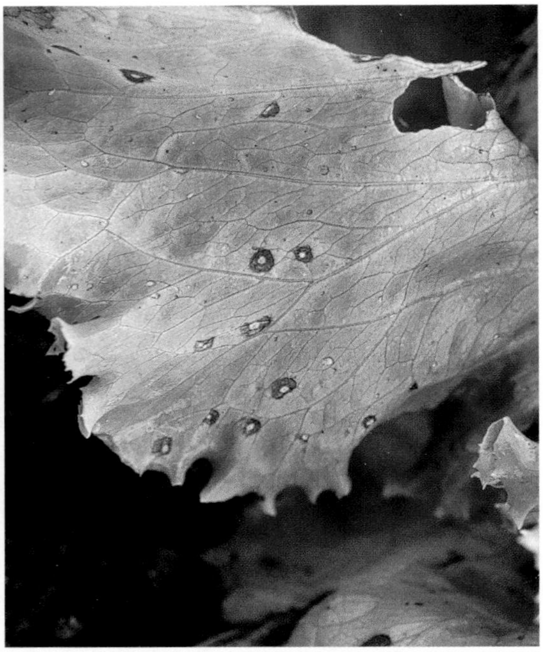

178

179

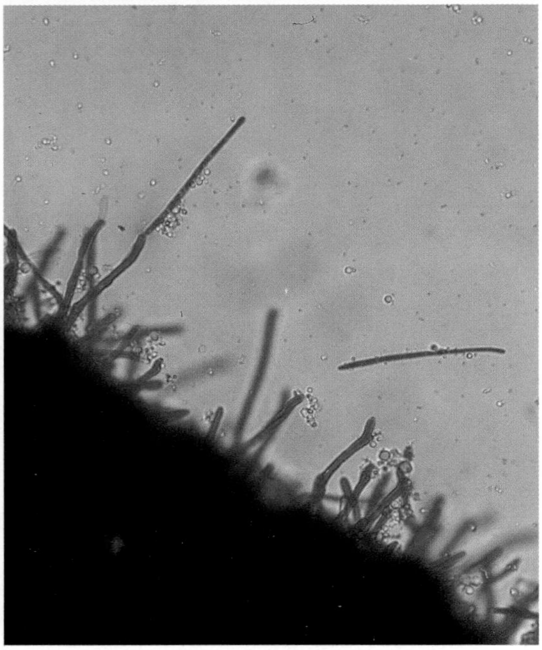

180

## *Cercospora longissima* ('Cercospora leaf spot')

**176** The majority of the lower leaves of this lettuce have dark brown spots located between the veins and along the edge of the lamina.
***C. longissima***

**177** The spots, which are damp at first, rapidly go brown while a central area remains light in colour.
***C. longissima***

**178** The spots spread out. Their shape is irregular, rather rounded to angular when delimited by the veins. A fairly limited yellow halo surrounds them.
***C. longissima***

**179** On necrotic tissues it is easy to observe tapered conidiophores, grouped together in a cluster, with filiform and multicellular conidia at their extremities.
***C. longissima***

**180** Eventually the spots converge and the tissues of some of these spots tear.
***C. longissima***

• *Cercospora longissima* (Cercosporiosis, 'Cercospora leaf spot')

Cercosporiosis is widespread throughout the world; it essentially causes damage in certain tropical production zones, where the hot, damp climate encourages its development. At present this disease does not seem to be rife in France except in overseas territories. It is not common in the US or Western Europe.

*C. longissima* causes small wet spots located on old leaves (**176**). They are brown in colour and surrounded by a pale green halo. They spread and form areas of brown damage, delimited by the veins (**177, 178** and **180**). These spots are covered by a greyish to fawn down consisting of numerous fungus fruiting bodies (**179**).

## Botrytis cinerea ('Botrytis leaf spot')

**181** Broad wet spots beginning to become necrotic are spread over two leaves. The concentric 'patterns' can be clearly distinguished.
***Botrytis cinerea***

**182** Subsequently these spots become necrotic and go brown. Eventually they split if climatic conditions become drier.
***B. cinerea***

• ***Botrytis cinerea* ('Botrytis leaf spot')**

Mainly attacking the bottom leaves of lettuces (see page 167), this fungus is capable of producing aerial contamination which takes the form of large spots on the leaves of the lettuce head (**181** and **182**). These rapidly expand if climatic conditions are favourable; they can cause rotting of a large proportion of the head while the lettuce is growing or while it is being stored (see page 140). Unobtrusive conidiophores carrying numerous conidia can sometimes be seen on the tissues.

Spots

• *Alternaria cichorii*
(Alternariosis, 'Alternaria leaf spot')

Several species of *Alternaria* (***Alternaria porri*** **(Ell.) Neerg. f. sp.** ***cichorii*** **(Nattr.) Schmidt., A.** ***dauci*** **(Kühn) Gr. and Sk. f.** ***endiviae*** **(Nattr.) Janezic, A.** ***cichorii***) have been associated with brown spots on lettuce leaves, in several countries (USA, Argentina, Japan, and so on), and in particular those bordering France (Italy, Germany, and so on). It actually seems as though, in the majority of cases, they can be brought together under the same species **A. *cichorii***.

This fungus causes spots located on the lamina and its periphery. These spots are dark brown, round to oval in shape, sometimes converging. They have concentric streaks and are progressively covered with conidiophores topped by very characteristic club-shaped pluricellular conidia (**185**). Eventually the tissues located at the centre of the spots may become lighter. This *Alternaria* seems capable of attacking the witloof chicory, and, to a lesser extent, the escarole and lettuce; the curly endive appears to be only slightly sensitive. Young lettuce leaves seem to be the most susceptible.

In France, attacks were observed on escarole in the Pyrénées-Orientales many years ago.

We should also point out that an ***Alternaria alternata*** was probably responsible for damage on old leaves of lettuces grown in fields in Egypt, in the region of El-Minia.

183

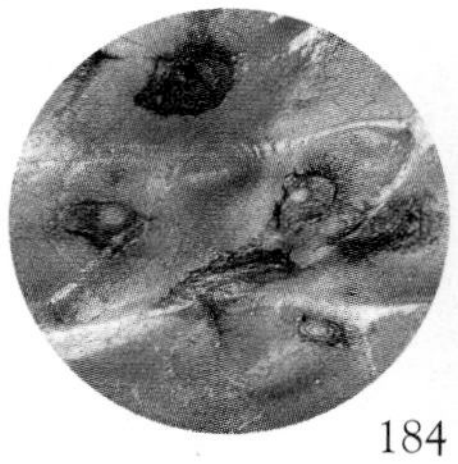

184

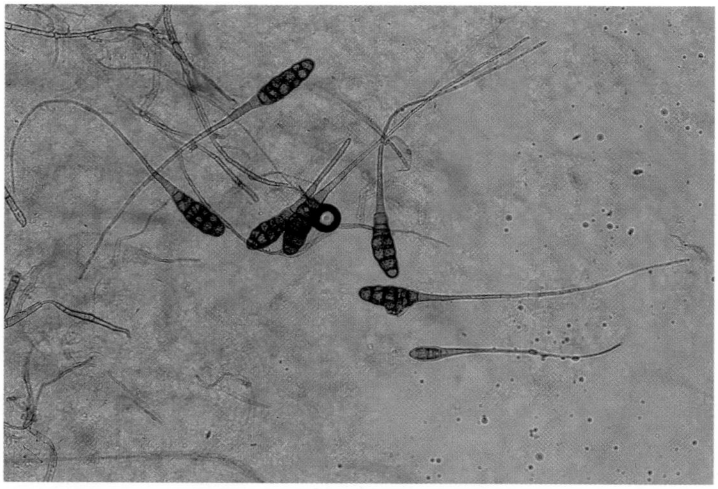

**185** Several multicellular, brown, club-shaped spores have transverse and longitudinal partitions. They extend into a hyaline tip.
***Alternaria cichorii***

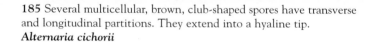

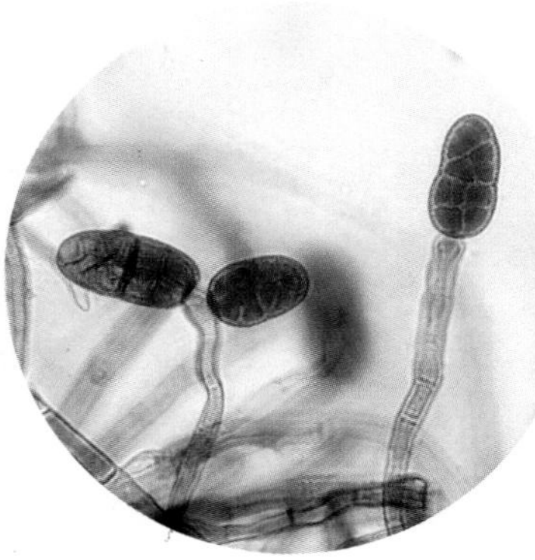

***Brown spots on leaves***

**In certain cases observation of the fruiting bodies present on the lamina will allow you to identify the fungus responsible. If you are in any doubt at all, consult a specialist laboratory.**

**186** Some multicellular, muriform conidia form at the ends of conidiophores swollen at their extremities. Subsequently these spores progressively go brown.
***Stemphylium botryosum*** **f. sp.** ***lactucum***

• *Stemphylium botryosum* f. sp. *lactucum* Wallr. (Stemphyliosis, 'Stemphylium leaf spot')

This disease of the leaves in the crown has been described in several European countries (England, Italy, Netherlands, Spain and Portugal), in Israel, South Africa, and the USA.

It causes spots which are initially small and damp and progressively form shapes varying from circular to oval, sometimes angular if they are restricted by veins. These spots have concentric areas which darken to varying degrees. Dark brown to black conidian-based matting covers them during damp periods. This consists of the conidiophores and the conidia of this fungus (photo 186). Ultimately the tissues located at the centre of spots become necrotic, dry out, then decompose. The leaves then have a number of perforations.

• *Myrothecium roridum* Tode: Fr

This fungus was described in Spain several years ago on lettuces (Trocadero variety) cultivated in fields in the Valencia region.

Initially it seems capable of causing small dark lesions around the perimeter of the leaves. These rapidly turn into damp circular spots. Subsequently they spread and become irregular. Their colour varies from chestnut to brown, or even black. They have some concentric patterns and are surrounded by a poorly defined chlorotic halo (fig. 7). The fruiting bodies of the fungus (its sporodochia) are formed on the tissues (**187** and **188**); they are arranged in a concentric or irregular way.

This is a facultative parasitic fungus, which selects as its host numerous cultivated or wild plants, found principally on red clover, coffee, cotton, and so on. On vegetables it has been observed on melon, tomato, and eggplant fruits. It has only been observed on lettuces in Spain, developing during hot, damp climatic periods which meet its climatic conditions.

**Figure 7: Appearance of spots on lettuce leaves caused by *Myrothecium roridum*.**

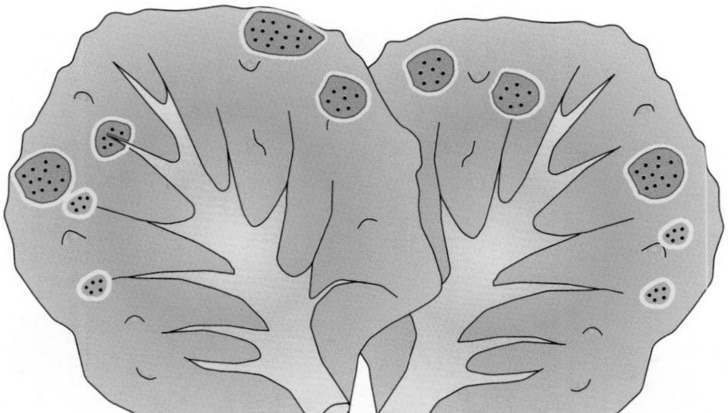

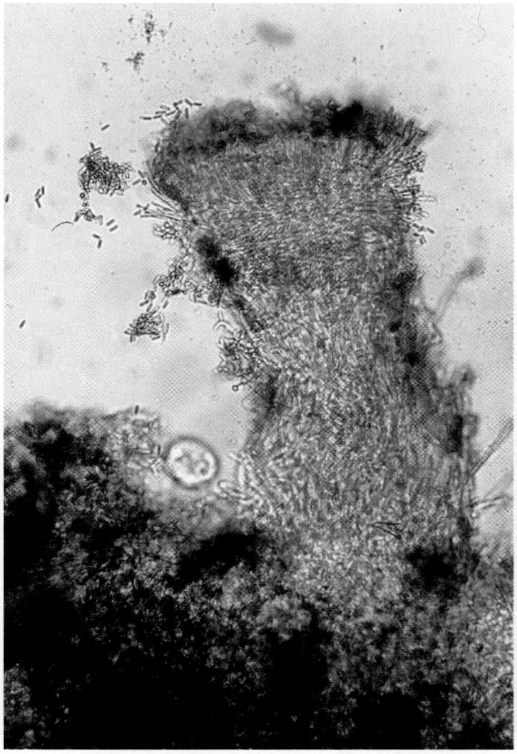

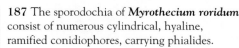

187 The sporodochia of ***Myrothecium roridum*** consist of numerous cylindrical, hyaline, ramified conidiophores, carrying phialides.

188 The phialides form cylindrical conidia, rounded at their extremities, hyaline and slightly olive-coloured. ***M. roridum***

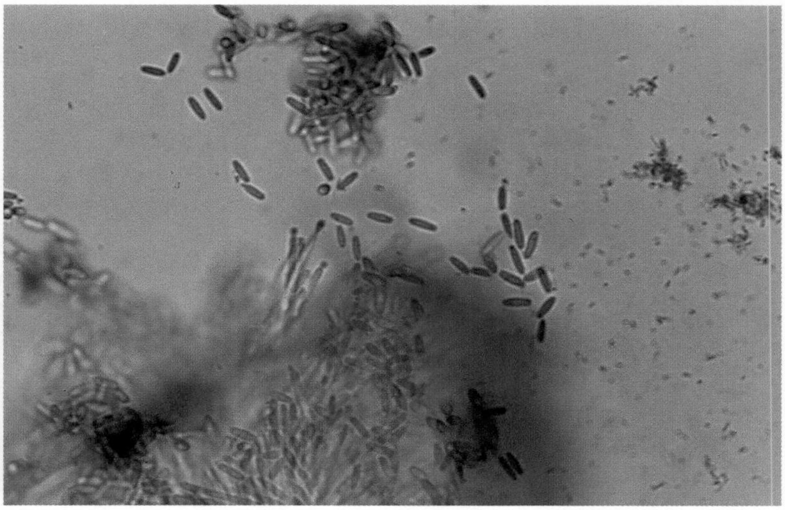

Spots

## Rust

**190** On these lettuce cotyledons, we can see that these spots are initially chlorotic before the appearance of uredosores. ***Puccinia* sp.**

### • Autoecious and heteroecious rust on l.s.v. ('Rust')

Several types of rust have been referred to in literature in relation to l.s.v. It seems that the same types of fungus have been described numerous times under different synonyms. Consequently it is difficult to give a precise opinion about rust and its effect on lettuce. For the sake of clarity we will group together all these types of rust into two categories: autoecious rust (fungus cycle completed on one single host) whose symptomatology is less obtrusive, and the heteroecious rusts (fungus cycle completed on 2 different hosts) which are listed on page 121.

The Italians have observed an autoecious rust, essentially on chicory, ***Puccinia hieracii* var. *hieracii* f. sp. *endiviae*** (Bellynck ex. Kickx) Boerema and Verhoeren (syn.: ***Puccinia cichorii*** D.C.). Initially this causes small pale yellow spots on chicory leaves. Subsequently these are covered with rust-coloured pustules (uredosores) which then go dark brown (teliosores) (**189** and **190**). This rust is probably encountered in various Mediterranean countries. Other types of rust have been reported on l.s.v., with names which are reminiscent of the above: ***Puccinia hieracii* (Roehl) Mart., *Puccinia hieracii* f. sp. *cichoriae*, *Puccinia endiviae***.

Another autoecious type of rust, ***Puccinia prenanthis* (Pers.) Lindroth**, has been described in Australia, affecting lettuces and chicory.

In France, an autoecious rust, possibly ***Puccinia hieracii* var. *hieracii* f. sp. *endiviae***, was encountered sporadically on lettuce several years ago. Currently only the symptoms of heteroecious rust are observed from time to time on lettuce (see page 121).

**189** Numerous spots, rust-coloured, are scattered over these lettuce leaves. ***Puccinia* sp.**

### • Tomato spotted wilt virus (TSWV)

This virus, transmitted by several species of thrips, initially causes spots and chlorotic damage to lettuce leaves. They then become brown and necrotic. The veins can be affected and become necrotic. Subsequently this damage extends and takes over a large proportion of the lamina (**191**). Other symptoms and information can be consulted in the section entitled 'Wilting, drying out, necrosis of leaves' page **147**.

**191** On these few leaves, necrotic lesions of a rust-brown colour, varying in extent, indicate the effects of TSWV.

• **Chemical injury**

You must be particularly vigilant when confronted with spots on l.s.v. leaves. Indeed you must be aware that their leaf tissues are tender and fragile, especially when l.s.v. are grown under protection. Consequently certain products (especially pesticides) cause problems of selectivity when used under certain conditions, and are considered to be phytotoxic for l.s.v. Various types of necrotic damage can then appear on the leaves (**192–194**). These can be confused with certain spots which we have described in this section. If you are in any doubt whatsoever, we recommend that you have your diagnosis confirmed by a specialist laboratory. If you require further information on types of chemical injury we suggest that you consult pages 39 and 83.

**192** A multitude of small beige to brown areas of damage, in various shapes, cover this lettuce leaf. These appeared following a fungicide treatment carried out incorrectly.
**Chemical injury**

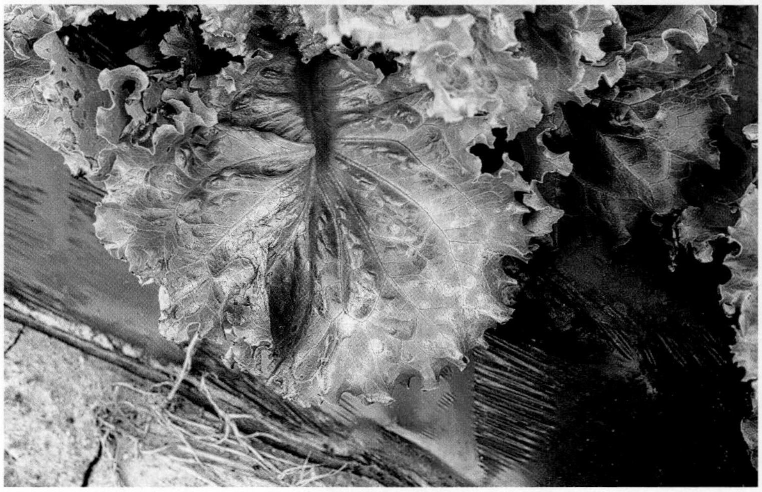

**193** These brown, necrotic spots, which converge in places, appeared after treatment with a herbicide intended to destroy adventitious roots growing between the rows.
**Chemical injury**

**194** Sprayed herbicide has fallen in a more or less regular way onto this young lettuce. Several types of irregular, brown inter-veinal damage are progressively appearing on one of its leaves.
**Chemical injury**

**Some examples of leaf spots caused by phytotoxicity ('Chemical injuries')**

**195** Several chlorotic spots are scattered over the lower leaves of this lettuce.
***Bremia lactucae***

**196** A multitude of elliptical orange coloured spots, varying in size, sometimes torn at the centre, are partially covering the lamina of this batavia.
***Microdochium panattonianum***

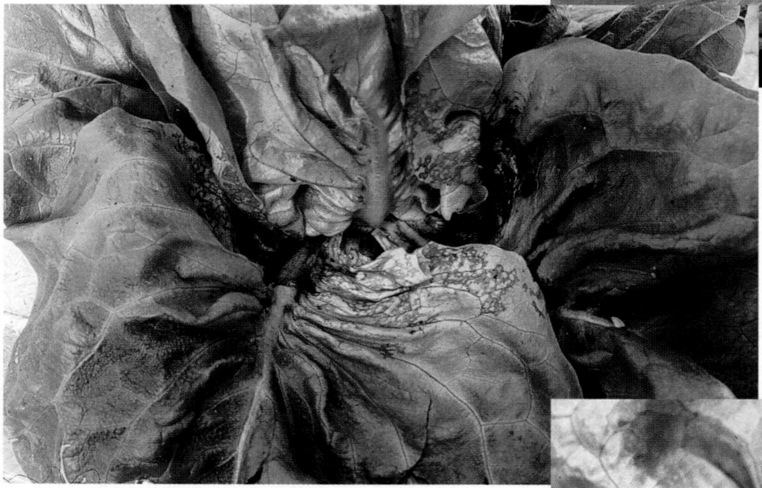

**197** Orange-coloured damage, varying in extent, has appeared on one sector of the lamina of this lettuce.
**Lettuce ring necrosis agent (LRNA)**

**198** Between the veins of this leaf we can distinguish islets of cells which have broken up, become corky and taken on an orange shade.
**Cold injury**

# Yellow, orange to light brown spots and damage on leaves

## Possible causes

- *Bremia lactucae* (fact file 1)
- *Microdochium panattonianum* (fact file 3)
- *Puccinia opizii*
- *Rhizoctonia solani* (fact file 7)
- Lettuce ring necrosis agent (LRNA) (Fact file 35)
- Other viruses responsible for spots, rings and/or yellow patterns (tobacco rattle virus (TRV); tobacco ring spot virus (TRSV); tomato spotted wilt virus (TSWV)); consult the section "Wilting, drying out, necrosis of leaves" page 147 (fact files 32 and 33).
- Damage due to predators (slugs, leaf miners, thrips, and so on)
- Cold injury
- Various types of chemical injury
- Vitrescence of the lamina

## Arguments in support of the diagnosis

- ***Bremia lactucae***
  **('Downy mildew')**

Lettuce mildew is a disease which is particularly feared by growers. Essentially it causes fairly broad light green to yellow spots to appear, both on young plants (**199–202**) and on older plants (**203–214**). These spots are frequently angular in appearance, since they are delimited by the secondary veins (**200**, **201**, **204**, and **206**). On the affected tissues on the lower face of the lamina we can also distinguish the presence of a white down (**201**, **206**, **210**, and **211**), which may be very dense, consisting of the sporangiophores and sporangia of the mildew (**215**). Ultimately these tissues become necrotic and take on a chestnut to brown colouration (**201** and **212**). The spots converge when damp climatic conditions encourage the disease. Eventually extensive sections of the lamina are affected, ultimately drying out and falling (**214**). Some leaves can be completely destroyed.

The oospores, the sexual reproductive form of this Chromist which is now classified with algae, are sometimes visible in damaged tissue (**213**).

- ***Microdochium panattonianum***
  **('Anthracnose')**

During cold, wet autumns and in the winter it is quite common to observe numerous spots on the lower leaves of lettuces (**220**). If we examine them more closely we can see that initially they are small in size, damp, and round in shape. They become chlorotic fairly quickly, taking on an orange colour before going brown. Ultimately they measure 3–5 mm in diameter (**216**, **217**, **221–223**). They are sometimes rather angular if delimited by the veins. The tissues go brown fairly quickly, become necrotic and fall. The lamina then has numerous perforations, giving the leaves a very characteristic riddled appearance (**218**). On the sides spots are more elongated and concave, and their colour is often darker (**224**). In the presence of high surrounding humidity, very unobtrusive masses of pink spores (the acervulus of the fungus bearing conidia) form around the perimeter of the lesions (**119**). Badly affected leaves quickly wilt and become completely necrotic.

The symptoms caused by this fungus have given rise to various English names for the disease: **'Shot Hole'**, **'Ring spot'** and sometimes even **'Rust'**.

199

200

201

202

## *Bremia lactucae* in the nursery and on young plants 'Downy mildew'

**199** This young lettuce displays several chlorotic spots on the lower leaves.
***B. lactucae***

**200** On this leaf we can clearly distinguish the possible development of spots of mildew on the young plant. They are initially pale green, and progressively take on a yellow hue. We can also note unobtrusive white down on the surface of the lamina. Their outline is generally delimited by the veins.
***B. lactucae***

**201** Well-developed spots are an intense yellow colour and ultimately become necrotic.
***B. lactucae***

**202** The spots extend and converge if conditions are particularly favourable. Ultimately broad sections of the lamina are affected.
***B. lactucae***

203

204

205

206

207

## *Bremia lactucae* on butterhead lettuce 'Downy mildew'

**203** The lower and intermediate leaves of this lettuce are covered with yellow and/or necrotic spots.
***B. lactucae***

**204** If we examine the spot more closely we can see that it is angular and elongated and thus seems to be delimited by the veins.
***B. lactucae***

**205** The yellow spots are more visible on the upper face of the lamina. They can be located in the lamina or on its periphery.
***B. lactucae***

**206** An unobtrusive white down, constituted by the sporangiophores of the mildew exiting through the stomata, sprinkles the spots. These fruiting bodies produce large quantities of sporangia.
***B. lactucae***

**207** On the lower face of the lamina the spots are more spread out and less prominent.
***B. lactucae***

208

209

210

211

## *Bremia lactucae* on batavia lettuce 'Downy mildew'

208 The symptoms of mildew are fairly comparable on batavia. Some of these lettuce leaves are scattered with pale yellow spots.
***B. lactucae***

209 Larger numbers of chlorotic spots are also delimited by the veins.
***B. lactucae***

210 Intense sporulation covers the spots on the lower face of the lamina.
***B. lactucae***

211 The numerous sporangiophores present are clearly visible.
***B. lactucae***

212

213

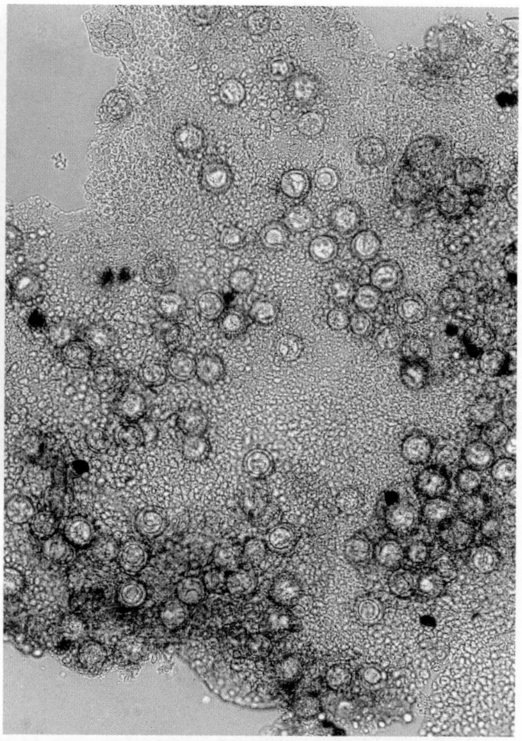

214

215

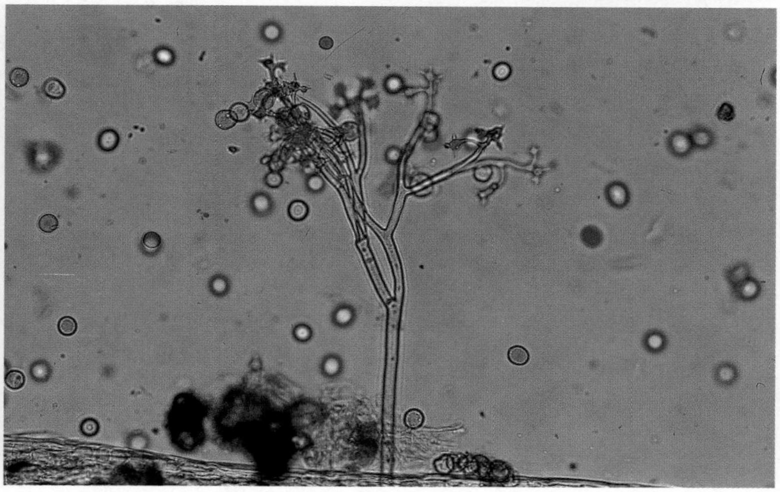

Spots

# *Bremia lactucae* on stem lettuce and fruiting bodies 'Downy mildew'

**212** On one of the lower leaves of this young stem lettuce we can distinguish a broad yellow necrotic spot indicating the start of an attack of mildew.
***B. lactucae***

**213** ***B. lactucae*** is a heterothallic fungus; two sexual types exist (B1 and B2). When they meet they form spherical oospores with a thick wall. These structures are occasionally encountered in nature on plant debris.

**214** The damaged tissues, still covered by white down, ultimately fall; some of these are riddled to varying extents.
***B. lactucae***

**215** The bushy sporangiophores of ***B. lactucae*** exit via the stomata. At the end of the sterigmata they have fairly spherical sporangia.

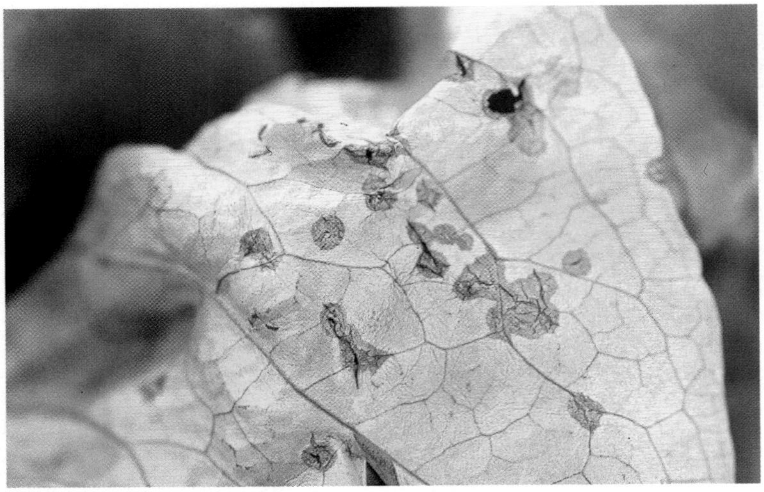

**216** Circular, damp spots, orange in colour, are characteristic of the onset of attacks of anthracnose on the lettuce lamina.
***M. panattonianum***

**217** In general they become lighter in their centre; the tissues quickly become necrotic, split and fall. Ultimately numerous holes sprinkle the leaves giving them a riddled appearance ('Shot-hole').
***M. panattonianum***

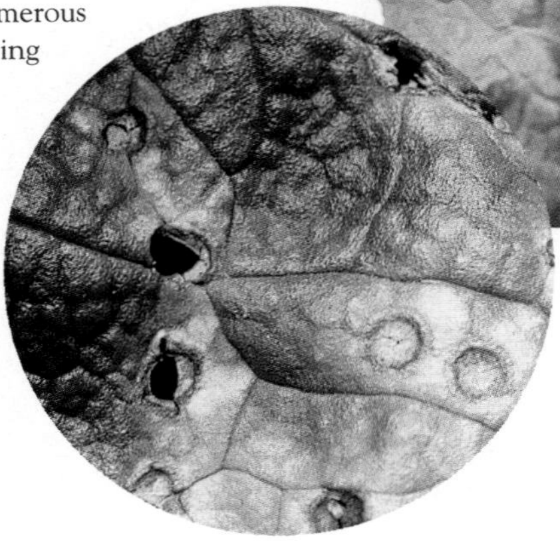

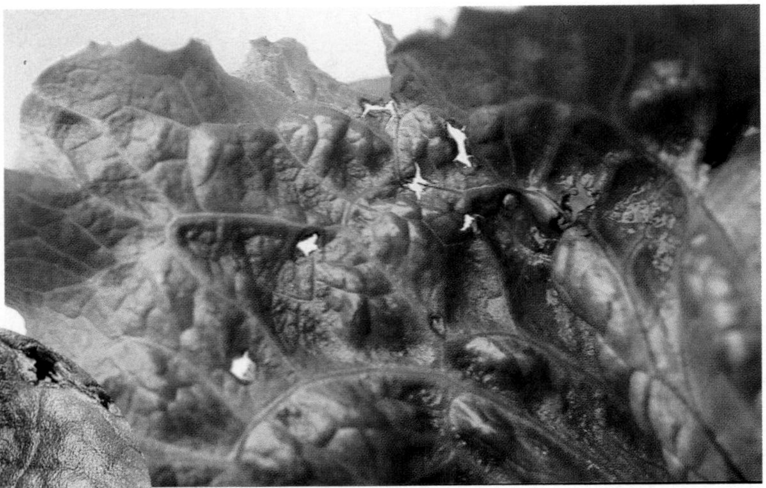

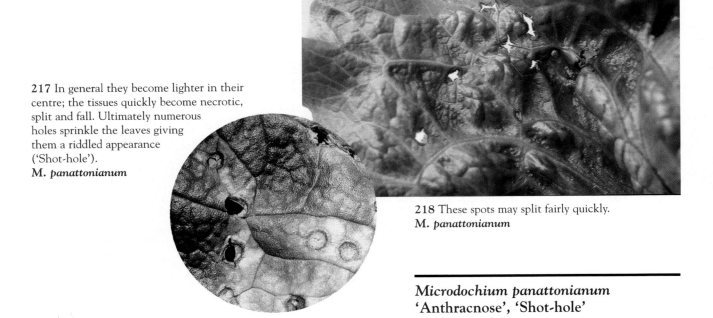

**218** These spots may split fairly quickly.
***M. panattonianum***

## *Microdochium panattonianum* 'Anthracnose', 'Shot-hole'

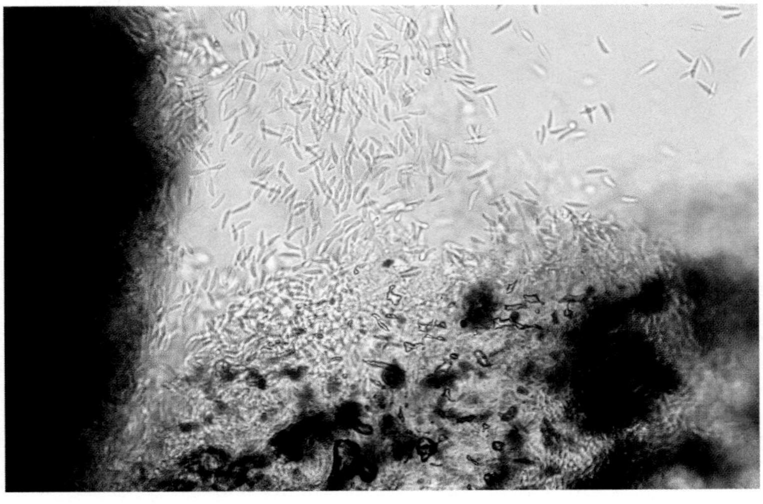

**219** Unobtrusive fruiting bodies of the fungus, acervula, form on the spots. They consist of large numbers of conidiophores grouped together; these generate hyaline, bi-cellular, cylindrical and slightly curved conidia.
***M. panattonianum***

220 

221 

222 

223 

224 

Spots

## *Microdochium panattonianum*
## 'Anthracnose', 'Shot-hole'

**220** Anthracnose mainly attacks the lower leaves of lettuces.
***M. panattonianum***

**221** The spots may be distributed over the whole lamina; in general they tend to be located at the base of the lamina.
***M. panattonianum***

**222** If conditions are damp, the spots take on a rust brown colour.
***M. panattonianum***

**223** In certain situations, broad sections of the lamina are damaged. This is due to the large number of spots and to the fact that they converge under damp conditions.
***M. panattonianum***

**224** These are more elongated on the veins and slightly concave. Under certain circumstances this involves actual small elliptical orange-coloured cankers.
***M. panattonianum***

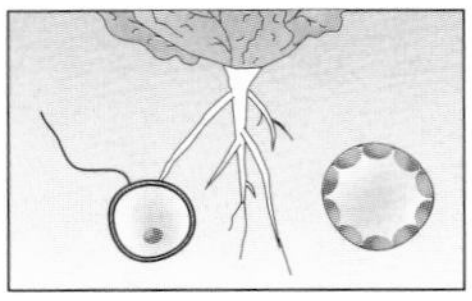

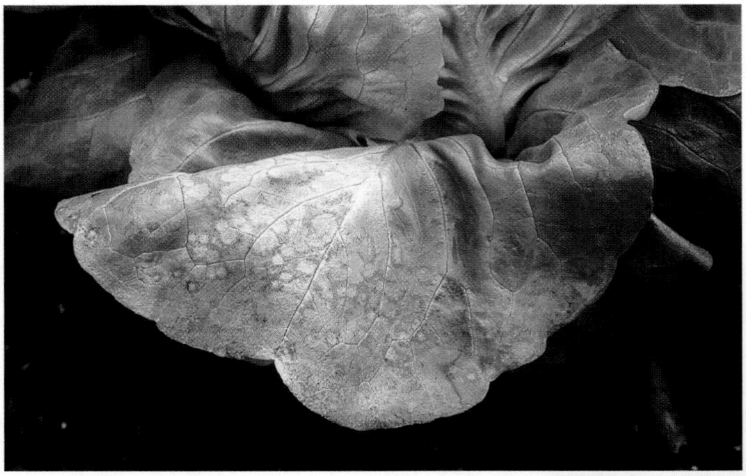

**225** Numerous transparent rings can be seen on the upper face of this lettuce leaf. Some of them are beginning to take on an orange colour.
**Lettuce ring necrosis agent (LRNA)**

**226** On the lower face of the lamina, the rings are more oily.
**LRNA**

**227** Subsequently, the rings visible on the lamina take on a brown colour.
**LRNA**

**228** Under the lamina, the more advanced damage takes on a very characteristic orange colour, which gives rise to the French name for this disease: 'orange spot disease'.
**LRNA**

**229** This damage can affect numerous leaves and be more or less widespread over the whole lamina. Portions of leaves affected quickly become necrotic and dry out.
**LRNA**

**• Lettuce ring necrosis agent (LRNA)**
(commonly known as 'orange spot disease')
This virus, which appeared during the 1980s in Europe (Netherlands, Belgium, and France), essentially attacks during winter. As its name indicates, it causes orange-coloured damage on the lower leaves, which can vary fairly substantial in size and shape. In fact, in many cases, it would be more appropriate to refer to rings rather than spots (**225–227**), therefore differentiating it from anthracnose. In addition, it is very rare for these spots and rings, although they sometimes cover the whole of the lamina of several leaves (**229** and **230**), to lead to total decomposition of the affected tissues, which then fall. The presence of 'riddled' leaves is not therefore relevant. Damage is sometimes present on the principal vein of the leaves; refer to pages 73 and 127.

This is a disease which is now fairly frequently observed under protection in various French areas of production. It is encouraged by lettuce monoculture and presents certain similarities with 'big vein' in terms of its method of transmission. Like the latter, it seems to be transmitted by a soil-based fungus which has totally adapted to aquatic life: *Olpidium brassicae* (see fact file 14). This can be easily observed in the cells of the root cortex (**231**). The virus responsible for this disease has not yet been clearly identified.

## *Lettuce ring necrotic agent* (LRNA)

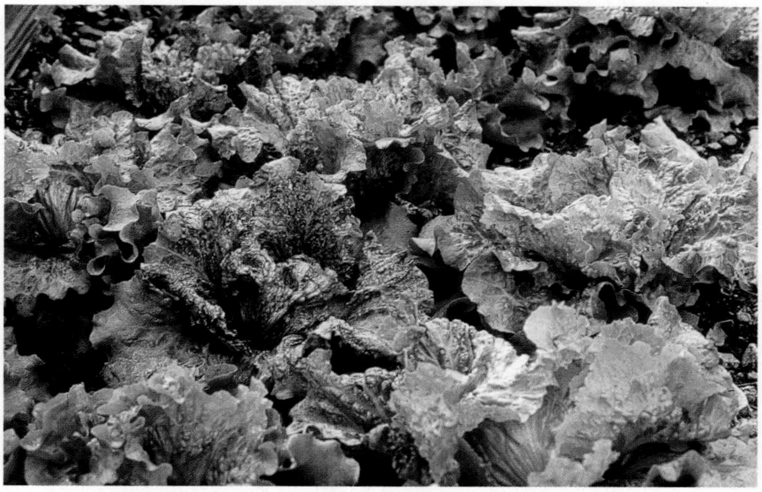

**230** Batavia lettuces are particularly sensitive; completely necrotic lettuces are found fairly frequently.
**LRNA**

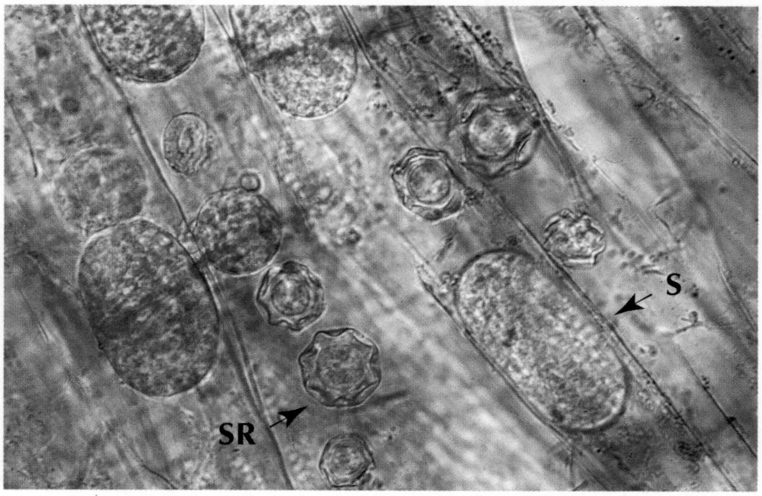


**231** In the cells of the root cortex of diseased plants, the presence of sporangia (S) and resting spores (DS) of the vector fungus, *Olpidium brassicae*, can easily be seen.
**LRNA**

*Spots*

## *Puccinia opizii*
## 'Rust'

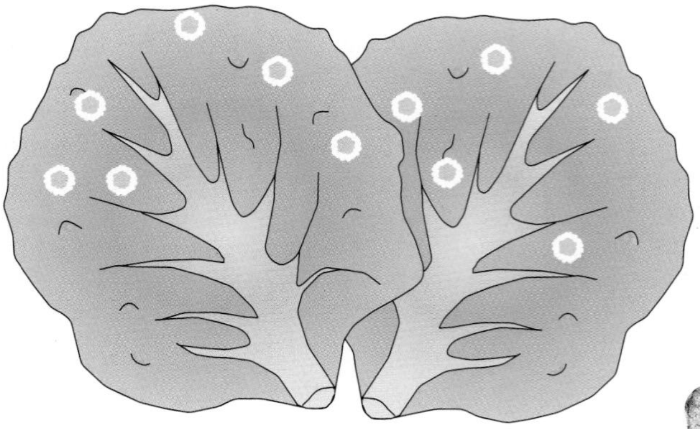

**Figure 8: The orange spots, ringed by a white halo, consist of the juxtaposition of between 50 and 200 aecidia.**

232 Some broad orange-coloured or necrotic spots can be seen on this leaf.
***P. opizii***

233 On the upper face of the lamina, we can see that these spots seem to consist of several orange-yellow juxtaposed structures. In fact, minuscule cupules can be identified on the lower face of the lamina. These are the aecidia of the fungus.
***P. opizii***

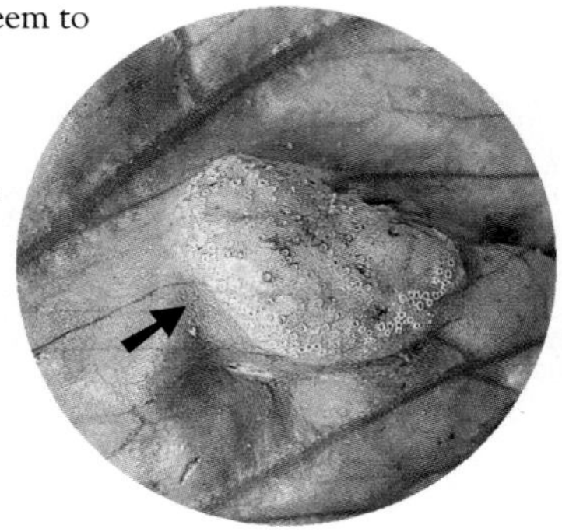

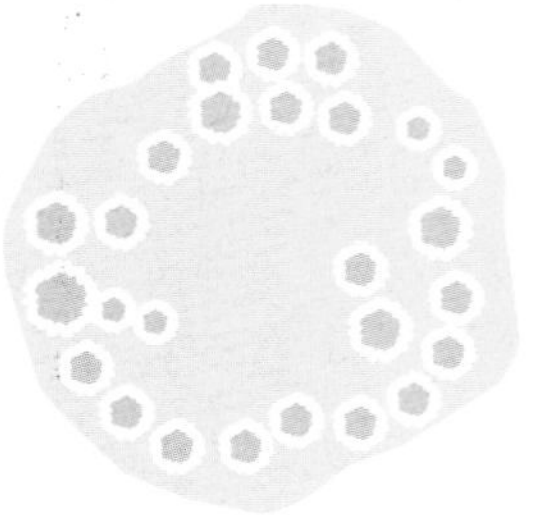

**Figure 9: Appearance of some aecidia on the lower face of the lamina, visible under a stereoscopic microscope.**

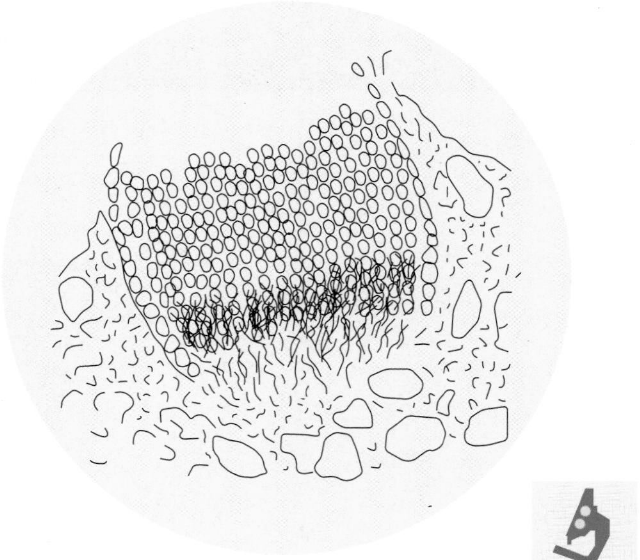

**Figure 10: Cross section view of an aecidium of *P. opizii* forming numerous globular aecidiospores on a lettuce leaf.**

**Table 14: Some characteristics for determining pathogenic agents and non-parasitic problems which cause orange-coloured spots and damage on leaves***

| Several characteristics | *Microdochium panattonianum* | Lettuce ring necrosis agent (LRNA) | Cold injury | Various types of chemical injury |
|---|---|---|---|---|
| **Frequency on lettuces** | Common. | Fairly common in France. | Occasional in winter. | Occasional, often following treatments. |
| **Stage of symptom appearance** | Any time in the nursery and after planting. | Often once the head has formed, during harvesting and trimming. | At all stages. | At any point during growing season, both in the nuersery and in the open field. |
| **Location of symptoms on the plant** | Mainly at the base of leaves, close to and all along the principal vein. | Mainly on the leaves of the crown, also close to and along the principal vein (see page 127). | Mainly on leaves surrounding head. | On leaves most exposed to pesticides during spraying. |
| **Location of diseased plants within the plot** | In groups, rapidly becoming widespread following rainfall or sprinkler irrigation. | Mainly in groups or scattered plants, especially in damper areas. | Fairly widespread or locally in the coldest parts of the plot or shelter. | Fairly widespread, variable distribution (see pages 39, 83 and 109). |

* Some uncertainty over diagnosis is possible in relation to necrotic damage caused by certain predators and *Rhizoctonia solani* on the lower part of the lamina and on the sides (see page 174).

- **Heteroecious rust ('Rust')**

***Puccinia opizii* Bub.** (heteroecious rust, whose cycle takes place on two hosts) is a fungus responsible for one of the types of lettuce rust (butterhead lettuce, oak leaf, iceberg) identified in the Netherlands, Germany, and Italy on lettuce, but also sporadically in France. The spots, which can be observed on the leaves, are fairly round and an orange-yellow colour (fig. 8 and **232**). On examining the lower face of the lamina more carefully, we can see that these consist of the juxtaposition of numerous aecidia (aecidiolum, aecium) which, on maturity, take the form of minuscule cupulae (figs 9 and 10). These, which are yellow to orange in colour (**233**), release large yellow spores (aecidiospores). The other stages of this rust (uredospores and teliospores) take place on an alternative host, the carex (*Carex muricata*) which invades swamps.

A heterocyclic rust, causing identical symptoms, has been reported in the USA and in Canada under the name of ***Puccinia dioicae* Magnus**. This is encouraged by hot, wet weather. There seem to be differences in susceptibility between varieties of lettuce.

In general, no intensive methods need to be implemented in order to combat this disease. In the rare nurseries where it may cause damage, treatment can be carried out with the fungicides traditionally used on lettuce, such as zineb and mancozeb which are effective on this type of rust. Crops should not be placed in proximity to zones where carex grows spontaneously. If this is the case, this plant must be removed (buried or destroyed) over a range of at least 200 metres. In Italy, crop rotation of at least 2 years is recommended.

A similar rust is probably rife in the USA: ***P. dioicae* Magnus**. It has several synonyms: ***P. extensicola* (Plowr.), *P. extensicola* (Plowr.) var. *hieraciata* (Schw.) Artur.**

## Cold injury

**234** On these leaves from the lettuce crown we can easily see elongated inter-veinal damage of an orange colour on the lower face of the lamina (leaf on the right). The latter is slightly blistered underneath (leaf on the left).
**Cold injury**

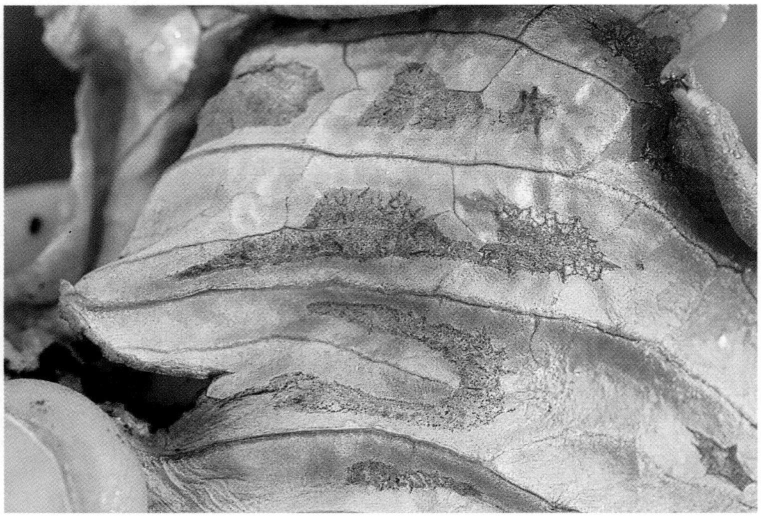

**235** The epidermis has split in certain places; orange lesions, which are slightly depressed, have developed.
**Cold injury**

**236** Places where the tissue has split can also be seen all along the principal vein.
**Cold injury**

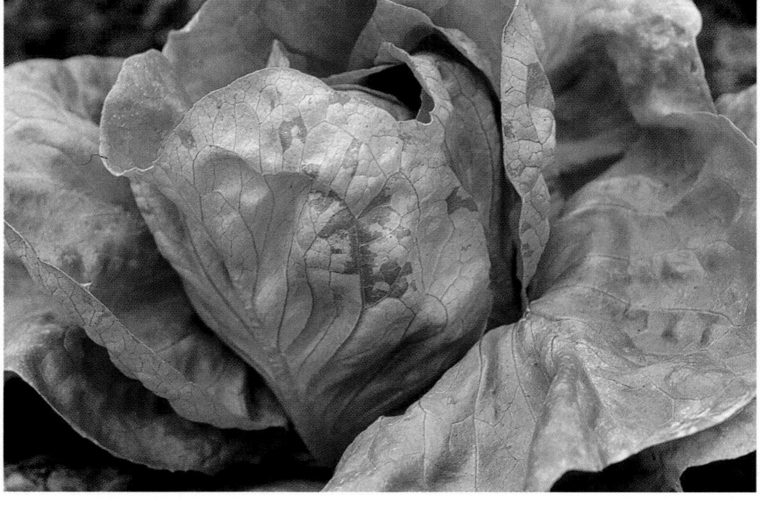

**237** Several islets of cells are transparent and damp in appearance.
**Vitrescence of the lamina**

## • Cold injury

The tender and succulent tissues of certain l.s.v. do badly in low temperatures. After a frost, it is quite frequent to see the appearance of elongated inter-veinal lesions, corresponding to the formation of ice on the epidermis. The upper face of the lamina can be wrinkled or blistered (**234**). Under the lamina the epidermis peels away from the underlying tissue; the affected portions then take on a silver-coloured appearance. Fairly frequently the epidermis breaks and the mesophyll tissues, which are better protected, take on an orange to brown colour (**235**). The fragmenting tissues may split along the principal vein (**236**).

It should be noted that during winter, in greenhouses and cold tunnels, lettuce leaves can have a lamina which is superficially 'blistered' with a duller appearance. Experts call this physiological disease lettuce **'toad skin'**.

Low temperatures and condensation of cold water on the leaves lie at the origins of these symptoms, which do not cause very much damage. High salinity can cause comparable damage.

## • Lamina vitrescence

This minor physiological affliction gives rise to oily transparent spots, delimited by the veins, appearing on the lamina of lettuces (**237**). These stains can be seen in greenhouses where the humidity of the air is excessive and/or the water content of the plants is too high. Within this context, evaporation from the plants is inadequate and the leaf tissues, engorged with water, create a more favourable medium for this manifestation. Comparable symptoms can be noted after harvesting if storage temperatures have been lowered too abruptly.

In the greenhouse, the symptoms of lamina vitrescence (glassiness) can be reversed if the climate within the shelter is modified by providing ventilation and heating. Differences in sensitivity between varieties and types of lettuce have been noted in the field.

## • Chemical injury

Orange-coloured leaf damage (**238** and **239**) are sometimes observed following treatment with pesticides. The very tender l.s.v. tissues, particularly those grown under protection, are particularly vulnerable. This situation is encountered from time to time, for example when fungicides based on phosetyl-Al are used.

If you require further information about chemical injury we suggest that you refer to the pages 39, 83 and 109.

## Chemical injury

**238** The lower face of this lettuce leaf presents numerous small orange to brown patches of necrosis distributed over the distal portion of the lamina; this part is the one most exposed to chemical treatments.
**Chemical injury**

**239** Some chlorotic lesions quickly becoming necrotic are developing close to the veins of this lettuce leaf.
**Chemical injury**

240 Damp orange brown lesions can be seen on the lamina and the principal vein of several lettuce leaves.
**Slug damage**

## Slug damage

241 On carefully examining the leaves, we can see that these lesions are due to bites from the slugs which are eating the lamina. The animals which have produced this damage are never far away.
**Slug damage**

## Leaf miner damage

242 Numerous puncture marks, slightly in relief, oily to orange in colour, indicate the action of leaf miners. These are minuscule flies, approximately 2 mm long.
**Leaf miner damage**

244

243 Rather sinuous deposits are present on old leaves.
**Leaf miner damage**

- **Damage due to predators (slugs, leaf miners, thrips)**

Several predators are responsible for the rather orange damage on l.s.v. laminae. We have deliberately included these types of damage under this heading because they may often be confused with other symptoms studied in this section.

This is the case with damage caused by certain **slugs** (*Deroceras reticulatum*, *D. laeve*, *Arion hortensis*) which feed very locally on the leaf tissues of the lamina and veins (**240**). Observation of these predators on or within the l.s.v. environment should allow you to confirm this hypothesis very quickly (**241**).

Various insects also cause minuscule orange damage, which can often be detrimental if they are numerous. **Leaf miners** produce numerous oily to orange holes, slightly in relief, after feeding (**242**). We can also distinguish sinuous deposits located mainly on old leaves (**243**). Several species of *Liriomyza* can cause these symptoms. In France it is essentially *Liriomyza huidobrensis* which appears on lettuces. In the USA it is primarily *L. langei*. Leaf miners are small flies of a few millimetres in length (around 2 mm), predominantly black in colour (**244**). Their minuscule yellow larvae move around within the thickness of the lamina producing the deposits.

**Thrips** cause minuscule patches of elongated orange necrosis, with a metallic tinge (**245** and **246**). These can be located on or between the veins. Although this causes little damage in France, thrips can cause serious damage in tropical areas, as is the case with *Thrips palmi*. Thrips are minuscule insects of approximately 1 mm in length, varying in colour (black, yellowish, amber yellow, and so on) which can sometimes be seen with the naked eye. The adults are fairly mobile. Some species (*Frankliniella occidentalis*) are vectors of a serious lettuce virus: tomato spotted wilt virus (TSWV). You can refer to the symptoms caused by this virus on pages 150 and 151.

## Damage caused by thrips

**245** The areas of the lamina where the thrips take their nourishment take on an orange colour with a metallic glint.
**Damage caused by thrips**

**246** In damaged areas of the leaf, we can distinguish elongated orange-coloured necrosis on the lamina, associated with the presence of thrips.
**Damage caused by thrips**

**247** A damp brown and black lesion extends over the principal vein and some secondary veins of this batavia lettuce leaf.
***Pseudomonas cichorii***

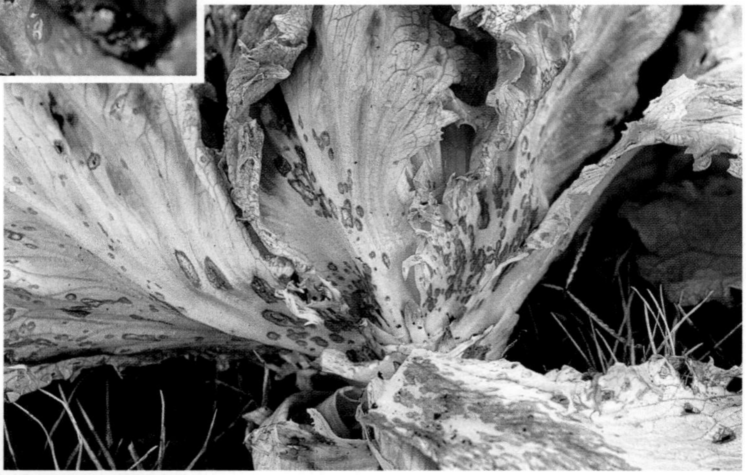

**248** A multitude of small elliptical spots, which may be depressed, cover the principal vein and lamina of this batavia lettuce.
***Michrodochium panattonianum***

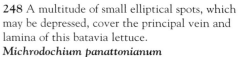

**249** On the vein of this curly endive we can clearly see an elongated area of damage, split in the centre.
**Latex spots**

**250** Numerous leaves of this lettuce have dried out all around the lamina edge.
**Dry marginal necrosis ('Tip burn')**

## Examples of damage situated on the principal vein or on the periphery of the lamina

# Spots and damage mainly located on the principal vein or on the leaf edges

## Possible causes

- *Microdochium panattonium* (fact file 3)
- *Rhizoctonia solani* (fact file 7)
- *Erwinia carotovora* subsp. *carotovora* (see photo 18)
- *Pseudomonas cichorii* (fact file 15)
- Other bacteria (*Pseudomonas marginalis* pv. *marginalis*, *P. viridiflora*, and so on) (fact file 17)
- Phytoplasma from the Aster yellows group (fact file 20) (refer to page 79)
- Lettuce ring necrosis agent (LNRA) (fact file 35)
- Other viruses (BWYV, INSV, and so on) (refer to pages 75 and 152)
- Marginal necrosis ('Tip burn')
- Excessive salinity
- Latex spots
- Other vein damage ('Pink rib'; 'Brown rib'; 'Russet spotting'; 'Brown stain'; various types of chemical injury)

A certain number of diseases still or in particular contexts, cause symptoms mainly located on the principal vein or around the periphery of the lamina. This location should certainly attract your attention. It is for this reason that we have gathered together, under this heading, the principal parasitic and non-parasitic diseases presenting this particular aspect.

## Supporting arguments

- **Lettuce ring necrosis agent (LRNA)**

The first symptoms of **orange spot disease** can occur on the periphery of certain leaves in the crown. Thus, a certain portion of the tissues located around the edge of the lamina become transparent or oily, go brown and become progressively necrotic (**251** and **252**). In certain cases the principal vein can also be affected. It then presents a fairly localized, orange to rust brown colour, which is quite characteristic.

The agent of this viral disease is transmitted by a soil-based fungus: *Olpidium brassicae*. If you are in any doubt, this disease also causes much more characteristic symptoms, which you can refer to on page 73 and 119.

- ***Rhizoctonia solani*** **(Bottom rot)**

When attacks from this soil fungus are particularly severe, a multitude of superficial or deep areas of damage can develop on the lettuce veins (**253**). Very often the neighbouring lamina is rotten and/or may have decomposed. If this has not occurred, the observer tends to focus his attention on the veins. This fungus is found very frequently in the soil in market gardens.

See pages 167, 169, 174–176 where all the symptoms this fungus causes on lettuce are described and illustrated.

- ***Microdochium panattonianum*** **('Anthracnose')**

This fungus, responsible for anthracnose in l.s.v., attacks the whole lamina. On the veins, the spots it causes have a particular appearance. They are always more elongated than on the lamina and very depressed (**254**). In certain cases we could describe them as cankerous spots.

Pages 111, 116, and 117 contain more information about this airborne fungus disease.

251

252
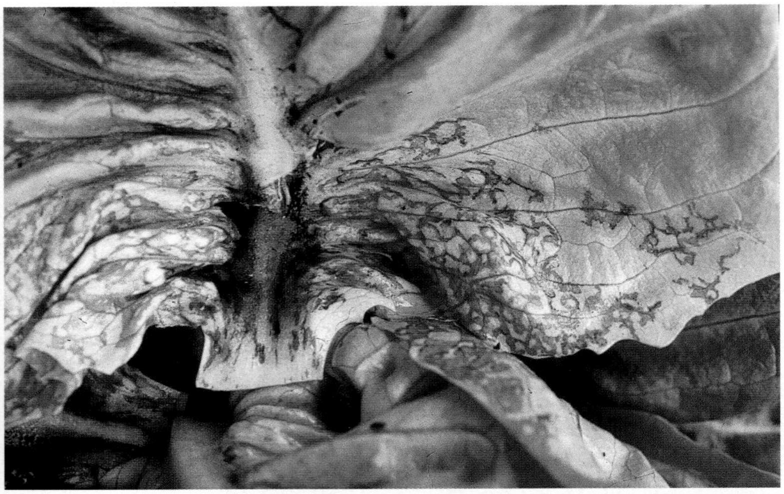

253
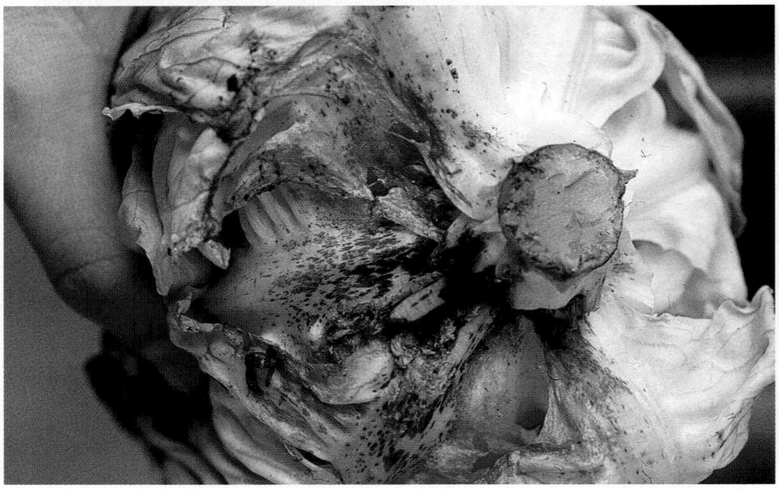
254

**Some parasitic diseases that sometimes affects the principal vein of lettuce leaves and which may lead to confusion**

251 Damp to oily areas of damage, orange to brown in colour, are clearly visible on the periphery of the chlorotic lamina of this lettuce.
**Lettuce ring necrosis agent (LRNA)**

252 We find the same type of damage at the base of the lamina and the principal vein of this lettuce leaf.
**LRNA**

253 Several veins of this lettuce are sprinkled, or even covered, by numerous small, elongated lesions of a rust-brown colour.
***Rhizoctonia solani***

254 ***Microdochium panattonianum*** is particularly aggressive on lettuce veins. Several concave, elongated cankerous lesions on this lettuce leaf clearly illustrate this fact.

- ***Pseudomonas cichorii* and other bacteria**

Damp brown damage is observed from time to time around the periphery of the lamina or on the principal vein of lettuce leaves (**255–257**). Several bacteria with different parasitic potential can be isolated on these lesions:

- pathogenic bacteria benefiting from climatic conditions which favour its establishment (***Pseudomonas cichorii*, and so on**);
- saprophytic, opportunist bacteria which take advantage of weakened plants, necrotic tissues (due for example to marginal necrosis), in order to colonize the leaves. They produce dark brown to black rotting, the extent of which depends on surrounding humidity (***P. marginalis* pv. *marginalis*, *P. viridiflava*, and so on**).

Additional information about this bacteria is provided on pages 97 and 99.

**255** Several leaves from the head of this escarole reveal peripheral blackening. ***Pseudomonas* spp.**

**256** Damp, transparent lesions blacken and progressively extend to the periphery of this lettuce leaf, particularly if damp conditions persist. These symptoms can occur following 'tip burn' damage, on which the bacteria can establish themselves. ***Pseudomonas* spp.**

**257 *Pseudomonas cichorii*,** in addition to producing spots on leaves, has the particular feature of blackening the principal vein of the l.s.v. on which it is a parasite, sometimes entirely. In certain cases, the bacterium spreads towards the lamina which quickly deteriorates and blackens.

258  259 

260  261 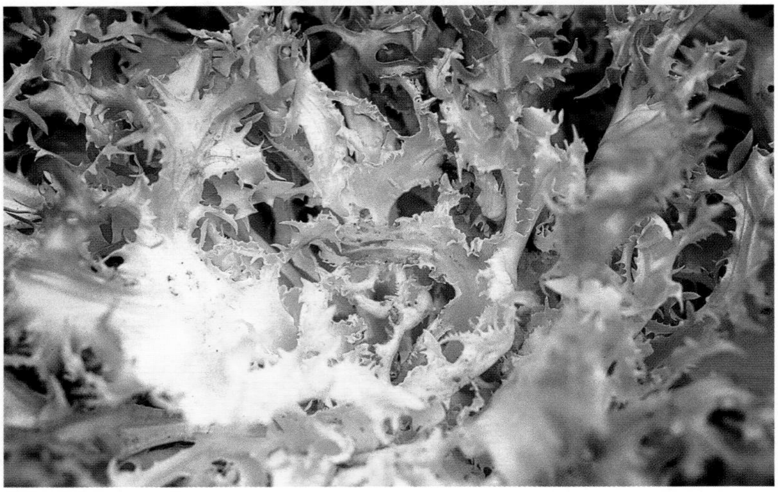

## Latex spots
### *An indeterminate disease*

**258** Two latex spots, orange pink to brown, are clearly visible on the principal vein on the lower face of this lettuce leaf.
**Latex spots**

**259** Islands of cells where the latex is accumulating finally break up, and the latex spreads out over the surrounding lamina.
**Latex spots**

**260** Secondary bacteria are rapidly colonizing the damaged tissue and increasing the browning effects observed.
**Latex spots**

**261** Latex can be released and may be found on certain young leaves close to areas of damage; they take on a fairly specific rust colour.
**Latex spots**

## • Latex spots

L.s.v. growers, especially those cultivating chicory in the south-east of France, are occasionally confronted with a new syndrome whose cause is not yet known.

The damage essentially manifests itself on adjacent veins and/or the lamina. Here we can see the presence of a few fairly small lesions, orange to light brown in colour (**258**). In fact these are pockets where latex accumulates before splitting (**259** and **260**). When this oxidizes, it takes on a rust brown colour and stains the surrounding leaf tissue (**261**).

The aetiology of the disease mainly concerns the possibility of phytoplasmosis. In fact, a phytoplasma from the Aster yellows group has been indicated in literature as being responsible for latex spots on leaf veins. This symptom is frequently accompanied by others, such as yellowing and reduction in the size of leaves and dwarfism of plants. None of these has been observed within the context of French crops. The laboratory investigations carried out in order to reveal a phytoplasma have been fruitless, failing to confirm the possibility of late attacks which only produce latex spots.

More exhaustive aetiological studies, incorporating research in the field, need to be carried out before we can properly understand and possibly identify a factor capable of explaining this syndrome.

## • Other damage affecting the veins and adjacent tissues

There have been American and British reports of other diseases, either non-parasitic or whose cause is uncertain, affecting the veins or adjacent tissues of lettuce leaves.

- 'Pink rib' mainly affects crisphead lettuce, of advanced maturity, and after harvesting. As its name indicates, the vein (rib) tissues of external leaves take on a fairly unusual pink colour. The cause of this disease is not known. Low oxygen content and high temperatures within storage premises increase the damage. Another problem, associated with $CO_2$ content in storage premises, has been identified after harvesting on the same type of lettuce, and this is known as 'brown stain'. Brown, oval shaped lesions, which are rather dry, appear at the base of the leaves.
- **'Rib blight', rib discolouration and 'brown rib'** involve the progressive blackening of tissues located on the principal and secondary veins, especially on the external leaves of the head. The origin of this disease is unknown.
- **'Russet spotting'** essentially occurs when lettuces which have undergone various types of stress in the field are stored. Small oval-shaped dry lesions, 2–4 mm in length, more or less cover the veins and the lamina, particularly on the lower part of the leaves. The presence of ethylene and temperatures of around 5°C encourage the appearance of these lesions.

Do not forget that certain herbicides can cause necrotic damage to appear on the principal vein and secondary veins of l.s.v. (**262**). See pages 39, 83 and 109 in order to obtain further information about chemical injury.

**262** Orange to brown lesions, located on the veins, can be a sign of the effects of certain herbicides on lettuces.
**Miscellaneous chemical injury**

## Marginal necrosis ('Tip burn')

**263** Marginal necrosis can affect the lower leaves as is the case with this lettuce.
**Marginal necrosis**

**264** Dark chestnut to brown spots, dry in consistency, are scattered around the perimeter of the lamina of the lower leaves; we can see that the lamina is also chlorotic in this area.
**Marginal necrosis**

**265** Sometimes the lamina can be entirely edged with brown, necrotic damage which is also slightly chlorotic.
**Marginal necrosis**

**266** Marginal necrosis can appear on the leaves surrounding the head and heart. Here, a few small unobtrusive transparent areas of damage, which are difficult to see, are scattered on the periphery of this leaf from the heart.
**Marginal necrosis**

**267** If conditions favourable to the appearance of this physiological disorder persist, a multitude of brown necrotic spots appear.
**Marginal necrosis**

• **Marginal necrosis ('Tip burn')**

For many years the physiological types of leaf necrosis observed around the edge of the laminae of lettuces have all been grouped together under the name of marginal necrosis or 'tip burn' (**263–272**). The symptomatology and aetiologies of these types of **physiological necrosis** have recently been more accurately described. In fact, two syndromes seem to appear:

- firstly a more traditional syndrome, characterized by small brown to black spots appearing around the edge of the laminae of **old leaves from the crown**. The small veins, present nearby, may also go brown. Subsequently, these localized lesions rapidly extend. Ultimately, brown areas, rather dry in appearance, progressively take over the interior of the lamina. This syndrome is known as **dry marginal necrosis or 'tip burn'** of lettuces. It may be associated with 'latex necrosis' which also produces dry damage around the periphery of the lamina, but is preceded by vessels bursting with local emission of this milky compound;
- secondly a syndrome grouping together various types of damp necrosis developing on the **leaves surrounding the head** and **the heart**. As previously, small veins located around the periphery of the lamina become transparent and take on a brown colour. The surrounding tissues go brown in a fairly widespread way and soft rotting begins. In places we can see minuscule spots and the browning of short sections of the veins. This type of damage is known as **damp marginal necrosis**.

We may wonder whether there is any point in making a distinction between these two types of 'tip burn', as their origin and many of the means used to avoid them are the same. These are summarized in *Table 15*. In addition, the distinction between dry or damp marginal necrosis is fairly subjective. In fact, depending on the climatic conditions occurring after their appearance, the damaged tissue will remain damp or will become so in the presence of high humidity and *vice versa* during dry periods.

Whatever the nature of these types of necrosis, they constitute an excellent nutritional base on which opportunist micro-organisms, such as *Botrytis cinerea* (**272**) and various bacteria, establish themselves. They lead to rotting of the lamina and the head, particularly if the prevailing humidity is high.

We should point out that l.s.v., and lettuce in particular, although moderately sensitive to salinity, may present comparable symptoms following abrupt increases in the **concentration of salt in the soil**.

**268** The veins located around the periphery of the leaves may become necrotic.
**Marginal necrosis**

**269** Subsequently, the tissues display widespread browning and give the impression that soft rotting is starting to occur.
**Marginal necrosis**

**270 Marginal necrosis** can result in the formation of broad fringes of necrotic tissue around the edge of the lamina; these will dry out if climatic conditions allow.

271 There are differences in sensitivity between types of lettuce. Varieties with fine and very serrated leaves are particularly susceptible.
**Marginal necrosis**

272 The damaged tissues constitute a nutritional base which is very conducive to the establishment of *Botrytis cinerea*, and to a lesser degree, *Sclerotinia sclerotiorum*.
**'Tip burn' + *Botrytis cinerea***

## Marginal necrosis ('Tip burn')

Table 15: Principal characteristics of marginal necrosis observed on lettuces

| | 'Tip burn' | Deep marginal necrosis |
|---|---|---|
| **Causes** | • Reduced water supply to leaves, resulting in reduced calcium flow.<br>• Rupture of latex bearing channels releasing toxic latex. | • Poor migration of calcium to young gorwing leaves following insufficient plant transpiration. |
| **Situations which favour marginal necrosis** | • Insufficient water supply to plants.<br>• Climatic conditions becoming hot and dry at an excessively rapid rate.<br>• Poor condition of root system (poorly established, damage associated with biotic or abiotic aggression).<br>• Inappropriate nitrogenous fertilizing.<br>• Excessive evaporation associated with wind or excessive ventilation under protection. | • Insufficient ventilation of shelters whose prevailing humidity is too high.<br>• Nitrogenous fertilizing which is out of balance.<br>• Prolonged period of overcast warm weather conducive to high humidity under shelter. |
| **Measures to be taken** | • Carry out an analysis of the soil before planting in order to ensure balanced fertilizing and to avoid excessive nitrogen and calcium deficiency.<br>• Prepare the ground carefully and encourage good root development of plants.<br>• Choose a variety which has a vigorous root system as this is often less susceptible.<br>• Ensure optimum water supply especially during dry climatic periods. | • Carry out soil analysis before planting in order to ensure balanced fertilizing and to avoid excessive nitrogen and calcium deficiency.<br>• Avoid susceptible varieties which have a small root system.<br>• Provide maximum ventilation under protection.<br>• Limit, or even halt irrigation during periods of high humidity.<br>• Encourage good plant functions as much as possible.<br>• Counteract abrupt rises in temperature by gentle spraying. |

- **Excessive salinity**

An abrupt rise in soil salt content can result in the necrosis of the periphery of the lamina as well as browning of the veins.

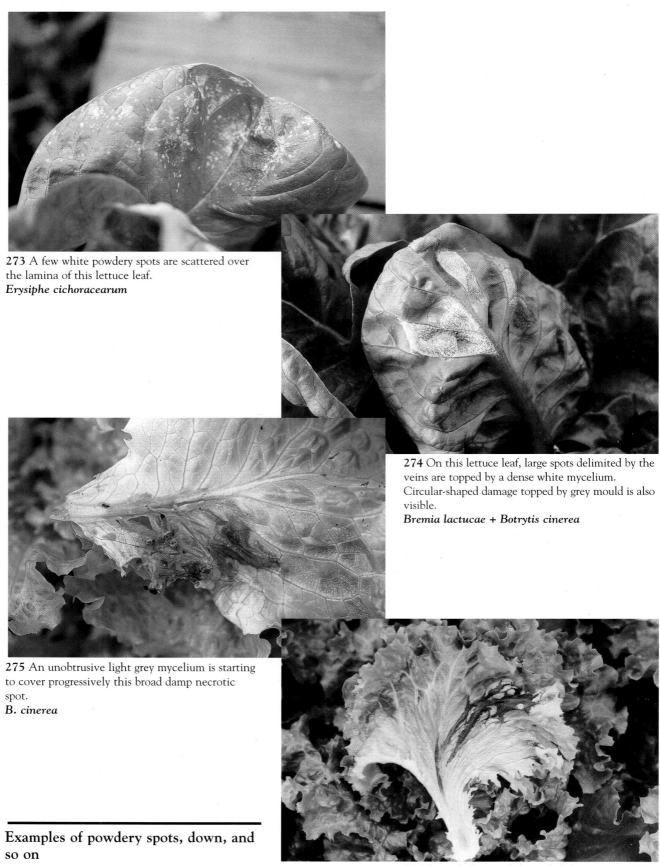

**273** A few white powdery spots are scattered over the lamina of this lettuce leaf.
***Erysiphe cichoracearum***

**274** On this lettuce leaf, large spots delimited by the veins are topped by a dense white mycelium. Circular-shaped damage topped by grey mould is also visible.
***Bremia lactucae* + *Botrytis cinerea***

**275** An unobtrusive light grey mycelium is starting to cover progressively this broad damp necrotic spot.
***B. cinerea***

**276** Several elongated damp dark brown areas of damage are extending from the veins.
***Pseudomonas cichorii***

## Examples of powdery spots, down, and so on

# Spots with powdery areas, matting, mould on leaves, damp spots leading to rotting of the head

## Possible causes

- *Botrytis cinerea* (fact file 5)
- *Bremia lactucae* (fact file 1)
- *Erysiphe cichoracearum* (fact file 2)
- *Sclerotinia sclerotiorum* (fact file 6)
- *Sclerotium rolfsii* (refer to page 178) (fact file 10)
- Various types of fungus (*Alternaria porri* f. sp. *cichorii*, *Cercospora longissima*, *Stemphylium botryosum* f. *lactucum*, and so on) (see page 105, 106, and so on) (fact file 4)
- *Erwinia carotovora* subsp. *carotovora* (fact file 18)
- *Pseudomonas cichorii* (fact file 15)
- *Pseudomonas marginalis* pv. *marginalis* (fact file 17)
- *Pseudomonas viridiflava*
- *Xanthomonas campestris* pv. *vitians* (fact file 16)
- Sooty mould associated with the presence of insects

## Arguments in support of the diagnosis

- **Erysiphe cichoracearum (Powdery mildew)**

Determining this mildew of l.s.v. does not generally pose any problem. In fact, the white powdery matting covering the upper face of the leaves (**277–279**) is a symptom which is sufficiently characteristic to eliminate any uncertainty in diagnosis. This white matting actually consists of a network colonizing the surface of the lamina, topped by numerous short conidiophores, each of which has several hyaline conidia in a chain (**280**). The presence of *Erysiphe cichoracearum* on the leaves ultimately leads to chlorosis and browning of the lamina. Minuscule globular cleistothecia (organs responsible for its sexual reproduction), located close to the veins, have been referred to in literature. In spite of the specific nature of these symptoms, some confusion sometimes occurs, especially when the leaves are spotted with white following sprinkler irrigation using very chalky water or chemical treatments which leave a deposit. In France, this obligate parasitic mould is fairly rare on l.s.v. It tends to be found on chicory, and more rarely on lettuce. Initially it colonizes the lower leaves. It seems clear that these leaves are more receptive and that this mould finds, within the plant cover, conditions (especially climatic ones such as poor light, higher humidity) particularly propitious to its development.

- ***Bremia lactucae* (Downy mildew)**

Lettuce mildew, in addition to producing fairly broad spots, light green to yellow in colour, fruits abundantly on these spots, essentially on the lower face of the lamina. A white down, varying in density, constituted by the sporangiophores and sporangia of the mildew covers the affected tissues (**281**). In certain situations of extreme humidity, it may sporulate abundantly on the leaves of the heart (**282**). If you require more information and details about this mould, see pages 111 and 115.

**277** Several small white powdery spots sprinkle the upper face of the lower leaves of this lettuce.
***E. cichoracearum***

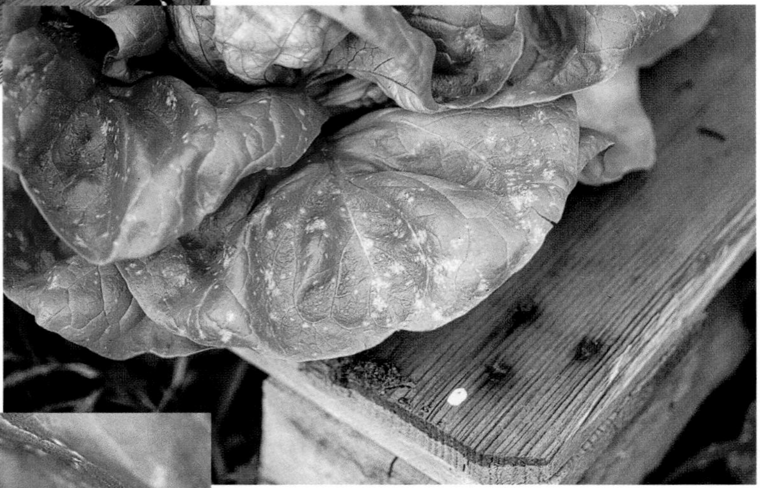

**278** The spots actually consist of the superficial mycelium of the mould and a profusion of conidiophores and conidia which form a chain.
***E. cichoracearum***

**279** This fungus, which is an obligate parasite, disturbs the functioning of the plant cells which die after a certain time. This results in fairly widespread chlorotic sections appearing on the lamina, which become greyish to light brown in colour.
***E. cichoracearum***

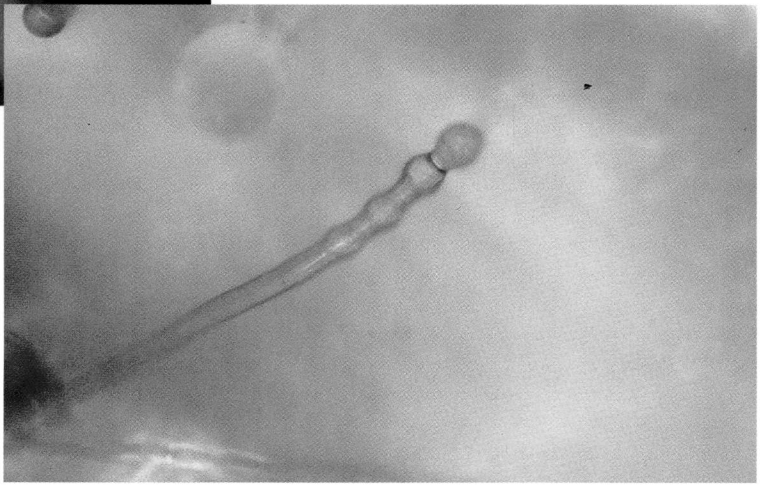

**280** Detail of a young conidiophore producing its first conidia.
***E. cichoracearum***

• *Botrytis cinerea* (Botrytis leaf spot)
*Sclerotinia sclerotiorum*
('Sclerotinia leaf spot')

**Various types of bacteria** *(Erwinia carotovora* subsp. *carotovora, Pseudomonas cichorii, Pseudomonas marginalis, Pseudomonas viridiflava, Xanthomonas campestris* pv. *vitians)*

These two types of fungi and these various types of bacteria behave with a certain degree of similarity on lettuces; they may even cohabit on the same areas of damage. They are micro-organisms which are often opportunistic and establish themselves on plants, sometimes causing serious damage in the presence of particular favourable conditions:

- prolonged period of damp, rainy weather;
- stagnation of water on the leaves, partic-ularly at the edge of the lamina;
- existence of miscellaneous types of damage;
- presence of senescent or necrotic tissue;
- plants growing very quickly, with succulent tissues.

They find it fairly easy to colonize areas of damage and senescent tissues which constitute entrances or nutritional bases which will facilitate their penetration and their development within the plants. With a wealth of effective enzymatic equipment, they break down the leaf tissues fairly quickly.

*Botrytis cinerea* and *Sclerotinia sclerotiorum* are easily identifiable because fairly quickly they produce a grey mould or white mycelium and very large sclerotia on the affected tissues. In order to encourage the appearance of these fruiting bodies, we recommend that you place a few samples of affected lettuces in a sealed box or a plastic bag containing filter paper soaked in water. This device will provide the humidity which is indispensable to the formation of these characteristic structures on plant tissues.

The damage caused by bacteria often takes on a fairly specific dark brown to black colouration. In this case no particular structures are observed.

We recommend that you consult *Table 16*, page 143.

282 Under certain conditions, the sporulation of the mildew can be abundant on both faces of the lamina, as is the case on these young leaves. ***B. lactucae***

281 A white down covers portions delimited by the veins on the lower face of the lamina. This consists of the very numerous sporangiophores and sporangia produced by the fungus. ***Bremia lactucae***

***Bremia lactucae***
**(Downy mildew)**

## *Botrytis cinerea* (Botrytis leaf spot)

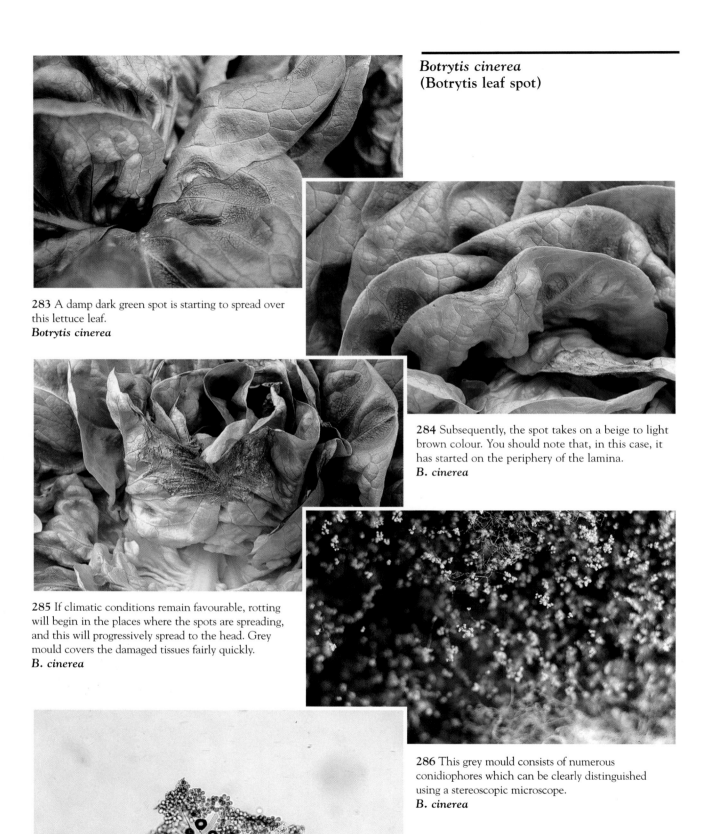

283 A damp dark green spot is starting to spread over this lettuce leaf.
***Botrytis cinerea***

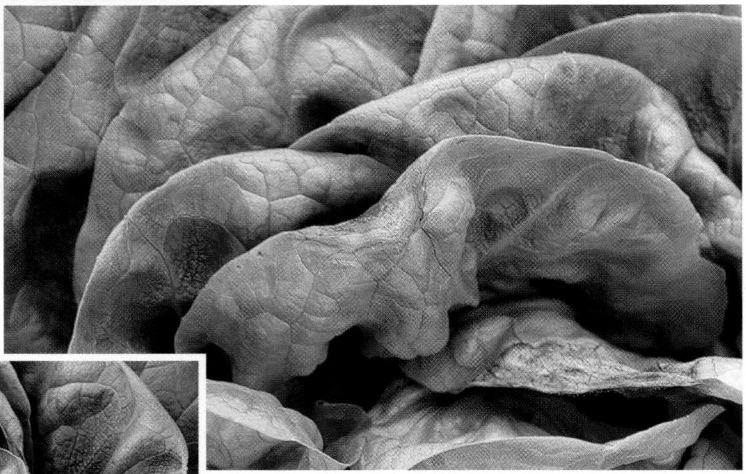

284 Subsequently, the spot takes on a beige to light brown colour. You should note that, in this case, it has started on the periphery of the lamina.
***B. cinerea***

285 If climatic conditions remain favourable, rotting will begin in the places where the spots are spreading, and this will progressively spread to the head. Grey mould covers the damaged tissues fairly quickly.
***B. cinerea***

286 This grey mould consists of numerous conidiophores which can be clearly distinguished using a stereoscopic microscope.
***B. cinerea***

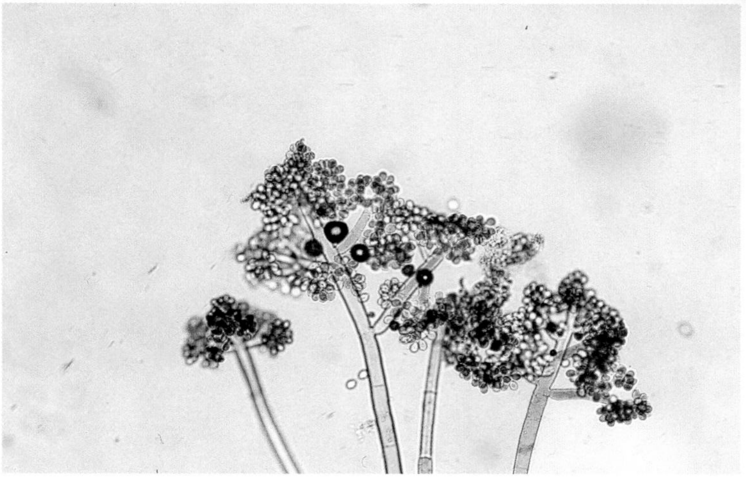

287 The conidiophores are ramified and have numerous conidia.
***B. cinerea***

288 A broad bluish spot is beginning at the edge of the lamina. This is associated with aerial contamination due to one or several ascospores which have come from the apothecia. A dense, cottony white mycelium is beginning to partially cover it.
***Sclerotinia sclerotiorum***

289 ***S. sclerotiorum*** has spread to leaves in the head which are completely rotten. Here and there we can distinguish the white mycelium and one to several large black sclerotia.

## *Sclerotinia sclerotiorum* ('Sclerotinia leaf spot')

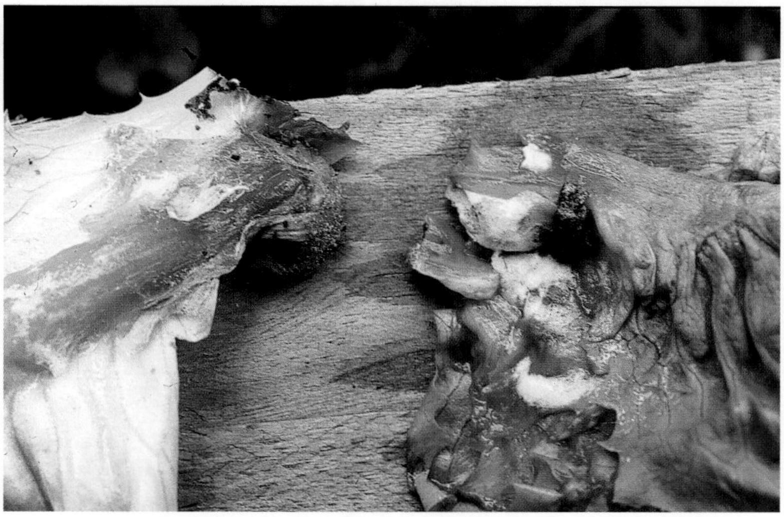

290 How to distinguish between the two types of fungi:
- ***Botrytis cinerea*** (on the left) produces a grey mould;
- ***Sclerotinia sclerotiorum*** (on the right) is characterized by its dense white mycelium and large sclerotia.

291

292

293

## Bacterial leaf rot

**291 Bacterial attacks** often produce damp dark brown to black spots which are sometimes surrounded by a yellow halo, which may be quite marked.

**292** If free water is present over a prolonged period it is fairly common for the leaves in the head and the head to rot completely.
**Various types of bacteria**

**293** In certain situations, the leaves decompose and the lettuces give the impression of melting.
**Various types of bacteria**

Table 16: Some criteria to differentiate between the pathogenic agents responsible for damp spots on leaves which develop into rot

| Certain characteristics | *Botrytis cinerea* | *Sclerotinia sclerotiorum* | Various types of bacteria |
|---|---|---|---|
| **Principal symptoms observed on lettuce leaves** | • Damp light brown to beige spots on leaves, often situated on the edge of the lamina (**283–285**); damp dark brown rotting of the head.<br>• Other symptoms: see pages 168 and 171. | • Bluish or light brown to beige spots on leaves (**288**); damp dark brown rotting of the head (**289**).<br>• Other symptoms: see page 173. | • Damp, black spots on leaves, necrotic to a greater or lesser degree, surrounded by a yellow halo (**291**); black, damp rotting of the head (**292** and **293**).<br>• Other symptoms: see pages 96, 98, and 100. |
| **Organs mainly attacked (in decreasing order of frequency of observation)** | Leaves in contact with the soil and crown, aerial leaves. | Leaves in contact with the soil and crown, aerial leaves. | All leaves, variable depending on the bacterium or bacteria present *Erwinia carotovora* subsp. *carotovora* prefers to attack the pith of the stem and the taproot. |
| **Frequency** | Very common. | Fairly common. | Fairly common. |
| **Presence of special structures on damaged tissues** | Grey mould consisting of numerous conidiophores and conidia (**286**, **287**, and **290**). | White mycelium and black sclerotia, sometimes apothecia within the plant's environment (**289** and **290**). | Bacteria which are difficult to see in the tissues; isolation on a culture medium is required. |

**294 Aphids** are insects with little mobility, 1.5–2.5 mm in length, whose colour varies according to species (green, black, pink, and so on). They are found in apterous (wingless) or winged forms.

**295 Whiteflies** are small winged insects, white in colour, from 1.2–1.5 mm in length. Their larvae are flat, with an oval and ciliated outline.

## Why does Sooty mould develop on lettuce?

The appearance of Sooty mould on lettuce leaves is always associated with the development of parasitic insects on the plants. Among the latter, aphids and whiteflies are those most commonly associated with the presence of this black mould on the lamina. The explanation is fairly simple: it is associated with the feeding behaviour of these insects. In fact aphids and whiteflies must take in large quantities of sap in order to satisfy their protein requirements. This obliges them to reject excessive sugar in the form of honeydew. This is consequently present in large quantities where these insects are living and it soils the surface of the parts of the lamina which have been invaded. This sugary honeydew is a real golden opportunity for several types of mould of the leaves which use it to feed on, progressively covering it with black mould (*Alternaria* spp., *Cladosporium* spp., *Capnodium* sp., and so on). This Sooty mould certainly has an adverse effect on photosynthesis, to varying extents.

**To rectify this, the populations of insect pests must be controlled.**

In addition to excreting the honeydew which lies at the origins of Sooty mould and damaging the quality of the lettuces harvested, **aphids** and **whiteflies** disturb the functions and growth of lettuces. If these insects are present in large numbers, it is fairly common to note a slowing down in plant growth.

Several species of aphids have been recorded as attacking lettuces: *Nasonovia ribisnigri*, *Myzus persicae*, *Macrosiphum euphorbiae*, *Hypermyzus lactucae*, and so on (**294**).

With regard to the whiteflies, *Trialeurodes vaporariorum* is certainly more common on lettuces than *Bemisia tabaci* (**295**).

These insects, and sometimes other species, are efficient vectors of several types of virus which are a very serious threat to lettuces. The principal types of virus are indicated in *Table 17*. Their symptoms are mainly grouped together under the headings 'Abnormalities in the growth of lettuces and/or the shape of leaves' and 'Abnormalities in the leaf colouration'.

Other insect pests, which are sometimes vectors of the virus, develop on l.s.v. leaves and cause rather different symptoms. The damage they cause is illustrated on pages 124, 125 and 205.

**Table 17: Principal types of virus transmitted by parasitic insects on l.s.v.**

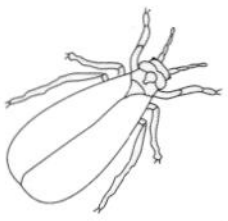

| Insect vectors | Principal types of virus transmitted to l.s.v. |
|---|---|
| **Aphids** **(*Nasonovia ribisnigri, Myzus persicae, Macrosiphum euphorbiae, Hypermyzus lactucae,* and so on)** | NP: lettuce mosaic virus (LMV), cucumber mosaic virus (CMV) Alfalfa mosaic virus (AMV). broad bean wilt virus (BBWV), dandelion yellow mosaic virus (DaYMV), endive necrotic mosaic virus (ENMV), turnip mosaic virus (TuMV). P: Beet Western yellows virus (BWYV). |
| **(*Aphis craccivora, Aphis gossypii, Acyrthosiphon solani,* and so on)** | NP: lettuce mottling virus (LMoV), P: beet yellow stunt virus (BYSV), lettuce necrotic yellows virus (LNYV), Sonchus yellow net virus (SYNV), Sowthistle yellow vein virus (SYVV). |
| **White fly** | P: beet pseudo-yellows virus (BPYV). |
| **(*Trialeurodes vaporariorum*)** **(*Bemisia tabaci, Bemisia argentifolii*)** | P : lettuce infectious yellows virus (LIYV), lettuce chlorosis virus (LCV). |
| **Thrips** **'*Thrips tabaci, Frankliniella occidentalis* and so on)** | P: tomato spotted wilt virus (TSWV), impatiens necrotic spot virus (INSV). |

NP: virus transmitted in a non-persistent way
P: virus transmitted in a semi-persistent or persistent way

**296** Lettuce beginning to wilt.
***Botrytis cinerea***

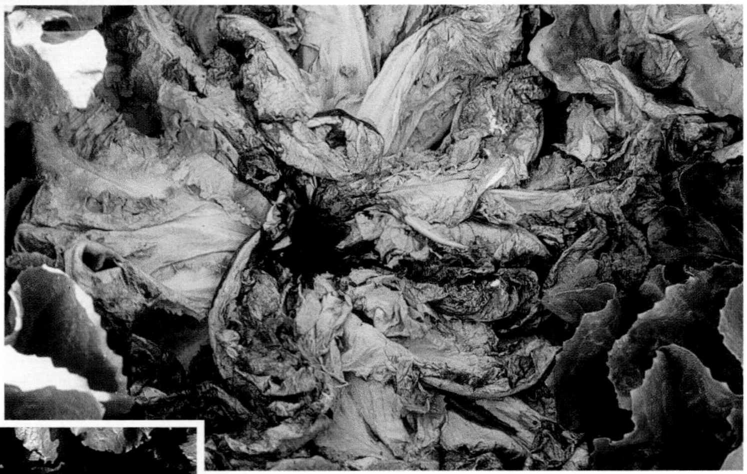

**297** Lettuce which has collapsed completely. Several of its leaves have dried out.
***Erwinia carotovora* subsp. *carotovora***

**298** Escarole, several of whose leaves have broad necrotic areas.
**Tomato spotted wilt virus (TSWV)**

**Example of wilting, drying out and leaf necrosis occurring on lettuces**

# Wilting, drying out, necrosis of leaves (may or may not be preceded or accompanied by yellowing)

## Possible causes

- Pathogenic agents or non-parasitic diseases responsible for damage to the roots, crown, and lower leaves of lettuces (consult section entitled 'Damage, abnormalities on leaves in contact with the soil and/or underground organs')
- Pathogenic agents responsible for leaf spots (advanced symptoms) (bacteriosis, *Bremia lactucae*, and so on; consult the section entitled 'Spots and damage on leaves')
- Various types of virus
  - Lettuce ring necrosis agent (LRNA) (fact file 35)
  - Tomato spotted wilt virus (TSWV) (fact file 32)
  - Turnip mosaic virus (TuMV) (fact file 27)
  - Lettuce mosaic virus (LMV) (fact file 21)
  - Cucumber mosaic virus (CMV) (fact file 22)
  - Tobacco streak virus (TSV) (fact file 32)
  - Tomato bushy stunt virus (TBSV)
  - Lettuce necrotic spot virus (LNSV) (fact file 33)
  - Impatiens necrotic spot virus (INSV) (fact file 32)

    Lettuce dieback (TBSV, LNSV) (see pages 153 and 365)
- Field mouse damage (see page 183)
- *Pemphigus bursarius* (see page 205)
- Other soil-based insect pests (*Agriotes* spp., *Agrotis* spp., *Hepialus* spp., *Melolontha melolontha*, *Tipula* spp., and so on) (see page 205)
- Root suffocation
- Heat damage
- Lightning damage
- Cold injury

Nutritional disorders (see page 82)

'Tip burn' (see page 133)

Various types of chemical injury

Atmospheric pollutants (PAN and ozone)

'Soft head'

## How to analyse and understand wilting

In a fairly general way, when lettuces have one or several leaves which are wilting and/or drying out, growers and experts, when confronted with these symptoms, tend to focus their attention solely on the leaves, where they are convinced that they will find the cause of the problem. In the majority of cases, the origins of wilting must be sought elsewhere. In fact the cause of this wilting will be found more commonly on the roots, the crown, and on the exterior or interior of the taproot and stem.

Various types of damage which occur on these organs actually disturb the functions of absorption (roots) and transport (crown, stem) of water and mineral elements in lettuces. The main results of this are lack of water in the plants which will wilt fairly rapidly (temporarily or irreversibly) and dry out partially or totally.

The appearance of wilting and the speed at which it develops essentially depend on three parameters:

- the stage of development of the plants;
- the nature and seriousness of the damage which has caused this. For example, at the onset of an attack, if the damage is not too extensive, wilting can be more or less reversible, since at night the leaves are subject to less evaporation, and once again become turgid;
- climatic conditions. High temperatures or windy spells increase evapotranspiration of plants, thus encouraging early wilting of lettuces. On the other hand, if there are conditions of low evapotranspiration, during a cold, damp spring for example, the plants may not wilt initially, in spite of plant health problems affecting the organs, as described previously (canker of the crown, root rotting, and so on). When they do wilt, it will often be too late to intervene.

In certain cases, wilting can be found in a particular location on the leaves and on the plants. This is characteristic of one or several diseases, as you will note subsequently, particularly when studying vascular diseases (page 211–215).

In fact, the foliage is a good indication of the state of hydration of the plants. Wilting and drying out of one or several leaves must be considered to stem from problems whose origins must be investigated by carefully examining all the organs which could have caused this.

In this section we have looked at several diseases which are particularly responsible for wilting, drying out or comparable symptoms on lettuces

## Examples of wilting developing on lettuce

**299** One lettuce is beginning to wilt in the middle of several lettuces which are adequately turgid.

**300** Fairly quickly, more and more leaves go yellow, while the plant continues to collapse.

**301** Now the lettuce is completely wilted and chlorotic. Soon it will die.

**302** Under certain situations, the pathogenic agent responsible for this wilting will be able to invade the plant, weakened in this way, completely. In this case the agent is ***Botrytis cinerea***.

303

304

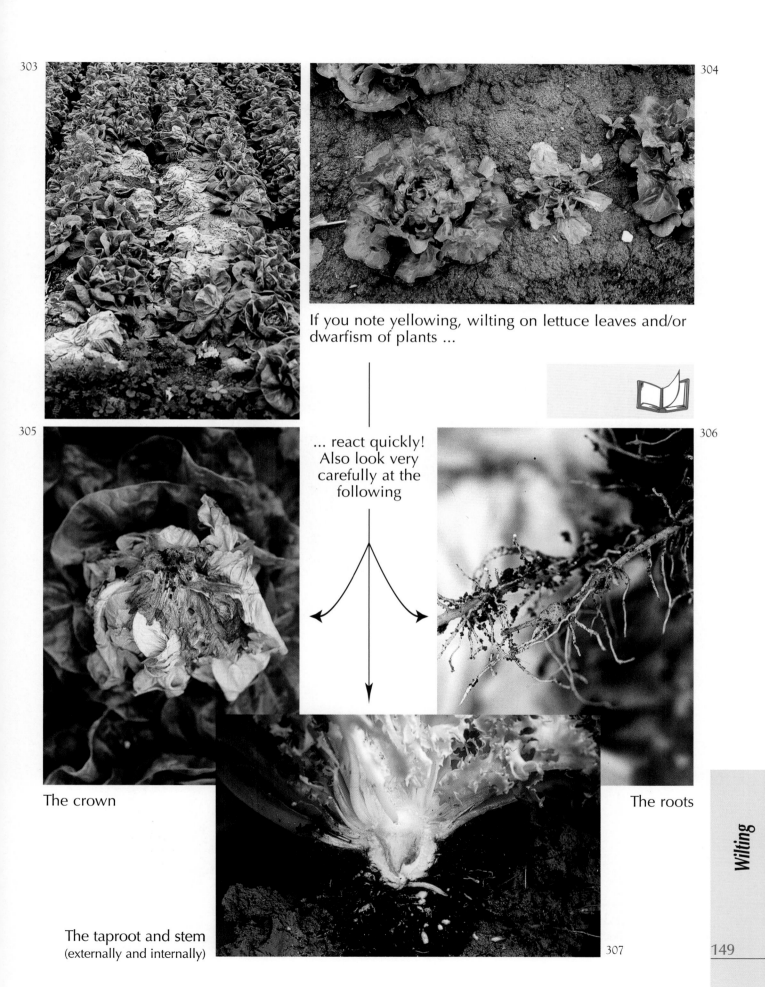

If you note yellowing, wilting on lettuce leaves and/or dwarfism of plants ...

... react quickly! Also look very carefully at the following

305

The crown

306

The roots

The taproot and stem (externally and internally)

307

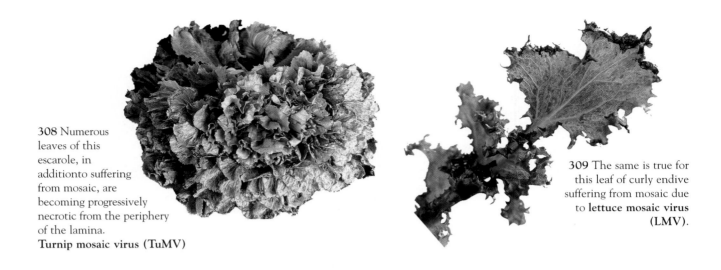

**308** Numerous leaves of this escarole, in additionto suffering from mosaic, are becoming progressively necrotic from the periphery of the lamina.
**Turnip mosaic virus (TuMV)**

**309** The same is true for this leaf of curly endive suffering from mosaic due to **lettuce mosaic virus (LMV).**

**310** Numerous chlorotic and necrotic rings are progressively covering the lamina; ultimately several leaves of this lettuce will end up by becoming yellow and more or less completely necrotic.
**Lettuce ring necrosis agent (LRNA)**

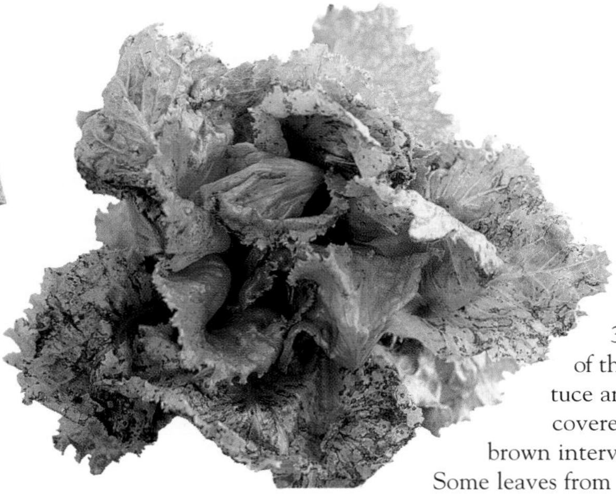

**311** The leaves of this batavia lettuce are more or less covered by necrotic, brown interveinal damage. Some leaves from the crown are beginning to dry out entirely.
**Complex of LMV + CMV**

**312** Broad chlorotic and necrotic lesions are more or less covering the leaves of this lettuce. Some leaves in the head are completely necrotic.
**Tomato spotted wilt virus (TSWV)**

## Examples of drying out and necrosis of leaves caused by the presence of one or several viruses in lettuces

## Diversity of symptoms caused by TSWV on lettuce leaves

**313** Early symptoms of **TSWV** on lettuce growing in the open field.

**314** At an advanced stage, numerous leaves are affected. They are chlorotic and have broad necrotic areas.
**Tomato spotted wilt virus (TSWV)**

**315** Chlorotic spots are becoming progressively necrotic.
**TSWV**

**316** Brown necrotic spots; necrosis and drying out of one section of the lamina.
**TSWV**

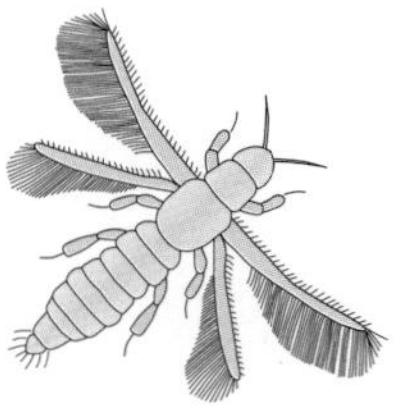

**317** Numerous minuscule areas of damage affecting the veins and surrounding tissues giving them a rather bronzed appearance.
**TSWV**

## Arguments in support of the diagnosis

**• Viruses and virus complexes**

A number of types of virus or certain strains of virus, alone (**308–310** and **312**) or as complex (**311**), are likely to cause wilting, drying out, and leaf necrosis. In the case of quite a few of these, other more characteristic symptoms are visible on the infected plants. In fact, the drying out and necrosis observed often closely correspond to the final stage of development of the viruses on the plants. If you are in any doubt whatsoever, we recommend that you consult the two sections entitled 'Abnormalities in leaf colouration' and 'Abnormalities in the growth of l.s.v. and/or the shape of their leaves'.

Two types of virus are likely to cause more systematic leaf necrosis, wilting and drying, with these being virtually the only symptoms. These are **tobacco streak virus (TSB)** and **tomato spotted wilt virus (TSWV)**. The latter, which is very polyphagous, has started to appear again over recent years on numerous floral and market garden plants in France. The introduction of a more effective vector in France (*Frankliniella occidentalis*), and the juxtaposition of market garden and floral crops susceptible to this virus have led to this situation. Lettuce is also affected and serious attacks take place at the end of spring, in the autumn or during winter under protection (**313** and **314**).

**Table 18: Viruses principally responsible for necrotic symptoms on lettuce leaves**

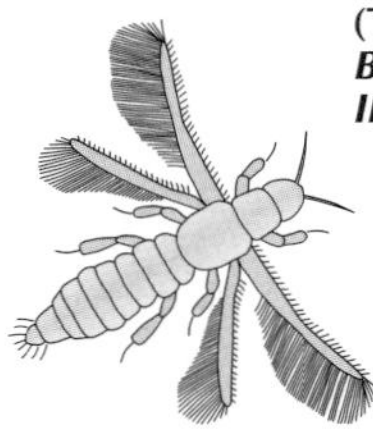

| Virus | Symptoms | Vector | Viral particle shape |
|---|---|---|---|
| **Tomato spotted wilt virus (TSWV)** ***Buniaviridae*** ***Tospovirus*** | Spots, chlorotic damage which becomes brown and necrotic on the leaves (**315**). The veins may be affected. This damage spreads and takes over a large proportion of the lamina. There is wilting and drying out of leaves, the latter sometimes having a bronzed appearance (**316** and **317**). | Thrips *Thrips tabaci, Frankliniella occidentalis*, and so on (See fact file 32). | 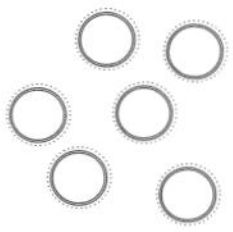 |
| **Tobacco streak virus (TSV)** ***Bromoviridae*** ***Ilarvirus*** | Numerous small necrotic spots on leaves. It is also possible to note concentric, chlorotic, and necrotic rings. Deformed, folded leaves give the plants a typical appearance. Growth of the latter may be slowed down or they may become entirely necrotic and die. | Several species of thrips *Thrips tabaci, Frankliniella* spp. It is also transmitted by seeds in beans and by several weeds such as *Datura stramonium, Chenopodium quinoa, Helilotus alba*, and so on. | 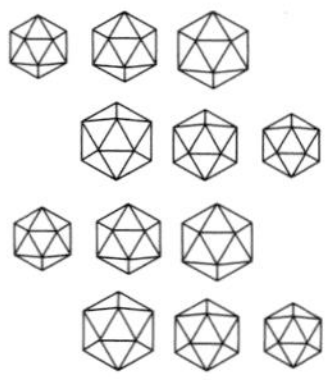 |
| **Impatiens necrotic spot virus (INSV)** ***Bromoviridae*** ***Tospovirus*** | Vein necrosis and concentric rings are visible on the lamina. The leaves may be deformed and the growth of the lettuces halted. | Transmitted by *Frankliniella occidentalis* and *Frankliniella fusca*. | 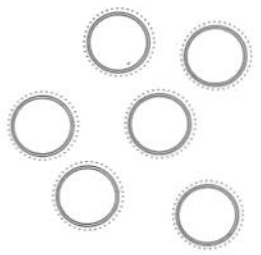 |
| **Lettuce necrotic spot virus (LNSV)** ***Comoviridae Nepovirus*** | Mosaic or mottling, concentric rings, arabesques and necrotic spots are visible on the leaves. | Unidentified vector. The disease seems to be associated with the presence of nematodes from the *Xiphinema* family. Transmission by seed has not been investigated. | 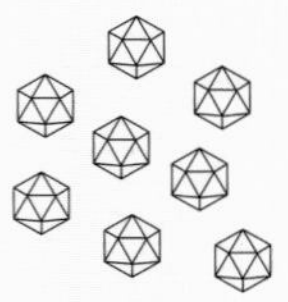 |
| **Tomato bushy stunt virus (TBSV)** ***Tombuviridae*** ***Tombuvirus*** | L.s.v. which are extremely stunted, presenting leaf necrosis. Progressive decline of plants. | Transmitted by soil and water, without any involvement of a vector. | 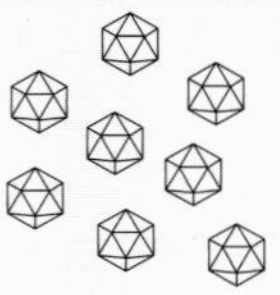 |

Some characteristics of these two viruses, as well as other ones responsible for leaf necrosis, are listed in *Table 18*.

## • Lettuce dieback

A disease of lettuce called dieback was identified in the Salinas Valley of California in the mid-1980s and early 1990s. It was shown to be caused by two tombusviruses, tomato bushy stunt virus (TBSV) and lettuce necrotic stunt virus (LNSV) (see p.365, **514**, **515**). It is most serious on romaine lettuce and on some butterhead and red and green leaf lettuces. It has similarities to a disease known as brown blight, which caused serious economic damage to the lettuce industry of California and Arizona in the early part of the 20th Century. Brown blight was eliminated through the use of resistant varieties of crisphead lettuce of the Imperial group and all subsequent varieties of the Great Lakes and Salinas types were also resistant. The cause of the disease was never identified. It appears that it has returned, afflicting lettuce types that were rarely grown at the earlier time.

Dieback is found on lettuce in the coastal valleys of California and the desert growing regions of the southwest (Imperial Valley in California, and Yuma area in Arizona). Therefore, it is important as these areas produce 98% of the lettuce grown in the US. Its greatest frequency is in recently flooded areas, although it has occurred on higher ground, perhaps in areas of high salinity.

Vector control is of no use since there is no known vector. Fumigation with methyl bromide or chloropicrin does not affect the incidence of the disease. Infected plants cannot be cured. Most modern crisphead varieties, some butterhead varieties, and some leaf varieties are resistant to the virus. No modern romaine varieties have resistance, but several romaine landraces are resistant. These are being used in breeding programmes which are in progress.

**318** Numerous lettuces are wilting and progressively withering away. They are located in an area of the plot where water has stagnated after a violent storm. **Root suffocation**

319 This lettuce has been subjected to temperatures well below 0°C. On thawing out, its lamina and leaves give the appearance of having been scalded.
**Frost damage**

320 After receiving spinklers of Gramoxone (paraquat) during herbicide treatment, this batavia lettuce displays several areas of leaf necrosis, varying in extent. You can see that some of these are located at the edge of the lamina where the product has been able to accumulate.
**Chemical injury**

321 This lettuce plant has not appreciated the effects of Mocap; the majority of its leaves, particularly the young ones, are going brown, wilting and starting to become necrotic.
**Chemical injury**

322 Some completely dried-out leaves show the effects of a mixture of pesticides, rather poorly tolerated by one of two varieties of oak leaf lettuce being grown (located on the right).
**Chemical injury**

## Examples of chemical injury caused by a herbicide progressively leading to drying out of lettuce leaves

### • Lightning damage

Damage has been reported on l.s.v. following storms. In this case growers see in their plots certain circular areas of l.s.v., varying in width, which give the impression of having been 'cooked'. These groups correspond to the point of impact of lightning. The leaves of plants which have been struck by lightning have very rapidly become necrotic. The seriousness of symptoms can be graded from the centre of the group, where the lettuces are particularly badly affected, to the exterior. Some plants are completely destroyed.

In this situation, the reaction of cause and effect is immediate; it is easy to make the association between the symptoms and a storm which occurred just before they appeared.

### • Root suffocation (Drowning)

Water stress manifests in several ways on crops of l.s.v. It can occur at a particular time, for example following heavy rain, burst irrigation pipes, water submerging a crop to a greater or lesser degree or accumulating in areas forming a basin. Plants which have suffered water stress of this kind on their roots wilt, often suddenly. Their laminae become rather yellow and subsequently dry out (**318**). If water retention has been prolonged and the soil is heavy, the plants may wilt and become completely necrotic.

**Water stress** can be unwittingly repeated in a salad crop. This is the case, for example, if a grower does not properly control plant irrigation and tends to supply too much water each time the crop is irrigated. This succession of drowning episodes results in the appearance of the symptoms described above. In this case, they may be preceded by reduced growth of the l.s.v. This situation will also be encountered, although more locally, on certain plots which are basin-shaped or whose soil is a heavier.

L.s.v. struggles if the root system is completely submerged over a long period. In this situation, they lack oxygen and therefore find themselves drowning. The consequence is the more or less significant destruction of the roots, which then causes wilting. This soil-based phenomenon is probably increased by the proliferation of anaerobic micro-organisms which probably contribute to the denaturation of the roots.

Generally speaking, significant damage is observed on poorly drained plots with heavy soil, over a period of hot weather which favours anaerobic micro-organisms and accelerates plant transpiration, and therefore wilting. On plots where drowning has occurred, the soil should be drained and l.s.v. planted on relatively high mounds.

### • Soft head

If varieties of lettuces are grown which are not well-suited to the production period chosen and/or to significant changes in temperature between day and night, this can account for partial wilting of the head in numerous types of lettuce.

This physiological problem mainly occurs during heat waves, both in the open field and under protection.

### • Heat injuries

During very hot periods, large areas of tissue which are going brown and progressively drying out can be observed on lettuce leaves. This is particularly the case on plants in full growth, especially those exposed to the rays of the sun. When their laminae, which are very succulent, are subjected to very low relative humidity and very high temperatures, they quickly suffer local dehydration, giving way to necrotic burns. This may also be referred to as sunstroke.

These conditions are also conducive to the appearance of peripheral drying out of leaves, referred to as 'tip burn' (see page 133).

### • Frost injuries

Lettuces are relatively tolerant of low temperatures. In spite of this, damage sometimes occurs at the end of the night, at temperatures of between approximately 0 and –1°C. We usually see the epidermis of the leaves peeling away after ice has formed on the cells of this tissue (see page 123). If temperatures are lower, all the tissues will be affected. When they thaw out the lettuces may look as if they have been scalded. They will be flaccid, and will look as though they are wilting (**319**).

### • Various types of chemical injury

Certain pesticides used on l.s.v. under incorrect conditions or at the wrong time can be responsible for symptoms which may be similar to wilting and/or drying out. The following symptoms are noted on affected lettuces:

- browning and rapid drying out of the lamina;
- necrosis and drying out of the periphery of the lamina of young and old leaves;
- the appearance of small white to brown areas of necrosis between the veins;
- wilting and drying out of the leaves on the shoot;
- inter-veinal drying out occurring on old leaves.

**320–322** show some of these symptoms. In order to help you to confirm this hypothesis, we suggest that you consult other symptoms and other information about chemical injury, on pages 39, 83, and 109.

- **Atmospheric pollutants**

The air normally consists of 78% nitrogen and 21% oxygen; most of the remaining 1% being water vapour and carbon dioxide gas. It can also contain pollutants which can be classified into two groups:

- **pollutants** from one or several **localized groups** affecting plants situated nearby (a factory, a volcano, and so on) (sulphur dioxide, nitrogen oxides, chlorine compounds, ammonia, ethylene, volatile herbicides such as 2,4-D, and dust);
- **general atmospheric pollutants** emitted by a multitude of different sources (car engines, and so on) and affecting the atmosphere over significant surface areas (ozone, peroxyacetyl nitrate (PAN)).

Gaseous effluents may remain as they are in the atmosphere, and we then refer to **direct pollutants**, or they may react with other normal or abnormal compounds in the air, resulting in **secondary pollutants**. The latter depend on mechanisms such as photochemical reactions, the formation of free radicals, oxidation, and so on, producing new compounds. These reactions are still poorly understood, but they are at the origins of oxidizing compounds (ozone) and photochemical compounds (smog).

Among these pollutants, two cause damage to lettuces (*Table 19*):

- **ozone** results in the appearance of numerous small areas of damage, more or less dark in colour, between the principal veins of the lamina of old leaves. These areas of damage finally go yellow and necrotic; affected leaves become chlorotic and prematurely senescent. This physiological phenomenon is still fairly rare. There are differences in sensitivity between varieties;
- **peroxyacetyl nitrate (PAN)** causes leaf discolouration affecting a significant number of plants; the leaves take on a bronzed to silvery metal colouring on the lower face of the lamina. This change in the colouration of the lamina is actually due to the absorption of PAN through the stomata. This gas then diffuses towards the cells of the mesophyll and destroys it. Pockets of air form between the lower epidermis and the palisade cells, and are at the origins of the metallic appearance of the leaves.

Damage appears in a few urban and suburban industrial areas, especially after periods of calm, sunny weather with thick fog. There are differences in sensitivity between varieties.

Several factors have an effect on the degree of severity of the damage caused by these pollutants. We can list the following: their own toxicity, their concentration, the duration of their emission, factors affecting their diffusion, sensitivity of plant species (direct pollutants) their stage of growth, the variety being grown, and growing conditions.

Table 19: Some characteristics of two significant pollutants in plant pathology

| | Ozone | peroxyacetyl nitrate (PAN) |
|---|---|---|
| **Plant functions affected** | • Loss of turgidity of guard cells.<br>• Change in membrane permeability.<br>• Reduction of gaseous exchange.<br>• Acceleration of starch hydrolysis. | Weakening of cells in the mesophyll whose place is filled up with air. |
| **Symptoms observed on plants** | • Small, depressed necrotic spots, varying from white to dark brown or even black or red in colour, depending on species.<br>• Precocious senescence of tissues, accompanied by chlorosis and necrosis.<br>• Loss of turgidity and whitening of tissues.<br>• Premature falling of leaves and fruits.<br>• Reduced growth of plants and yield. | The lower face of the leaves has a shiny, or even metallic appearance, and is silvery to bronze in colour (sylver leaf). |
| **Sources** | Atmospheric pollutant capable of causing most damage. Nitrogen oxides and hydrocarbons are the forerunners. | Petrol vapour +/– $O_3$ or $NO_2$ = PAN. Photochemical compounds which are toxic to plants. Damage is observed in very sunny regions where there are higher emissions of hydrocarbons associated with road traffic, close to large towns and in the presence of fog. It is now being recorded in northern Europe. |

# Damage and abnormalities of leaves in contact with the soil and/or underground organs

For the purposes of clarity, the symptoms of the diseases covered in this part of the book have been split into three sub-sections:

- Damage to leaves, in contact with the soil, and the crown (page 163)
- Damage and abnormalities of roots (page 185)
- Internal and/or external and abnormalities of the taproot and stem (page 207)

We have deliberately included damage which occurs on old leaves in contact with the soil, the crown, the roots, and the taproot in the same section. There are numerous reasons for this:

- first of all the lettuce is a fairly squat plant whose base leaves, with very short internodes, are in contact with the soil. It is therefore difficult to dissociate the different types of organs; and pathogenic agents certainly cannot do this. In fact, types of fungus which attack the roots easily spread to the crown and the proximal part of the lower leaves. In the same way, fungi which tend to predominate in the crown sometimes spill over on to the taproot and the roots;
- in addition, diagnosis of diseases affecting these organs is particularly difficult because they frequently cause very similar symptoms. Risks of confusion are therefore very significant and diagnosis cannot be carried out in the same simple, unstructured way proposed in other sections.

We advise you to be particularly observant and to follow carefully the advice we recommend throughout this section.

323 The crown of this lettuce has been completely destroyed by damp rot which began on several leaves in contact with the soil.
***Botrytis cinerea***

324 All the roots of this lettuce have taken on a uniform reddish-brown colour.
***Pratylenchus penetrans***

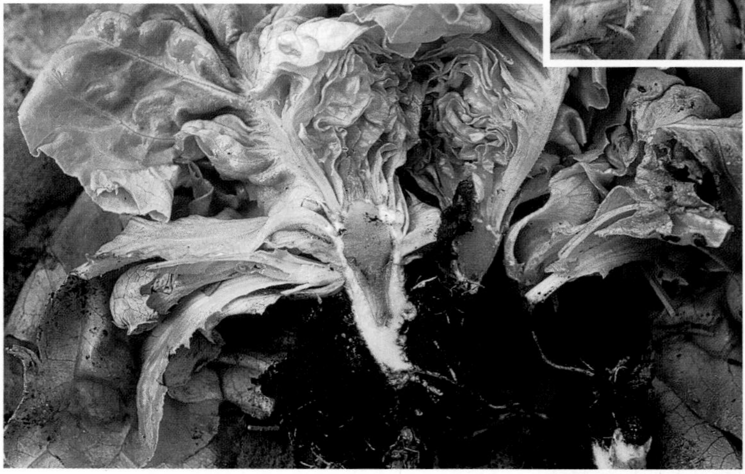

325 When this young lettuce is cut longitudinally, we can see necrotic damage located in the vessels and sometimes in adjacent tissues.
***Pythium tracheiphilum***

326 A cross section of this lettuce reveals that the pith is beginning to be invaded by an ***Erwinia* sp.**

## Examples of damage occurring to leaves in contact with the soil, the taproot, and roots

**Observation guide and examples of damage occurring to leaves in contact with the soil, the crown, and the roots.**

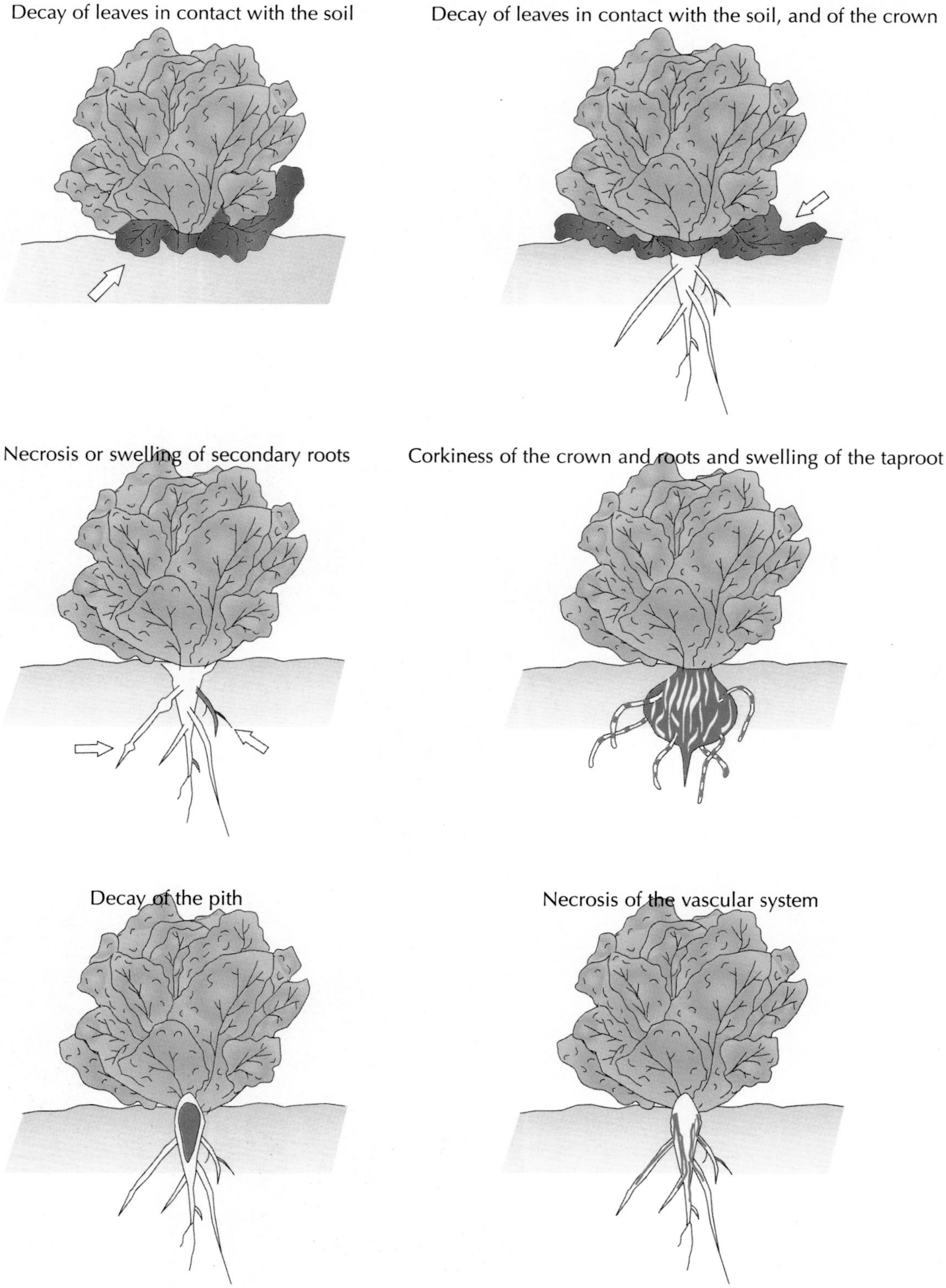


**Figure 11: Nature and possible location of damage and abnormalities affecting leaves in contact with the soil or underground organs in lettuce varieties.**

## Principal fungi attacking leaves in contact with the soil and the crown

327 Damp rot, reddish to dark brown in colour, has established itself on a few lettuce leaves in contact with the soil.
***Botrytis cinerea***

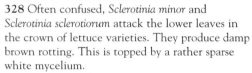

328 Often confused, *Sclerotinia minor* and *Sclerotinia sclerotiorum* attack the lower leaves in the crown of lettuce varieties. They produce damp brown rotting. This is topped by a rather sparse white mycelium.
***Sclerotinia minor***

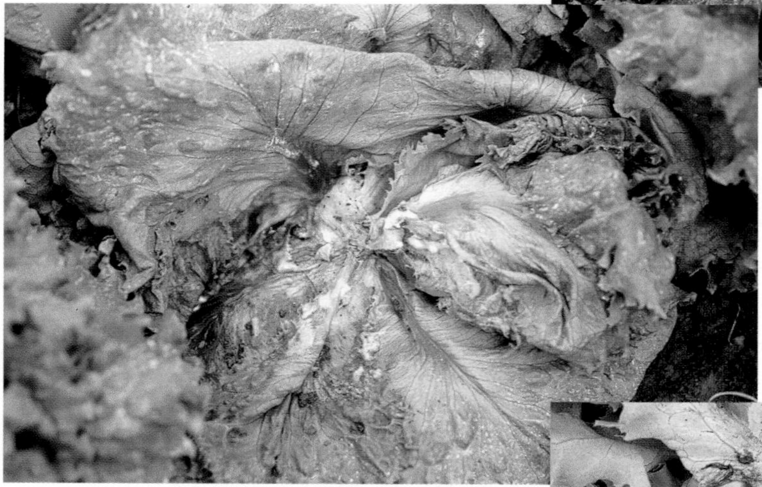

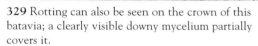

329 Rotting can also be seen on the crown of this batavia; a clearly visible downy mycelium partially covers it.
***Sclerotinia sclerotiorum***

330 Fairly dry brown rotting has appeared on several leaves of this lettuce next to the principal vein.
***Rhizoctonia solan***

Diagnosis of these four mycoses is sometimes difficult and confusion is possible; they may also attack at the same time or one after another on the same plant.

# Damage to leaves, in contact with the soil, and the crown

## Possible causes

- *Botrytis cinerea* (fact file 5)
- *Microdochium panattonianum* (fact file 3)
- *Mycocentrospora acerina* (fact file 4)
- *Phoma exigua* Desmaz. (pages 181 and 365)
- *Pythium spp.* (fact file 8)
- *Rhizoctonia solani* (fact file 7)
- *Sclerotinia minor*, *Sclerotinia sclerotiorum* (fact file 6)
- Other fungi acting as parasites on the roots and/or crown (*Athelia rolfsii*, *Phymatotrichopsis omnivora*) (fact file 10)
- *Erwinia carotovora* subsp. *carotovora* (fact file 18)
- Lettuce ring necrosis agent (LRNA) (fact file 35)
- 'Brittle crown'
- Vole damage

## Arguments in support of the diagnosis

The principal fungi which attack the lower leaves and crown of l.s.v. varieties behave in virtually the same way (**fig. 12**). They may even cohabit on the same areas of damage. They are opportunistic micro-organisms which will often take advantage of particular conditions to establish themselves on plants and sometimes cause serious damage:

- prolonged period of damp, rainy weather; excessive irrigation;
- stagnant free water on leaves and on the soil, especially around the edge of the lamina;
- presence of senescent tissues, particularly old leaves at the bottom in contact with the soil;
- plants growing very vigorously, with succulent tissue;
- plants at maturity, as harvest-time approaches (**fig. 13**).

They easily colonize areas of damage and senescent tissues which constitute entrance ways or nutrient bases. These facilitate their penetration and development in the plants. With their impressive enzymatic equipment these fungi break down the tissues they have invaded fairly quickly. In general, they take over the crown and girdle it, and we then see wilting (**fig. 14**). We suggest that you consult page 137.

In general they are easy to identify, particularly *Botrytis cinerea* and the two *Sclerotinia* spp. (*Tables 20* and *21*). In fact, they produce storage organs or fairly characteristic fruiting bodies fairly quickly on the affected tissues. **To encourage these to appear, we suggest that you place a few samples in a sealed box or plastic bag containing an absorbent paper soaked in water. This provides the humidity essential for the formation of these structures.**

These fungi produce other symptoms on l.s.v., particularly on the leaves in the head. Consequently we recommend that you consult the heading 'Spots with powdery areas, matting, mould on leaves, damp spots leading to rotting of the head', page 137.

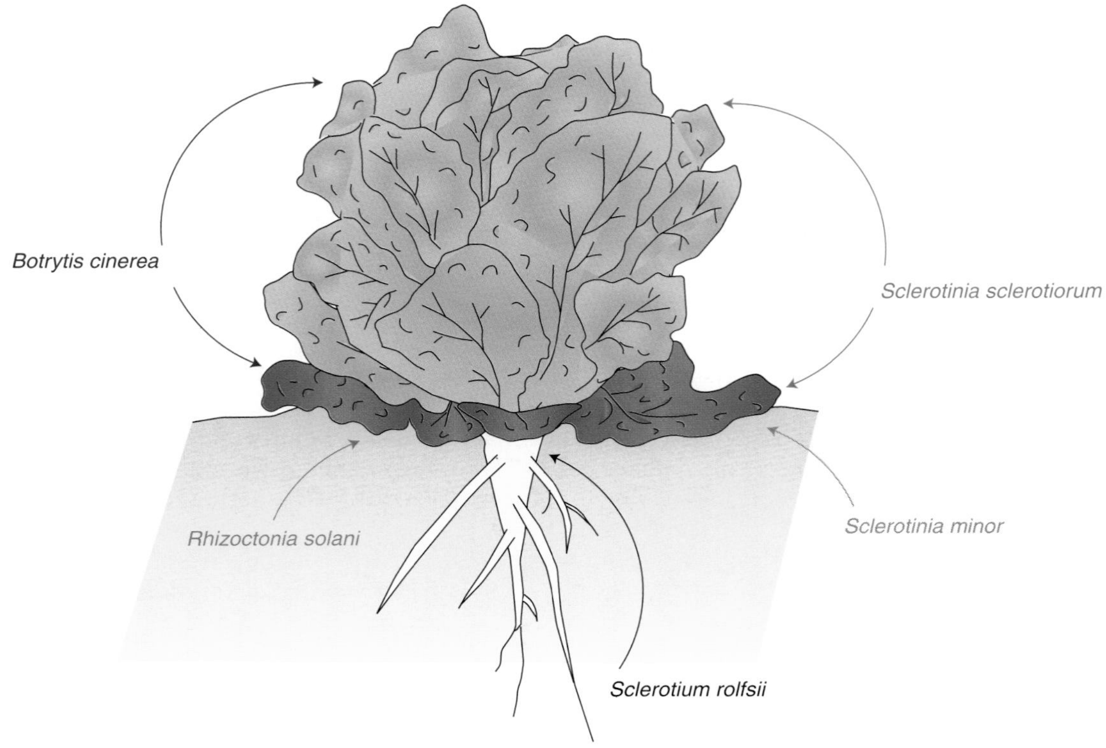


**Figure 12: Principal fungi responsible for basal rotting on l.s.v.**

**Figure 13: Development of l.s.v and the process showing how risks of basal rotting increase.**

1 The young plants are erect, small, and well ventilated.

2 Numerous leaves have formed, some of which are coming into contact with the soil.

3 Now the plants are covering the ground completely, and from this point onwards risks of basal rotting are significant, for various reasons:

- the light available to lower leaves is restricted;
- lower leaves rapidly become chlorotic, senescent and, therefore, very vulnerable;
- it is more difficult for fungicides to access these areas;
- the surrounding microclimate is damper.

**Figure 14: Probable development of basal rotting on l.s.v.**

• ***Botrytis cinerea* (Grey mould)**

This fungus frequently attacks lettuce, as well as numerous other vegetable crops. Very often, cold damp climatic conditions accompany its spread. It has significant saprophytic potential which allows it to colonize numerous senescent or damaged plant tissues. During growth, these tissues are affected:

- dehiscent cotyledons in the nursery (if sowing has been dense, *Botrytis cinerea* can form a sterile, greyish white mycelium web, similar to a 'spider's web', covering the plantlets which quickly collapse);
- damaged tissues of moribund plantlets attacked by a damping-off agent;
- damage to leaves, young or old, chlorotic and senescent, often in contact with the soil.

These nutrient bases, when in a damp environment, allow this opportunistic fungus to establish itself and rapidly colonize the tissues. These rapidly become flaccid and damp, then reddish, brown to dark brown in colour (**331**, **332**, **335–337**). Rotting rapidly affects numerous leaves, spreading to the crown (**338**). If conditions are particularly favourable, the crown may be completely destroyed, resulting in yellowing and wilting of the plant (**339**). Rotted tissues are generally covered by a very characteristic **grey mould** and, more rarely, few black sclerotia (**337**, **343**, **366**, and **368**).

*B. cinerea* also attacks leaves in the head on which it causes spots (see page 105) and damp rot (**340–342**) (see page 139). This particular symptom may be observed after harvesting and while lettuce is being stored.

• ***Sclerotinia minor* and *Sclerotinia sclerotiorum* ('Drop')**

These two fungi are common in plots of cultivated l.s.v., particularly in the open field. They cause very similar symptoms, so we will examine them together. They are responsible for damp damage affecting the parts of plants in contact with the soil, particularly senescent leaves. This damage develops very quickly into rot which becomes generalized to the leaf zones close to the ground (**333**, **334**, and **348**). Numerous petioles as well as the crown are invaded, resulting in virtually complete wilting of plants (**344**). Subsequently, rot spreads to all the leaf tissue which decomposes and collapses (**346**).

On certain portions of affected tissues, a rather downy white mycelium may form (**345** and **349**). On the mycelium, which may vary in density, we can see the structures which allow us to differentiate between these two species of *Sclerotinia*:

- groups of small, irregular black sclerotia for *S. minor* (**347** and **369**);
- a few large, black, fairly elongated sclerotia for *S. sclerotiorum* (**350** and **367**).

The perfect shape of these fungi can sometimes be seen on the surface of the ground, particularly in the case of *S. sclerotiorum*. This is revealed by the formation of small 'trumpets' or apothecia (**364**), whose ascospores may be at the origin of aerial contamination. In this case, rotting sets up initially on one leaf, and then progressively spreads to the whole crown (**352**).

• **Rhizoctonia solani (Bottom rot)**

Although lettuce seedlings are susceptible to *R. solani*, we rarely see the damping-off which is a sign of attacks of this fungus in the nursery. The use of quality substrates is definitely the explanation for this. In more extensive growing systems, seeds are sometimes sown in contaminated soil or in plugs placed at ground level; then we may note failed emergence and the damping-off of seedlings in localized groups.

It occurs more frequently during the growing season, at an advanced point in head formation. The nature and appearance of the damage may vary depending on the type of l.s.v. and climatic conditions.

In certain particularly badly infected soils and/or if climatic conditions are favourable, these fungi may attack young plants, well before formation of the head.

**331** Early attacks of these fungi on young plants, although not initially fatal, greatly reduce plant growth.

**332** Necrotic damage, beige to light brown in colour, has developed on a young leaf in contact with the soil. Subsequently it has spread to several leaves and the crown of this young lettuce. Grey mould is visible on certain portions of the damaged tissue.
***Botrytis cinerea***

**333** All the leaves of this young lettuce are in the process of decomposing. We can see a fairly dense, although not downy, mycelium network.
***Sclerotinia minor***

**334** ***Sclerotinia sclerotiorum*** has established itself on the leaf in contact with the soil which is now decomposing. Subsequently it has invaded several leaves and the crown. We can distinguish its characteristic white downy mycelium.

## Attacks on young lettuce

Initially on lettuce we see fairly localized lesions, varying in number, on the lamina and principal vein of the lower leaves in contact with the soil (**353**, **355**, and **357**). Initially they are dry in consistency, and are rusty to dark brown in colour. They develop at varying speeds depending on climatic conditions. They can spread to the point of forming concave cankerous spots on the principal vein (**354**). Fairly damp rotting quickly establishes itself on the principal veins of several leaves as well as on the lamina adjacent to this. It progressively spreads to the taproot via the leaf insertion zone. Ultimately, the first leaves affected, which are the most external ones, wilt. Their lamina quickly decomposes under the joint action of *R. solani* and **secondary bacteria** which increase the rotting which has already begun (**356**). ***Botrytis cinerea*** can also colonize damaged tissues in a secondary way (see following paragraph).

On escarole or frisée endive, we find more less the same symptoms. Lesions are browner to black in colour (**358**, **360**, and **362**). In addition *R. solani* gives the impression of temporarily sparing the principal vein (**359**).

The presence of *R. solani* on damaged tissues can be confirmed by visual observation of its rare **sclerotia** (**365**) or its characteristic **brown mycelium** using a stereoscopic and/or ordinary microscope (**361** and **363**). Its mycelium is not easy to see because it breaks down rapidly on damaged tissues.

**Table 20: A few characteristics of the principal micro-organisms responsible for attacks on lettuce seeds and plantlets**

| **Some characteristics** | ***Pythium* spp.** | ***Rhizoctonia solani*** | ***Botrytis cinerea*** | ***Sclerotinia sclerotiorum* or *Sclerotinia minor*** |
|---|---|---|---|---|
| **Principal symptoms observed** | Absence of germination, widespread root browning, constricting of tissues present on the surface of the soil, damping-off. | Absence of germination, widespread root browning, constricting of tissues present on the surface of the soil, damping-off. | Absence of germination, wide-spread root browning, constricting of tissues present on the surface of the soil, presence of a 'web' consisting of greyish white mycelium covering plantlets, damping-off. | Rotting of leaves in contact with the soil, rotting of the crown, death of plantlets. |
| **Frequency of attacks** | Fairly frequent. | Fairly frequent. | Fairly frequent. | More rare. |
| **Structures allowing identification on and/or in damaged tissues** | Oospores and chlamydospores visible in root cells (**394**). | Compartmentalized brown epiphytic mycelium, showing restricted branching (**361**, **363** and **395**). | Greyish white mycelium web, grey mould consisting of numerous conidiophores and conidia (**336**). | White mycelium, large black sclerotia, sometimes apothecia near the plant (**367**). Sparse white mycelium with small black sclerotia (**369**). |

335 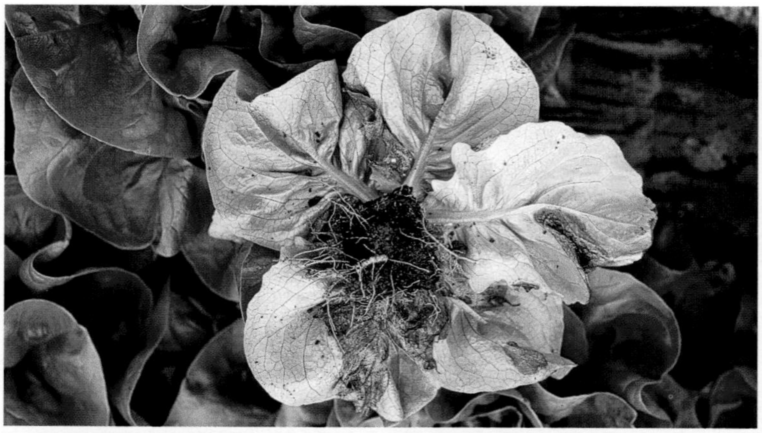

336 

337 

338 

339 

## Attacks of *Botrytis cinerea* (Grey mould) on leaves in contact with the soil

**335** A few broad damp spots are visible on the leaves in contact with the soil.
***B. cinerea***

**336** ***B. cinerea*** rapidly colonizes lower leaves, covering them with a characteristic grey mould.

**337** Rotting is now widespread on other leaves; it has completely destroyed the crown. We can see grey mould as well as a few small black sclerotia on the tissues. Some portions of the leaf sometimes react to the infection and take on a reddish hue, especially in the taproot.
***B. cinerea***

**338** After attacking the lower leaves, ***B. cinerea*** spreads to the interior of the taproot, at crown level, and all tissues are damaged, including the vascular system.

**339** Once the crown has been girdled, the lettuce suddenly wilts.
***B. cinerea***

**340** A damp brown lesion is developing around the periphery of the blade.
***B. cinerea***

**341** It rapidly progresses to the lamina then spreads to other leaves which in turn, rot.
***B. cinerea***

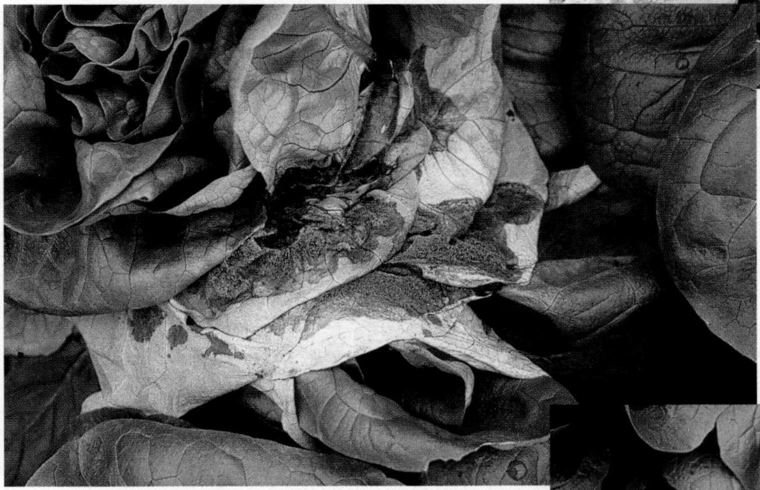

**342** ***B. cinerea*** can also establish itself in the head of plants; damaged tissue, initially damp and brown, is progressively covered by the characteristic grey mould. Leaves may go yellow because of damage to petioles.

**343** Ultimately, whichever leaves have been attacked, the lettuce may rot entirely and be completely covered by grey mould.
***B. cinerea***

## Attacks of *Botrytis cinerea* (Grey mould) on leaves in the head

**344** This lettuce, whose crown has been girdled, has suddenly collapsed. The fungus has now invaded a significant part of the head and some leaves are completely liquefied. We can see mycelial masses and a fairly sparse white mycelium.
***S. minor***

**345** Damaged tissues are first of all damp and transparent; they decompose fairly rapidly.
***S. minor***

**346** Plant tissues are now completely decomposed. We can only see a few principal veins on the leaves, as well as the disseminated sclerotia of the fungus.
***S. minor***

**347** Unlike Sclerotinia sclerotiorum, the sclerotia are irregular in shape and smaller in size.
***S. minor***

## Attacks of *Sclerotinia minor* ('Drop') on leaves in contact with the soil

## Attacks of *Sclerotinia sclerotiorum* ('Drop') on leaves in contact with soil and on leaves of the head

348 The head of this lettuce is very easy to pull up. Its crown and the principal vein of several of its lower leaves have been invaded by damp rot, light brown in colour.
***S. sclerotiorum***

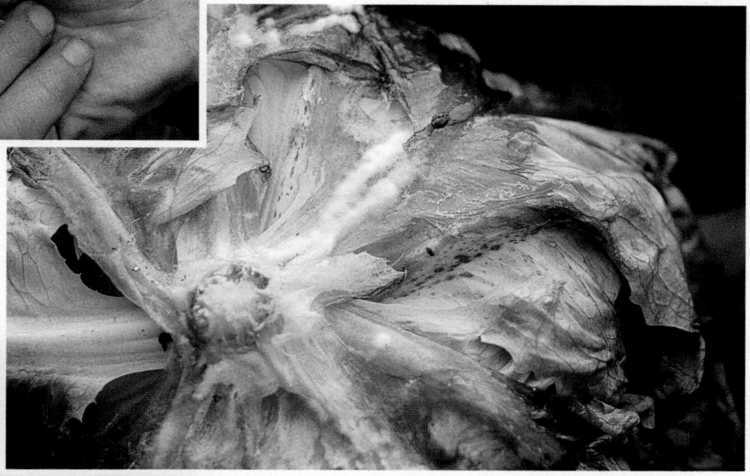

349 A dense white mycelium is progressively covering the partially decomposed tissues.
***S. sclerotiorum***

350 Fairly large structures, which are initially white, are forming on the parts which were colonized first.
***S. sclerotiorum***

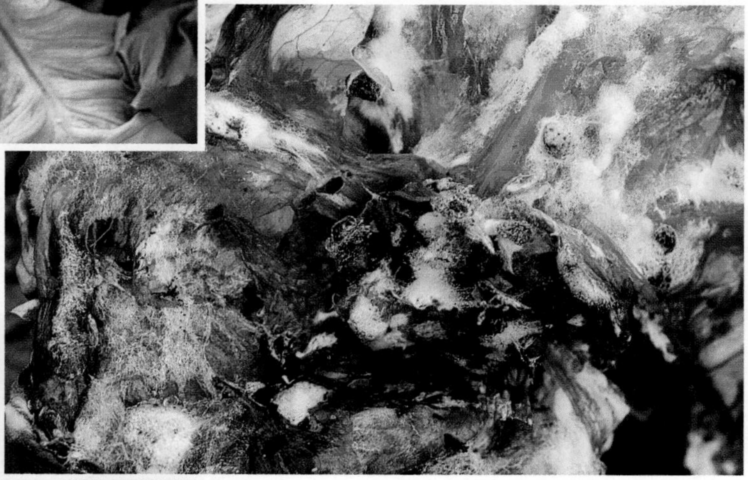

351 These sclerotia are forming and will quickly become black. They are hard in consistency.
***S. sclerotiorum***

352 Ascospores produced by the apothecia may cause aerial contamination resulting in damp brown rotting of the head, as has been the case with this endive.
***S. sclerotiorum***

353

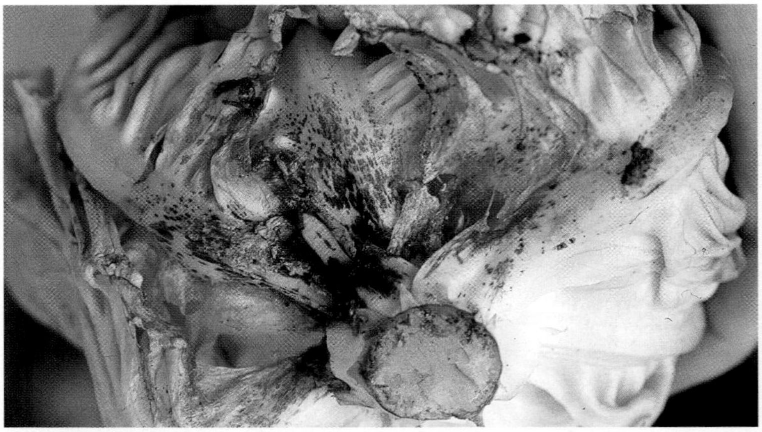

354

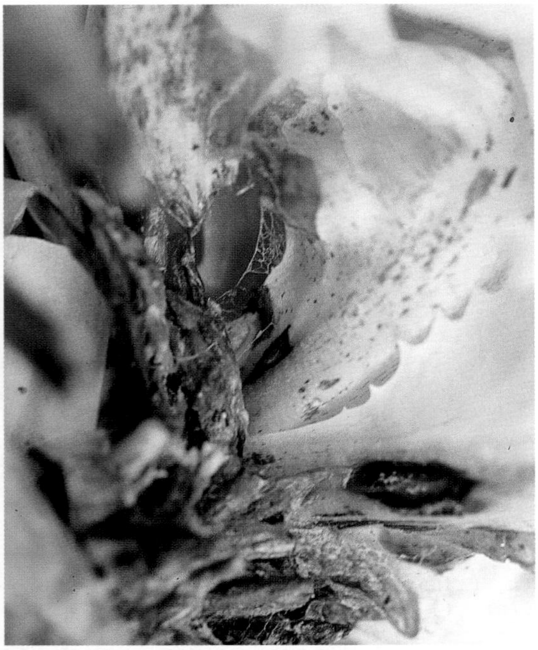

355

356

357

## *Rhizoctonia solani* ('Bottom rot') on lettuce and escarole

**353** Numerous small elongated lesions, reddish to rust in colour, and rather dry, scatter on the lamina and base of the principal vein of several leaves. Some portions of the lamina have decomposed.
***R. solani***

**354** On certain veins we sometimes note concave, dark brown, cankerous areas of damage. We may also notice an unobtrusive mycelium web.
***R. solani***

**355** Under certain situations, lesions are more extensive; broad necrotic areas, sometimes damp, are causing partial damage to the lamina.
***R. solani***

**356** A few leaves, whose principal vein base is damaged, wilt and decompose progressively.
***R. solani***

**357** In the presence of humidity and secondary bacteria, rotting is damper and dark brown to black in colour.
***R. solani***

358

359

360

361
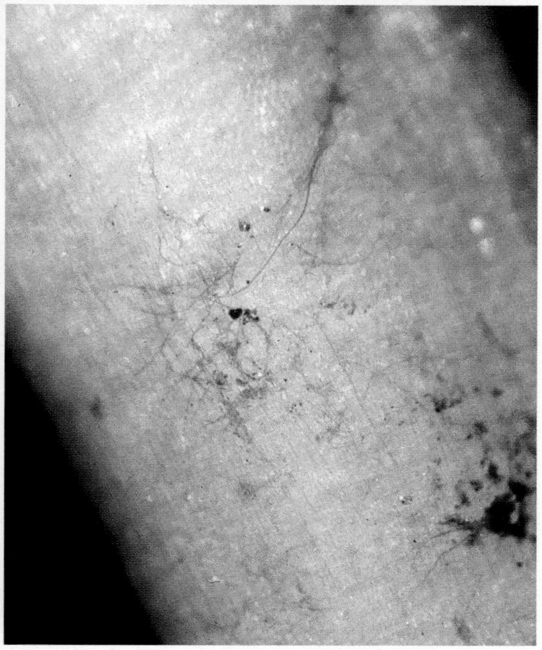

362

## *Rhizoctonia solani* ('Bottom rot') on endive

**358** Brownish black rotting is affecting several leaves of this frisée endive.
***R. solani***

**359** On this endive, it is initially the fragile, fairly filiform lamina, which is damaged, with the principal vein apparently remaining intact. Diseased tissues are initially beige in colour.
***R. solani***

**360** Fairly quickly, rotting turns black and then spreads to the principal veins. Ultimately these may go completely black; this symptom must not be attributed to *Pseudomonas cichorii*.
***R. solani***

**361** Careful observation of healthy tissues, close to diseased tissues, may reveal the characteristic brown mycelium of ***R. solani***.

**362** The young leaves from the head, which are particularly susceptible, literally collapse under the effects of this wet black rot.
***R. solani***

**Table 21: Some characteristics of the principal fungi affecting the lower leaves in contact with the soil and crown of lettuce**

| Some characteristics | *Botrytis cinerea* | *Sclerotinia minor* | *Sclerotinia sclerotiorum* | *Rhizoctonia solani* |
|---|---|---|---|---|
| **Location of symptoms:** | | | | |
| Crown | + | + | + | + |
| Leaves in contact with soil | ++ | ++ | ++ | + |
| Taproot | + | ++ | +/– | + |
| A few roots | – | +/– | +/– | +/– |
| **Extent of wilting observed** | Entire plants | Entire plants | Entire plants | Old external leaves |
| **Stage of symptom appearance:** | | | | |
| After planting | + | +/– | +/– | + |
| Growth phase | + | +/– | +/– | +/– |
| Close to harvest time | ++ | ++ | ++ | ++ |
| **Frequency in crops** | Very frequent | Frequent | Frequent | Very frequent |
| **Presence of special structures on or in lesions** | Greyish mycelium, not very dense, and grey mould consisting of numerous conidiophores and conidia (**366**). Rare black, flat and rather round sclerotia, 2–5 mm in diameter, can sometimes be seen (**368**). | White downy mycelium, more scattered, with small sclerotia of 0.5–2 mm in diameter or (**369**). The apothecia are small, measuring 1.5–2 mm in diameter. They are very rare. | A white downy mycelium (**349**) and large irregular, black sclerotia, of several millimetres in length (**367**). Sometimes we can see the presence of apothecia, 4–6 mm in diameter, forming on the sclerotia present on the soil (**364**). | Brown compartmentalized mycelium, showing restricted branching. The mycelium of this fungus is hard to see on the damaged tissues (**363**) as well as its sclerotia (**365**). |

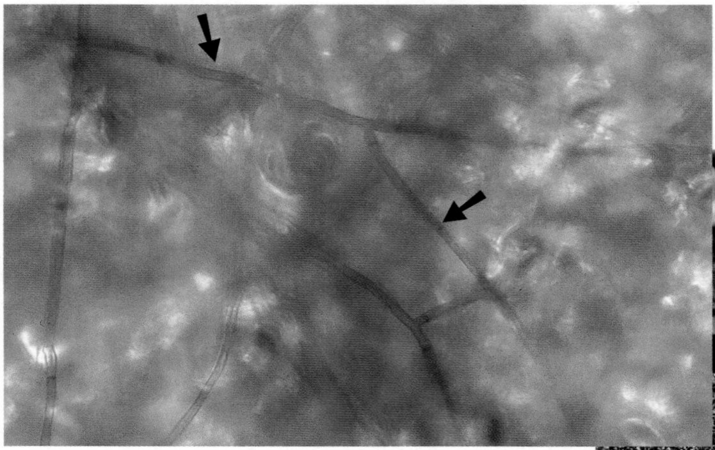

**363** On tissues colonized by ***R. solani***, using a stereoscopic microscope, the characteristic compartmentalized brown mycelium of this fungus can sometimes be seen.

**364** Small fragile trumpets, beige to brown in colour, form in clusters on the surface of the soil; these are the apothecia of ***S. sclerotiorum***.

365

366

367

368

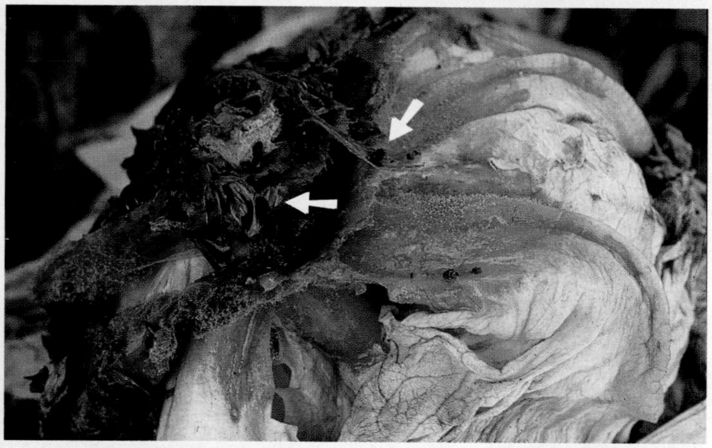

369

## Structures that differentiate the principal fungi that attack leaves in contact with the soil and the crown

**365** Sparse mycelial masses with sclerotia, brown and not dense, indicating the presence of ***Rhizoctonia solani***.

**366** ***Botrytis cinerea*** produces an abundant typical grey mould on affected tissue.

**367** ***Sclerotinia sclerotiorum*** produces large irregular sclerotia, 2.5–10 mm long. They are white when first formed and become progressively blacker.

**368** Black sclerotia, more or less round are rarely found in attacks of ***B. cinerea***.

**369** A few small sclerotia, more or less clustered, of 0.5–2 mm indicate the presence of ***Sclerotinia minor***.

370  371

372 

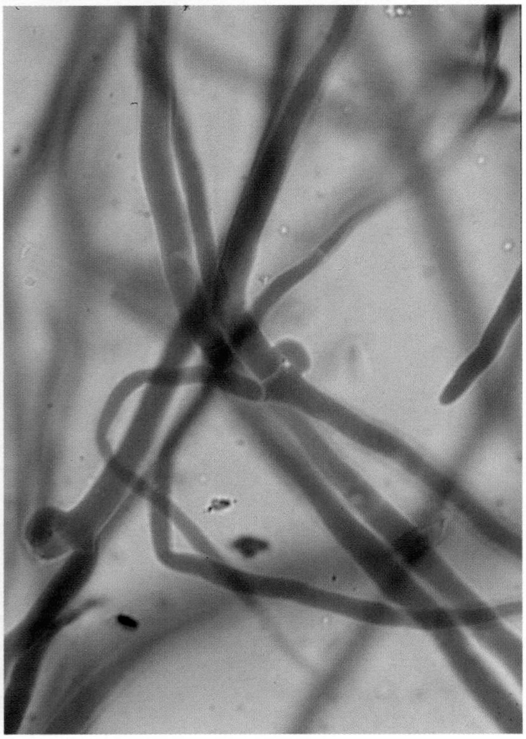 373

## *Sclerotium rolfsii* ('Stem rot')

**370** Rotting has established itself on several leaves and on the crown of this young lettuce which is in the process of collapsing.
***S. rolfsii***

**371** A dense white radiating mycelium characteristic of the presence of this fungus.
***S. rolfsii***

**372** Circular sclerotia (0.5–1.5 mm in diameter) are forming inside the mycelium; their colour changes from white to a shade of dark brown, of varying degrees of darkness.
***S. rolfsii***

**373** We can easily distinguish the loops of anastomosis of the Basidiomycete on its mycelium.
***S. rolfsii***

Table 22: Symptoms and principal characteristics of other pathogenic fungi recorded on the lower leaves, crown, and roots of lettuce

| | ***Athelia rolfsii*** **(Curzi) Tu & Kimbrough** ***Sclerotium rolfsii*** **Sacc.** | ***Phymatotrichopsis omnivora*** **(Duggar) Henneb. =** ***Phymatotrichum omnivorum*** **Duggar** |
|---|---|---|
| **Symptoms** | Damp lesions develop on leaves in contact with the soil. Rotting becomes established and progressively takes over the crown, certain roots, and the base of petioles (**370**); it spreads to the head. Affected tissues turn various shades of brown. We can also see wilting, yellowing of leaves as well as collapse of plants followed by their progressive decomposition. | Rotting of the root system associated with its colonization by the mycelium of this fungus. It establishes itself first of all in the cortex, then spreads to the vessels. It disturbs sap movements. Leaves quickly become yellow and take on a bronze hue. Plants wilt suddenly and die. |
| **Presence of particular structures on or in damaged tissues** | White mycelium, sometimes abundant, covering the base of plants and soil (**371** and **373**) (fig. 15). Numerous smooth and rather spherical small sclerotia (1–3 mm). First of all they are white, then go progressively brown as they age (**372**). | Downy mycelium and mycelium strands present on the roots, in the soil around the roots, and in the proximity of the soil surface. Isolated or massed small sclerotia, fairly spherical in shape, 1–2 mm in diameter, reddish-brown once they have matured (fig. 15). |
| **Principal characteristics** | Attacking primarily in hot regions, this fungus affects a very large number of plants. On lettuce, it is rarely very serious. High temperatures and significant humidity encourage its development (also consult fact file 10). It is present in some soils in the south-west of France (Pau region, Basque country). It does not seem to attack lettuce. | This fungus tends to predominate in semi-desert regions. It has been detected fairly recently in orchards of avocados and mangoes in semi-tropical zones. It is very polyphagous, affecting around 2,000 cultivated dicotyledons (cotton, alfalfa, vines, apple trees, and so on) or not. It is very rare on lettuce; it has mainly been recorded in the southern USA. |

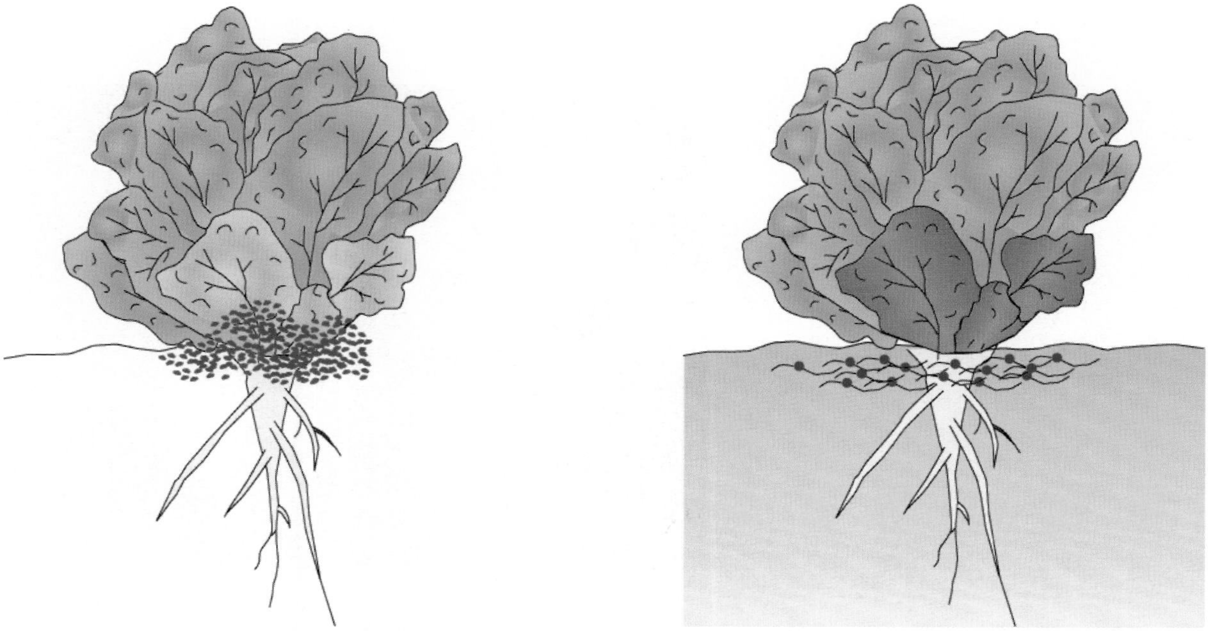

Figure 15: Appearance of lettuce colonized by *Athelia rolfsii* (on the left) and *Phymatotrichopsis omnivora* (on the right).

### • *Pythium* spp.

Several species of *Pythium* are capable of colonizing and destroying the lower leaves of l.s.v., under damp crop-growing conditions which have included splashing of water and contaminated particles of soil. For example, *Pythium uncinulatum*, *Pythium aphanidermatum*, *Pythium tracheiphilum* (see page 187) cause damp rot on leaves, particularly those in contact with the soil, in Holland, India, and the USA.

These symptoms have only been observed infrequently in France on l.s.v. We must remember that they may have been wrongly attributed to other pathogenic agents studied previously. In the course of a particularly damp autumn, we have observed broad, damp, brown lesions on the leaves of the external crown (**374**). These have been attributed to the presence of *Pythium* sp. whose numerous decorated oogonia in the damaged tissues can be seen (**375**).

**374** Several leaves of this lettuce have broad damp areas, soft in consistency and dark brown in colour. In some places, the tissues look as if theyhave melted. Rotting is in the process of progressing into the heart of the head.
***Pythium* sp.**

## *Pythium* sp.
## ('*Pythium* rot')

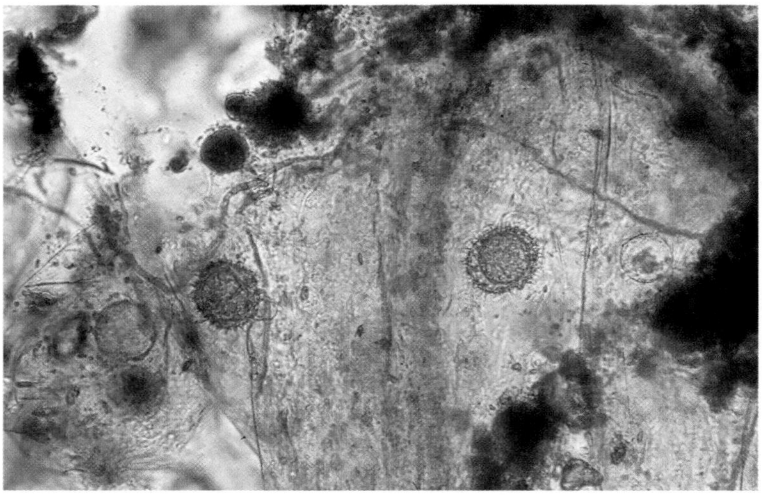

**375** Observation under a light microscope often reveals the presence of the characteristic structures of *Pythium:* from the mycelium oogonia (decorated in this case), of the sporangia.
***Pythium* sp.**

- ***Phoma exigua* Desmaz. (Phomal basal rot) (see also page 365)**

This is a newly identified fungal disease expressed as a dry rot of the crown of lettuce plants. It was first discovered in 2000 in a field of romaine lettuce in Santa Cruz County, California and has subsequently been seen in Monterey County and Santa Maria, all coastal lettuce districts. It has also been found on iceberg and leaf lettuces, but is most commonly found on romaine. The organism was identified as *Phoma exigua* in 2002.

Phoma basal rot is shown as a hard dry black rot, usually on one side of the crown area. A cross section of the crown reveals that the rot penetrates to the centre of the crown. It affects growth on one side of the plant, so that the leaves are smaller and shorter than those on the other side.

The origin, aetiology, epidemiology, and possible controls are not yet well known.

- **Other micro-organisms which may cause confusion in diagnosis**

As you can see on **376** and **378**, lettuce ring necrosis and anthracnose, caused by lettuce ring necrosis agent (LRNA) and *Microdochium panattonium* respectively, have caused spots and necrosis which sometimes cover broad sections of the lower leaves of lettuce. Their symptoms may sometimes lead to confusion with those caused by the pathogenic agents covered in this chapter. The same is true for *Erwinia carotovora* subsp. *carotovora* (**377**). Here the internal rotting of the lettuce stem may spread to the veins of the lower leaves and spill over on to the adjacent lamina, leading to a degree of confusion.

We therefore advise you to concentrate on the chapters entitled 'Spots and damage on leaves' in particular the section 'Yellow, orange to light brown spots and damage on leaves') 'Internal and/or external damage and abnormalities of the taproot and stem' on pages 111 and 207, respectively.

**376** Damp, orange areas of damage, becoming necrotic, are characteristic of the presence of **lettuce ring necrosis agent (LRNA)**. Once again, confusion in diagnosis is possible, although this type of damage always remains dry, unlike other types of rotting.

**377** On this lettuce, the principal veins of several lower leaves have rotted, to varying degrees.
***Erwinia carotovora* subsp. *carotovora***

**378** ***Microdochium panattonianum*** causes oily to orange coloured spots on the lower leaves. If these are numerous and convergent, the damage may be confused with that caused by basal rotting agents.

### • 'Brittle crown'

Young lettuce grown in the open field is particularly susceptible to the wind. Its crown is particularly stressed if the wind blows continuously. Plant tissues are subjected to very local mechanical stress, rubbing against the surrounding soil and establishing scarred, corky layers. These do not allow the tissues to develop normally, as they have lost their elasticity. Their crown is thus progressively constricted and breaks easily once plants have reached a certain height (**379–380**) (fig. 16). We can assume that these corky and sometimes spongy tissues have been invaded by secondary micro-organisms increasing the phenomenon of corkiness and sometimes giving the tissues a browner colour.

It is difficult to avoid this very natural problem. The advice is to choose plots which are not exposed to the wind.

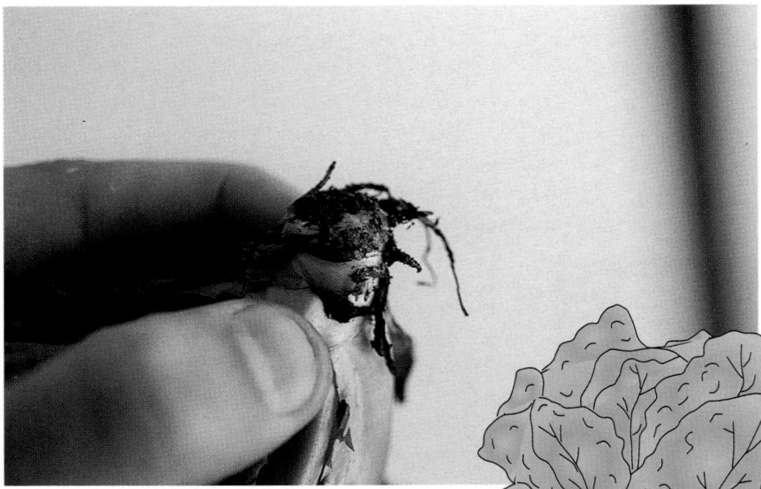

**379** The corky, constricted crown of this young lettuce has broken easily in the course of various cultural operations. The other part of the plant has remained in the soil.
**'Brittle crown'**

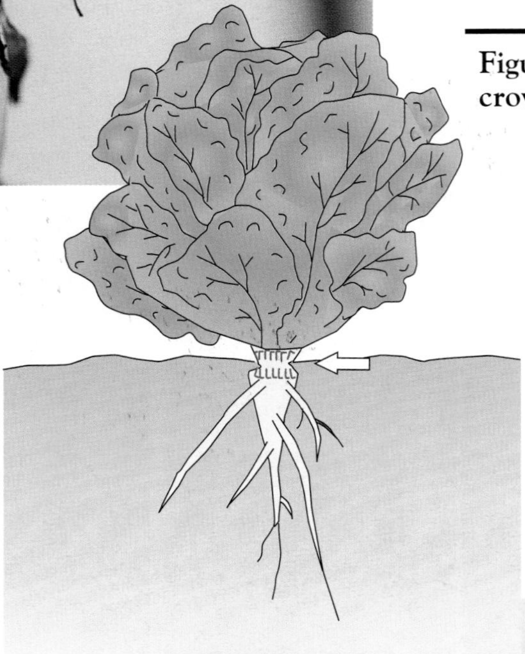

**Figure 16: Aspects of the 'Brittle crown' symptom on lettuce.**

**380** Tissues are only slightly affected. They are corky and brownish in colour, spreading to some degree into the interior.
**'Brittle crown'**

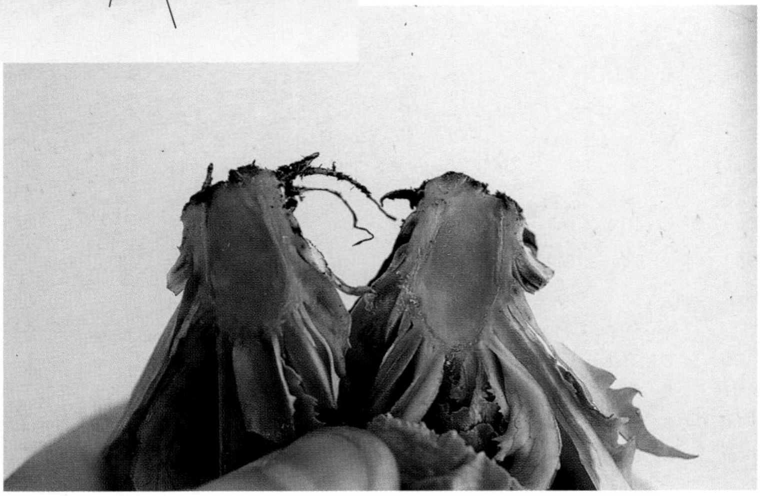

**'Brittle crown'**

## • Vole damage

These rodents, with essentially underground habits, feed on the roots of numerous vegetable and non-vegetable plants. L.s.v. may be affected by them, particularly if grown in tunnels. Voles gnaw lettuce roots up to the crown. Plants quickly and suddenly wilt (**381**). When the plants are turned over, we can see that the roots have completely disappeared and stem has been gnawed away (**383**). When a tunnel comes to the surface this is an indication that these rodents have passed by (**382**). Several species of vole may be responsible for damage to lettuce, particularly the Provençal vole *Pitymys duodecimcostatus*.

**381** Several lettuces in small groups have suddenly wilted.
**Vole damage**

**382** A vole tunnel is emerging right in the place where the roots should have been.

**383** The root system of the lettuce has disappeared. On the stem we can see traces of numerous bites.
**Vole damage**

**Vole damage**

**384** Numerous rootlets are reddish brown in colour.
**Ammonium toxicity**

**385** A few brown corky bands can be seen on several leaves.
**Corky root – *Rhizomonas suberifaciens***

**386** Some roots and rootlets are excessively swollen, forming very characteristic root knots.
***Meloidogyne* sp.**

# Damage and abnormalities of roots

## Symptoms studied

- Yellowing, browning, blackening of roots
- Miscellaneous damage and abnormalities of the roots (corky root, swelling, galls, feeding damage, and so on)

## Possible causes

- *Olpidium brassicae* (virus vector) (fact file 14)
- *Phymatotrichopsis omnivora* (see also page 179)
- *Pyrenochaeta lycopersici* (fact file 10)
- Various Pythiums (*Pythium* spp., *Phytophthora* spp.) (fact file 8)
- *Plasmopara lactucae-radicis*
- *Rhizoctonia solani* (fact file 7)
- *Thielaviopsis basicola* (fact file 9)
- *Agrobacterium tumefaciens*
- *Rhizomonas suberifaciens* (fact file 19)
- *Meloidogyne* spp. (root knot nematodes) (fact file 36)
- *Nacobbus* spp. (false gall nematodes)
- *Pratylenchus* spp. (lesion nematodes) (fact file 37)
- Other root infecting nematodes (*Longidorus* spp., *Rotylenchus robustus*, *Merlineus brevidens*, *Tylenchorhynchus claytoni*, and so on)
- *Pemphigus bursarius* (root aphid)
- Other soil pests (*Agriotes* spp., *Agrotis* spp., *Hepialus* spp., *Melolontha melolontha*, *Tipula* spp., and so on)
- Allelopathy
- Root suffocation
- Damage due to decomposition of organic material
- Non-parasitic swollen rot
- 'Acid soil'
- Various types of chemical injury
- Ammonia toxicity

## Very difficult diagnosis

The root system of leafy salad vegetables is the poorly known part of the plant; many practitioners have some difficulty in assessing their state since they are often not in good condition for observation. We suggest you seek help.

First of all you must **carefully recover the root system**; you must avoid pulling them up roughly and damaging them as the most fragile (but also the most interesting for diagnosis) will remain in the soil. It is very important to **wash them well with water** to remove soil particles that frequently hide certain symptoms. Now you can examine the roots very carefully and using a lens if you have one.

If the crown and the principal vein base of the leaves are affected, observe the root system as well and consult the chapter 'Internal and/or external damage and abnormalities of the taproot and stem'.

## *Pythium* spp. (Pythium root rot)

**387** The whole root system of this lettuce shows a dark brown colouration.
***Pythium* sp.**

**388** This young plant attacked by *Rhizoctonia solani* is not prematurely dead. The root system is rather reduced; many roots and rootlets have not formed or are missing after turning brown and necrotic.
***R. solani***

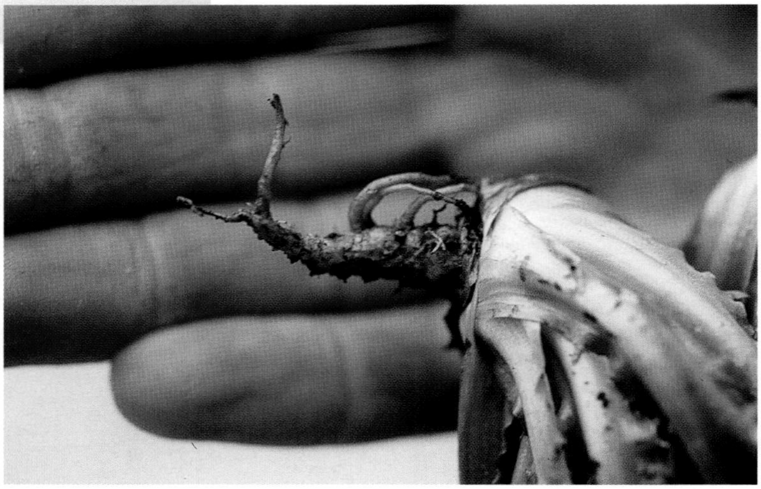

## *Rhizoctonia solani* ('Rhizoctonia root rot')

### *Ask some questions!*

Sometimes, in simply answering the following questions, you can provide an explanation to your problem (each affirmative response to one of the questions will confirm the possible action of the incriminating cause):

- Have you watered the plants too much?
- Are the diseased plants situated in a place where they receive too much moisture?
- Have you planted during a time or in a soil that is still cold and wet?
- Has the irrigation of the nursery or plantation been excessive?
- Has the application of fertilizer before planting or during cultivation been too large?

# Yellowing, browning, blackening of roots

## Possible causes

- *Olpidium brassicae* (virus vector) (fact file 14)
- *Phymatotrichopsis omnivora* (fact file 10)
- *Plasmopara lactucae-radicis*
- Various types of Pythiums (*Pythium* spp., *Phytophthora* spp) (fact file 8)
- *Rhizoctonia solani* (fact file 7)
- *Thielaviopsis basicola* (fact file 9)
- *Rhizomonas suberifaciens* (parasitic 'Corky root') (see page 197) (fact file 19)
- *Pratylenchus* spp. (semi-endomigratory nematodes) (fact file 37)
- Other nematodes predominantly on the roots (*Longidorus* spp., *Rotylenchus robustus*, *Merlineus brevidens*, *Tylenchorhynchus omnivorus*, and so on)
- Root suffocation
- Damage associated with an activator of the decomposition of organic matter
- Non-parasitic corky root, ammonium toxicity (see page 197)
- Various types of chemical injury (certain herbicides can cause necrosis of the roots and/or reduce or prevent their formation; see page 83)

The majority of diseases and predators affecting the root system of l.s.v. cause widespread yellowing, browning (localized or widespread), necrosis, and the disappearance of numerous rootlets and sometimes even the roots. In more serious cases, their root system is completely destroyed and the vessels located in the taproot may become yellow and slightly brown. Damage sometimes spreads to the crown and leaves (consult the heading 'Damage to leaves, in contact with the soil, and the crown').

Sometimes these diseases may attack at the same time or, more specifically, some of them may make the plants more susceptible to other diseases. For example attacks of Pythiums and *Rhizomonas suberifaciens* appear more serious in heavy, poorly-drained soil. Drowned plants are more receptive to these pathogenic agents.



## Arguments in support of the diagnosis

- ***Pythium* spp. ('Damping-off', 'Pythium root rot')**

In a fairly general way, *Pythium* spp. especially attacks seeds and young plantlets after sowing in the nursery, pre- and post-emergence. It is capable of inhibiting germination, rapidly colonizing succulent, tender and non-lignified young tissues, the roots and the crown. Subsequently it is fairly common to see plantlets, located in the same place (in groups), which wilt, go yellow, dry out, and disappear fairly rapidly (**fig. 17**). This is a classic syndrome which is called **'damping-off'**. In addition to *Pythium* spp., other pathogenic agents can be responsible for this damping-off; and some of their characteristics are indicated in table 20. As you will see, the symptoms they cause on the roots are fairly similar and so it is quite difficult to differentiate between them with the naked eye. They affect both the roots and the part of the stem located close to the soil or the substrate, thus constricting the latter. In certain situations, the effects of these types of fungus on l.s.v. persist throughout the season. Sometimes we notice reduced growth of plants, to a greater or lesser degree, or the plant abruptly wilting. In all cases, when the plants are pulled up, we note the effects of *Pythium* spp. on the roots: more or less widespread browning of the roots, disappearance of numerous rootlets (**387**) and sometimes vascular damage, and damage to the crown. For example in California, ***P. uncinulatum*** is responsible for the poor development of the root system of l.s.v. as well as for a deterioration of the tips of the roots. In addition, in Holland this same species is associated with damage to the lower leaves of l.s.v., just like ***P. aphanidermatum***, in India, which colonizes leaves in contact with the soil and produces damp translucent rotting which may spread to the entire head (see page 180).

**Table 23: Some characteristics of the principal micro-organisms responsible for attacks on seeds and young lettuce plants**

| Some characteristics | *Pythium* spp. | *Rhizoctonia solani* | *Botrytis cinerea* | *Sclerotinia sclerotiorum* or *Sclerotinia minor* |
|---|---|---|---|---|
| **Principal symptoms observed** | Absence of germination, widespread root browning, constricting of tissues present on the surface of the soil, damping-off. | Absence of germination, widespread root browning, constricting of tissues present on the surface of the soil, damping-off. | Absence of germination, widespread root browning, constricting of tissues present on the surface of the soil, presence of a 'web' consisting of greyish white mycelium covering the plantlets, damping-off. | Rotting of leaves in contact with the soil, rotting of the crown, death of plantlets. |
| **Frequency of attacks** | Fairly frequent. | Fairly frequent. | Fairly frequent. | More rare. |
| **Structures allowing them to be identified on and/or in damaged tissue** | Oospores and chlamydospores visible in the root cells (**394**). | Partitioned brown epiphytic mycelium showing constricted ramifications. | Greyish white mycelium frame-work, grey mould consisting of numerous conidiophores and conidia (**366**). | White mycelium and large black sclerotia, sometimes apothecia in the plant environment (**367**). White mycelium scattered with small black sclerotia (**369**). |

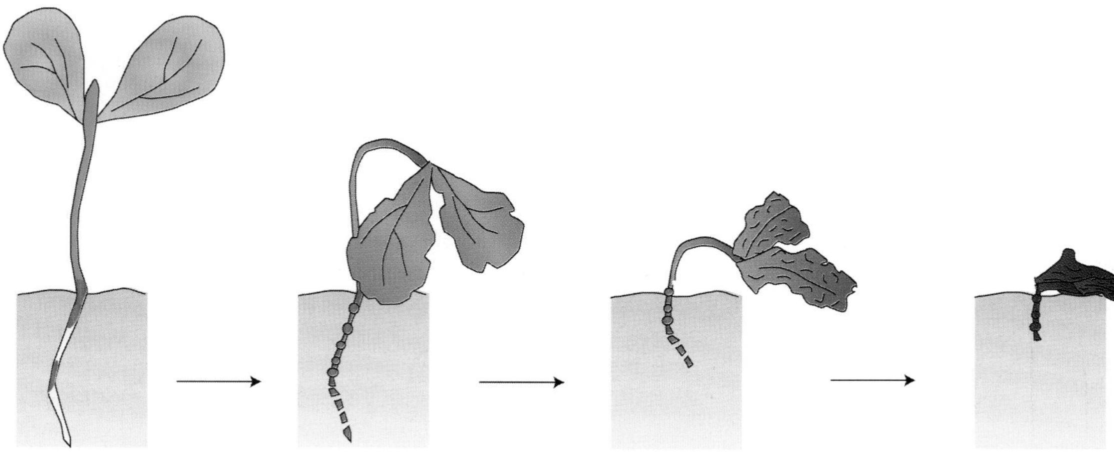

**Figure 17: Development of damping-off on a young lettuce plantlet.**

Several other species of *Pythium* spp. are likely to attack l.s.v. They have only been accurately identified in a fairly restricted number of situations worldwide. Some of these are ***P. dissotocum***, ***P. irregulare***, ***P. megalocantum***, ***P. polymastum***, ***P. spinosum***, ***P. sylvaticum***.

A 'systemic' *Pythium*, ***P. tracheiphilum*** also attacks l.s.v. We recommend that you consult the section entitled 'Internal and/or external damage and abnormalities of the taproot and stem'.

Only the observation of oospores (sometimes other structures such as sporangia or chlamydospores) on and in the affected tissues, or carrying out isolation on artificial media, will allow you to diagnose the action of this type of fungus with certainty (**394**).

- ***Rhizoctonia solani*** **('Bottom rot')**

This fungus is mainly predominant in lower leaves of l.s.v. which are in contact with the soil. It may attack young plantlets causing damping-off. The results are the appearance of dry damage, light brown to chocolate brown in colour, on the part of the stem or taproot located on the surface of the substrate or the soil. Attacks of *R. solani* may occur subsequently on plantlets. This may essentially affect the roots, causing them to go brown. A large number of the latter may disappear (**388**). In this case, the presence of its characteristic brown mycelium (**395**) will help you to identify it without too much difficulty.

- ***Thielaviopsis basicola*** **('Black root rot')**

This soil fungus is responsible for black root rot on numerous plants, particularly in the market garden (bean, eggplant, and so on). It is particularly responsible for damage to witloof chicory, damaging and deforming the roots. It has rarely been reported on other types of l.s.v. Attacks have been recorded on lettuce in Queensland, Australia. Affected lettuces grow poorly, with small heads. Yields are low.

In France, although the fungus is present in numerous types of soil, no damage had been observed on lettuces to date. Recently we have recorded *T. basicola* on curly endive growing poorly (**389**) in the open field in the Eastern Pyrenees. Numerous roots of these plants were brown to black in colour (**390–393**); and the largest of these had damage and a corky root.

As this fungus is not well known on l.s.v., no hypothesis has been issued about its action on these plants, and therefore none has been retained. Only microscopic observations of fragments of damaged roots allow us to observe the chlamydospores typical of *T. basicola* on and in the cells of the cortex (**396**).

389 

390 

391 

392 

393 

## *Thielaviopsis basicola* ('Black root rot')

**389** These four plants of curly endive, attacked to greater or lesser degree by ***T. basicola***, are very different in size, in spite of being planted on the same date.

**390** Their root system has been damaged to a greater or lesser degree, which explains the poor growth of the plants.
***T. basicola***

**391** The root systems are fairly varied. We can see corky root and cracking.
***T. basicola***

**392** ***T. basicola*** is also responsible for dark brown to black rotting affecting numerous roots; this damage lies at the origins of the English name for the disease: 'Black root rot'.

**393** Damp, brown damage is also visible. The rotting deep inside the tissues causes the roots to rupture when the plants are pulled up.
***T. basicola***

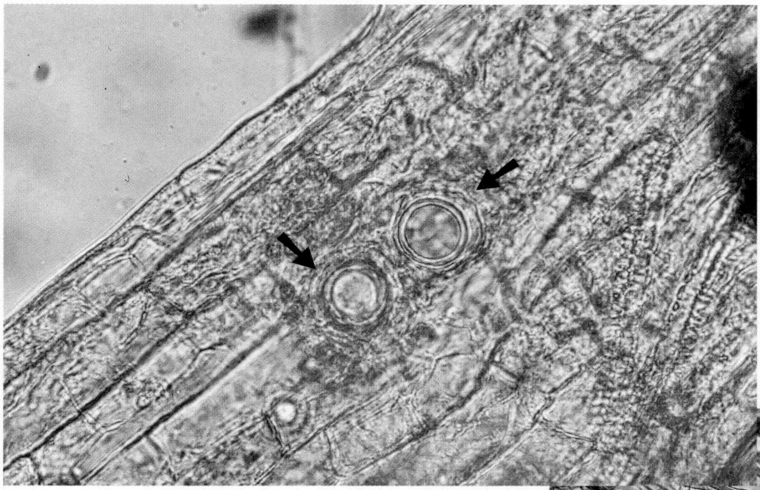

**394** Round oospores, with a thick wall, are present on or in the root tissues; they frequently show parasitism by ***Pythium* spp.**

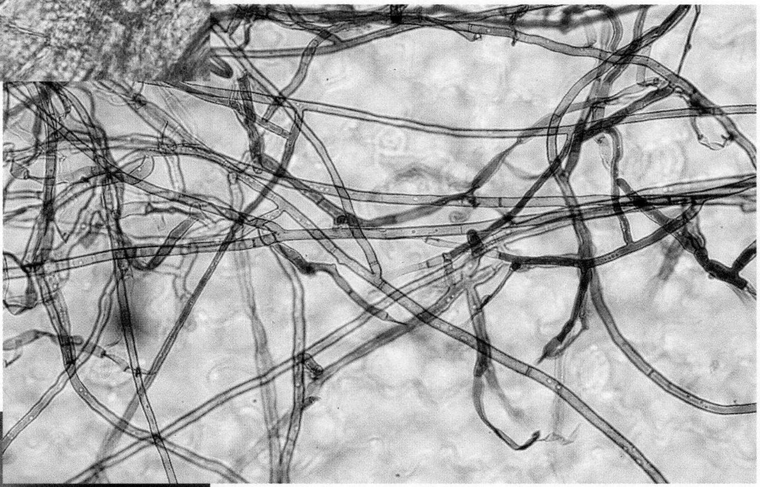

**395** ***Rhizoctonia solani*** is characterized by a brown, partitioned mycelium, which is constricted at the base of its branches.

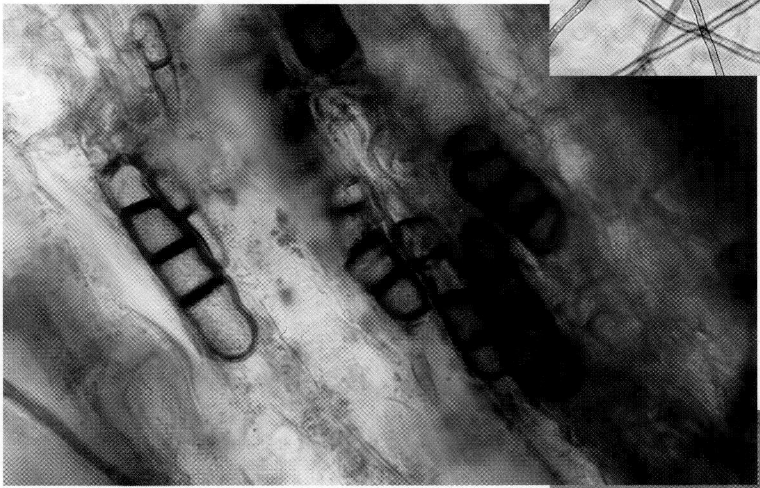

**396** Brown chlamydospores of ***Thielaviopsis basicola***, in chains, are clearly visible on this root.

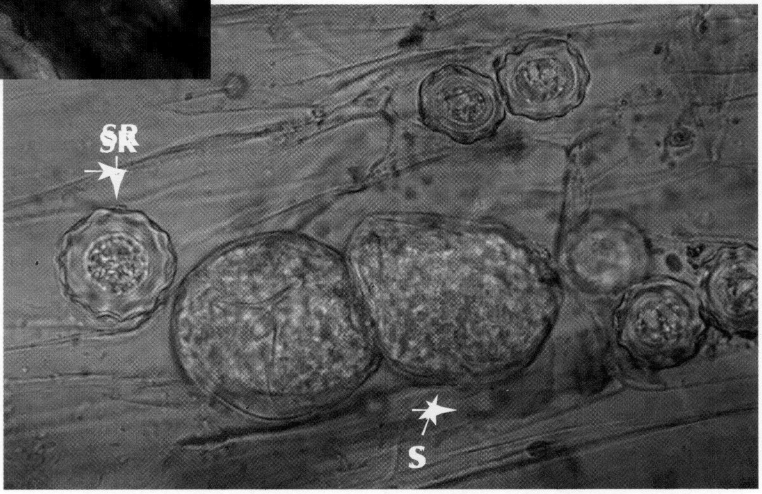


**397** **Sporangia** (S) and 'chlamydospores' (resting spores) (SR) of ***Olpidium brassicae*** are contained within several cells.

## Principal types of fungus which predominate in the root system of l.s.v. plantlets and adult plants

• *Olpidium brassicae*

As far as we know this aquatic fungus is not pathogenic to lettuce. However we frequently see its fruiting bodies (sporangia and chlamydospores or resting spores) in the cells of the epidermis and cortex of l.s.v. roots (**397**), whether grown in or out of soil.

The presence of this fungus, an obligate parasite, in the roots does not seem to have an adverse effect on plant development. However it is not entirely harmless since it is the vector of two very damaging viral diseases: big vein disease and orange spot disease (see pages 63, 73, and 119).

It is obvious that the conditions of the medium, particularly soil temperature and oxygenation, the substrate or nutrient solution, and more specifically the state of the roots, must have an influence on the behaviour of this fungus. We know that it is totally suited to aquatic life. Like *Pythium* spp. and other aquatic fungi it has mobile zoospores which allow it to spread easily in water.

**• *Phytophthora* spp. ('*Phytophthora* rot and root-rot')**

*Phytophthora* spp. are fairly rare in l.s.v. Two species have been recorded on this plant:

- *Phytophthora porri* which causes stem rot mainly located in the crown, but ultimately destroys the head. The strain responsible seems to have a fairly low thermal optimum explaining the fact that attacks occur mainly in winter, in poorly-drained soil in southern Australia;
- *Phytophthora cryptogea*, responsible for browning and rotting of roots of lettuces grown out of soil in southern California. In France this fungus is rife among endives, if they are grown in beds covered by soil.

**• *Plasmopara lactucae-radicis* ('Root downy mildew')**

*P. lactucae-radicis* is a fungus which was discovered fairly recently in some hydroponic crops of lettuce grown in Virginia. This fungus mainly develops on the roots, which it colonizes and on which it fructifies abundantly. It probably possesses parasitic features similar to those of sunflower mildew, *Plasmopara helianthi* f. *helianthi*. *P. lactucae-radicis* colonizes the interior of the roots; it is responsible for brown root necrosis varying in severity. This certainly causes the lower yields observed in infected greenhouses. In addition, it forms numerous sporangiophores on the roots, which produce a multitude of sporangia (**fig. 18**). The latter ultimately release mobile zoospores which spread easily in the nutrient solutions used in hydroponic cultures.

In the few areas where it is rife, the observation of sporangiophores and sporangia on diseased roots and aplerotic oospores in the cortical tissues, can confirm the presence of this root mildew which affects lettuce.

• *Phymatrichopsis omnivora* (Dugger) Hennebert
(*Phymatotrichum omnivorum*) (Shear) Dugger ('Texas root rot')

This very polyphagous fungus, whose thermal optimum is high (28–30°C) appears mostly in semi-desert zones. It has only been reported as affecting l.s.v. in the USA, where it is responsible for sudden wilting while crops are growing, caused by rotting of the roots and mycelial colonization of the vessels.

Damage is mainly observed in alkaline soils, low in organic material. Acid soils, whose pH is close to 4.7, do not seem to favour the development of this fungus or the production of sclerotia ensuring its ongoing presence. These sclerotia also cause the initial contamination. They are produced by the mycelium in the soil, often on and close to the roots. We find them in the soil, 15–60 cm under the surface, going down to over 2 metres in depth, according to certain authors. They can survive easily for several years. Conidiospores also form on its mycelium (**fig. 19**).

Several measures are recommended in order to limit the damage caused by this fungus:

- avoid planting l.s.v. in plots where particularly susceptible crops have been grown, notably alfalfa and cotton;
- get rid of as many diseased plants as possible, especially the root system;
- establish a fairly long period for crop rotation of at least 4 years, introducing monocotyledons which are not susceptible. Pea crops are also likely to reduce the presence of this fungus in the soil significantly, especially if numerous plants are buried in it (see page 179).



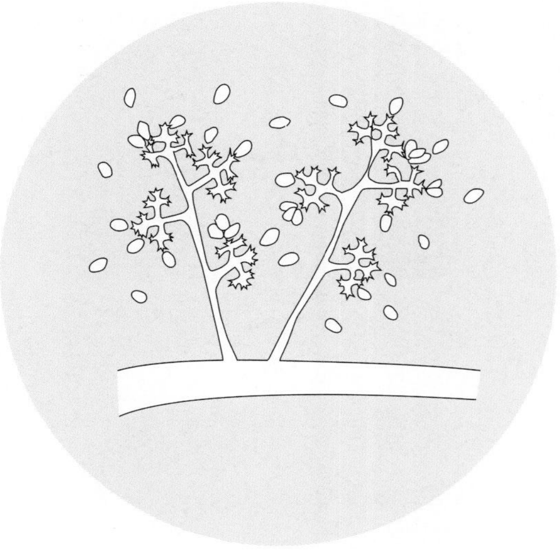

**Figure 18:** ***Plasmopara lactucae-radicis*** forms branched sporangia on the roots of infected lettuce. They are extended by short sterigmata on whose extremities elliptical sporangia develop.

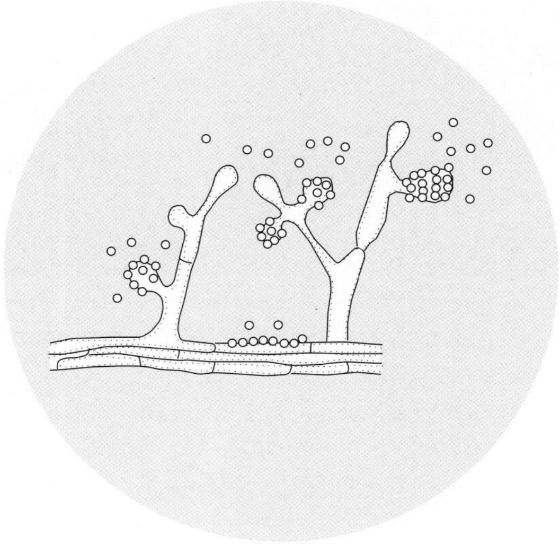

**Figure 19:** ***Phymatotrichopsis omnivora*** produces conidiophores which originate laterally on the mycelium. Their terminal part is swollen and they produce globular to oval-shaped conidia, often forming a mass.

398

399

400

401

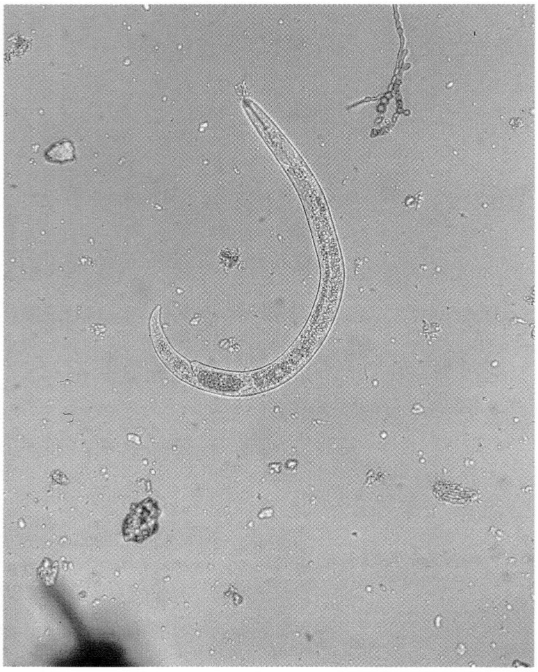

402

## *Pratylenchus* spp. ('Lesion nematodes')

**398** These two lettuce plants affected by nematode parasites are chlorotic and small in size.
***Pratylenchus penetrans***

**399** Attacks often occur early, as was the case with this root system whose oldest parts are completely rotten.
***P. penetrans***

**400** Roots infected by endomigratory parasitic nematodes are covered with numerous lesions, which are initially damp and yellowish to reddish-brown in colour. Ultimately the whole of the root system is affected, which is what has happened on these three plants.
***P. penetrans***

**401** *Pratylenchus* are fairly short nematodes with a clearly visible buccal stylet.
***P. penetrans***

**402** Roots often have no rootlets; the cortex may have deteriorated completely.
***P. penetrans***

• ***Pratylenchus*** **spp. ('Lesion nematodes')**
After penetrating the root cortex and progressively destroying numerous cells, these phytophagous, migratory endoparasitic nematodes (**fig. 20**) cause abundant reddish-brown to black lesions (**399**, **400**, and **402**). The colour of the tissues is due to the release of phenolic substances, following the action of the nematode hydrolytic enzymes. This root damage adversely affects lettuce growth and plants are often small in size with chlorotic leaves (**398**). Adults have a very marked stylet. Larvae and eggs are often visible in the cells of the root cortex (**401**), which will help you to confirm your diagnosis.

***P. penetrans*** is the principal species recorded on l.s.v. in various countries worldwide, particularly the USA and Canada. *P. crenatus* has also been recorded on l.s.v. (see *Tables 24* and *25*). Waterlogged soils, low temperatures, and unsatisfactory fertilization probably exacerbate the effects of this nematode.

In France the incidence of these nematodes on lettuce is unknown. We have seen serious attacks on lettuce in the Eastern Pyrenees. The fact that we do not know what damage they cause to lettuce certainly leads to confusion in diagnosis.

• **Ammonium toxicity (non-parasitic corky root)**
This toxicity causes reddening of the root cortex, and sometimes the stele and taproot of l.s.v. (**403**). Plant growth is reduced. Plants also wilt over time, are dark green in colour and have peripheral V-shaped patches of yellowing on the lamina, which become necrotic (**404**). We may see these symptoms if nitrogenous fertilizer has been used incorrectly or in the presence of an organic matter decomposition activator. Root symptoms can be generalized or localized in the area of the root system corresponding to the location of the fertilizer.

**403** The cortex of certain roots is brownish-red in colour. Numerous rootlets are also affected.
**Ammonium toxicity**

**404** Burns on the roots have caused reduced growth of this lettuce. The lamina of several leaves is becoming progressively yellow around the periphery and will ultimately become brown and necrotic.
**Ammonium toxicity**

405

406

407

408

## Parasitic corky root (*Rhizomonas suberifaciens*)

**405** This root has yellowish lesions which are progressively becoming brown.
**Corky root - *R. suberifaciens***

**406** In certain affected zones the roots, brown and corky on the surface, have burst longitudinally to varying extents.
**Corky root - *R. suberifaciens***

**407** The root system of this lettuce has numerous brown lesions which sometimes encircle a significant length of certain roots.
**Corky root - *R. suberifaciens***

**408** We find the same type of symptom on the taproot where surface corkiness and splitting can be observed.
**Corky root - *R. suberifaciens***

*We advise you to refer to the damage caused by* Thielaviopsis basicola *on lettuce roots on page 190.
To some extent this resembles damage caused by the diseases responsible for 'corky root', parasitic or not.*

# Miscellaneous damage and abnormalities of the roots (corky root, swelling, root knots, feeding damage, and so on)

## Possible causes

- *Pyrenochaeta lycopersici* (fact file 10)
- *Thielaviopsis basicola* (fact file 9) (see page 189)
- *Agrobacterium tumefaciens*
- *Rhizomonas suberifaciens* (parasitic corky root) (fact file 3)
- *Meloidogyne* spp. (root knot nematodes) (fact file 36)
- *Nacobbus* spp. (false root knot nematodes)
- Other nematodes
- *Pemphigus bursarius* (root aphid)
- Other soil-based pests (*Agriotes* spp., *Agrotis* spp., *Hepialus* spp., *Melolontha melolontha*, *Tipula* spp., and so on)
- Allelopathy
- Non-parasitic corky root
- Acid soil



## Arguments in support of the diagnosis

- **Non-parasitic corky root**
  ***Rhizomonas suberifaciens***
  **('Corky root')**

The presence of corky roots showing varying degrees of hypertrophy on lettuce has long been attributed to a non-parasitic disease. In fact the situation is more complex and the cause of these symptoms may vary depending on growing contexts.

Some authors have suggested the effects of **phytotoxins** released into the soil when green plant waste is decomposing under very humid conditions. Little information has been recorded about this situation.

These symptoms are more frequently associated with over-zealous application of **nitrogenous fertilizer** which leads to excessive release of ammoniacal nitrogen and nitrites into the soil. We can also note rotting of lateral roots and a red to brown coloured stele. This indicates **non-parasitic corky root**, which occurs much less frequently if we use reasonable quantities of nitrogenous fertilizer or nitrogen in the form of ammonium sulphate or nitrate, urea, or calcium nitrate (also see page 195).

More recently, a bacterium which is extremely difficult to isolate, ***Rhizomonas suberifaciens***, has been associated with this type of symptom in plots where several lettuce crops have been grown in close succession. Initially this bacterium is capable of causing yellow lesions on the taproot and on the large lateral roots. The lesions become brownish in colour and quickly encircle these roots which sometimes swell. Subsequently we notice significant surface corkiness of the tissues which split longitudinally (**405–407**). The taproot may be particularly badly affected, sometimes showing excessive hypertrophy (**408**) (see page 209). This disease is known as parasitic corky root or corky root.

It is quite clear that in some plots *R. suberifaciens* and excessive nitrogenous fertilizer may be jointly responsible. In addition we do not make any systematic distinction between parasitic and non-parasitic corky root as their symptoms, as well as the methods for controlling them, are fairly comparable.

Roots have yellowish lesions which rapidly become brown. In places we see the same surface corkiness and longitudinal splitting, as on the taproot. They also tend to swell. This causes their surface to become rough and cracked; we can distinguish grooves and corky ridges on the root system.

The taproot is often very badly affected. It is extremely corky, hypertrophic, and becomes brittle.

• **Allelopathy**

Previous crops of certain plants may prove to encourage or, on the other hand, inhibit lettuce growth. These effects are due to the production of chemical substances, exuded from the roots or from decomposed crop debris, which have passed into the soil. These can be responsible for the following:

- inhibition of seed germination;
- reduction in lettuce growth, dwarfism, and yellowing of leaves;
- various types of root damage (poor establishment, reduction in the number and length of roots, necrosis located at the shoot and in root portions in contact with plant waste, and so on).

Previously grown crops likely to cause a phenomenon of this kind include: asparagus, broad bean, broccoli, celery, lettuce, vetch, barley, wheat, and rice. The quantity of toxic residue produced and therefore the extent of the damage also depend on the type of soil, its humidity, the quantity of oxygen present, and the length of time the plant debris has been decomposing before the lettuce was planted.

In numerous situations allowing the plant debris to decompose for an adequate period can prevent the occurrence of this phenomenon.

• **Acid soil**

You should be aware that in certain excessively acid soils salad crops, particularly lettuce, may grow less successfully and roots may be a dull grey in colour.

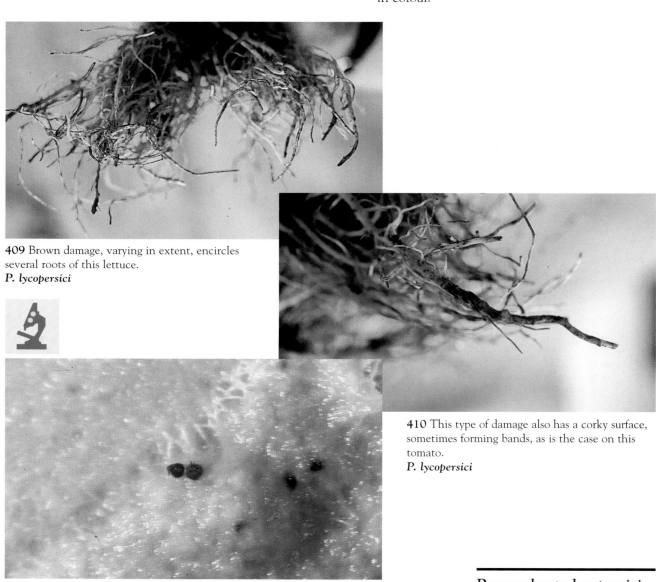

**409** Brown damage, varying in extent, encircles several roots of this lettuce.
***P. lycopersici***

**410** This type of damage also has a corky surface, sometimes forming bands, as is the case on this tomato.
***P. lycopersici***

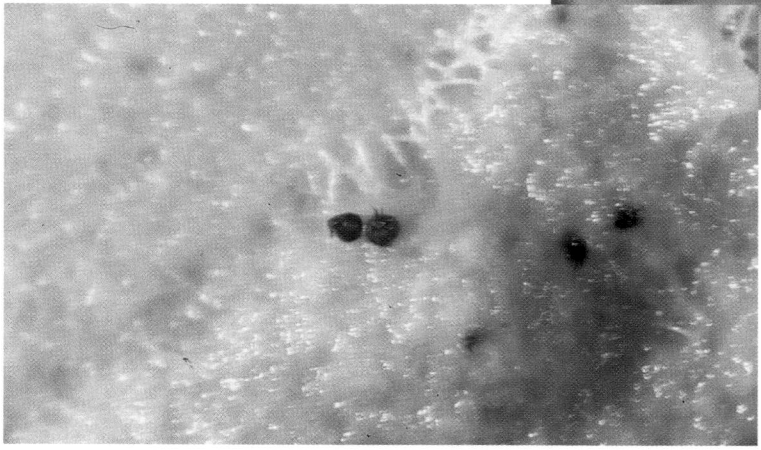

**411** On rare occasions ***P. lycopersici*** forms pycnidia on the roots, which have characteristic brown bristles located on the ostiole.

## *Pyrenochaeta lycopersici*

- *Pyrenochaeta lycopersici* **('*Pyrenochaeta* corky root')**

This fungus, responsible for tomato corky root also attacks other members of the Solanaceae and various Cucurbitaceae members. It also develops on the roots of lettuce involved in rotation with these crops.

As with all of its hosts, *P. lycopersici* causes brown damage, with a corky surface, which encircles the roots (**409–411**). Its effect on lettuce is only occasionally significant, causing reduction in plant vigour and therefore in the size of lettuces when they come to be harvested.

To our knowledge it has only been recorded on lettuce in England and France, where it is occurs in a few plots on which several rotations of susceptible plants, particularly tomato, have taken place.

- ***Meloidogyne* spp. ('Root knot nematodes')**

Root knot nematodes (**fig. 20**) affect numerous vegetable crops used in rotation with salad plants (tomato, eggplant, melon, and so on). It is not therefore surprising to note their effects on crops. The root damage caused by these nematodes, consisting of knots and swellings (**412–414**), is very characteristic and easy to identify. If the knots are cut in half we can see one or several round, whitish masses, corresponding to swollen adult females (**415** and **416**). This root damage causes delayed growth in affected plants, and sometimes wilting and decline.

Several species of *Meloidogyne* have been recorded on lettuce (*Tables 24* and *25*): ***M. incognita, M. javanica, M. arenaria, M. hapla***. The knots from the latter species are probably smaller and rounder than those of other species.

We ought to point out that ***Agrobacterium tumefaciens*** has only been reported once on lettuce, in Brazil in the state of Guanabara. Its attacks were frequently combined with those of *Meloidogyne* sp.

- ***Nacobbus* spp. ('False root-knot nematodes')**

These sedentary or migratory endoparasitic nematodes (fig. 20) cause various types of damage on lettuce roots:

- necrosis associated with cell destruction and the formation of cortex cavities;
- root knots which are morphologically very comparable to those caused by *Meloidogyne* sp. and in particular *M. hapla*. Initially the females establish a nutritional site within the root, close to the vascular system. They take their nourishment to the detriment of the cells, become much larger and cause small round root knots to form. Their eggs are expelled and grouped together in a gelatinous matrix. They constitute the most effective form of survival of these nematodes in decomposing plant debris. Several species have been reported on l.s.v. (*Table 25*): ***N. aberrans, N. batatiformis, N. serendipiticus***.

These very polyphagous nematodes affect both cultivated plants and weeds. Their attacks on lettuce are infrequent.

Confusion in diagnosis is possible with *Meloidogyne* sp., essentially in production zones where these two types of nematodes are commonly in combination. This confusion does not normally cause a problem as the methods of combating them are identical.

Numerous other nematodes have been recorded less commonly on lettuce. A list of these, together with information about some of them, can be consulted in *Tables 24* and *25*.

## *Meloidogyne* spp. ('Root-knot nematode')

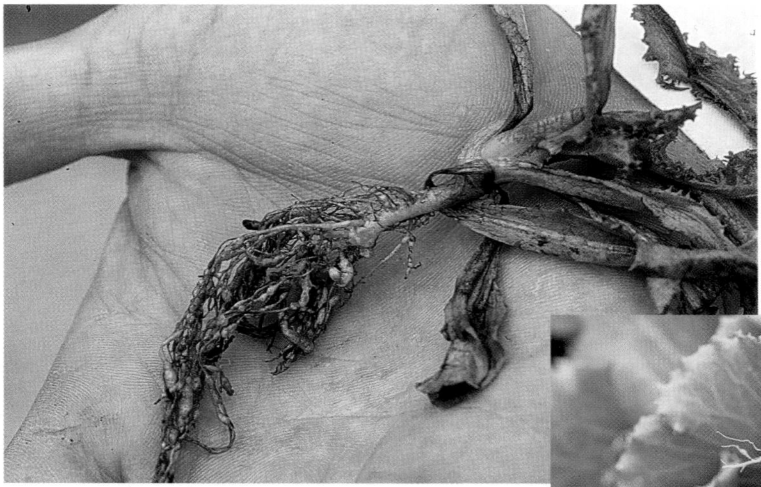

**412** The root system of this chicory plantlet is sprinkled with root knots and swellings which are often isolated; their colour varies from dirty white to brown. ***Meloidogyne* sp.**

**413** We find the same symptoms on adult plants. Root knots mainly form on the portions of roots present outside the main root mass. ***Meloidogyne* sp.**

**414** When the inoculum rate of the soil is high, we can see series of root knots and swellings establishing themselves all along the roots. ***Meloidogyne* sp.**

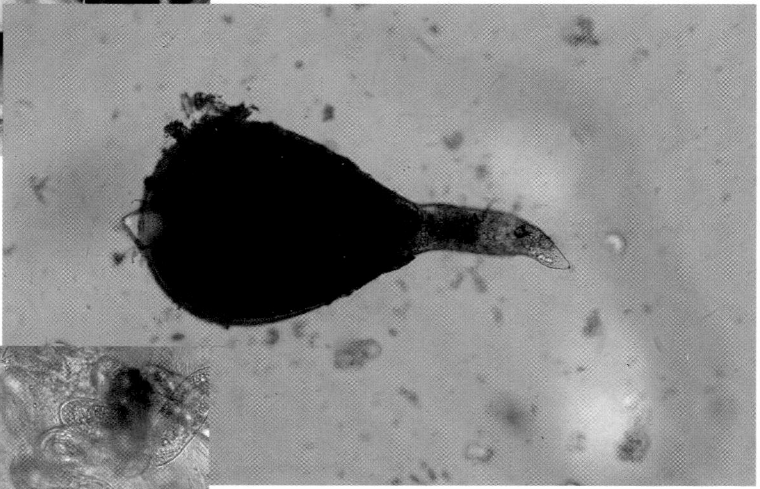

**415** These root knots contain one or several swollen adult females. Sometimes we can see the female stylet when a root knot is cut out. ***Meloidogyne* sp.**

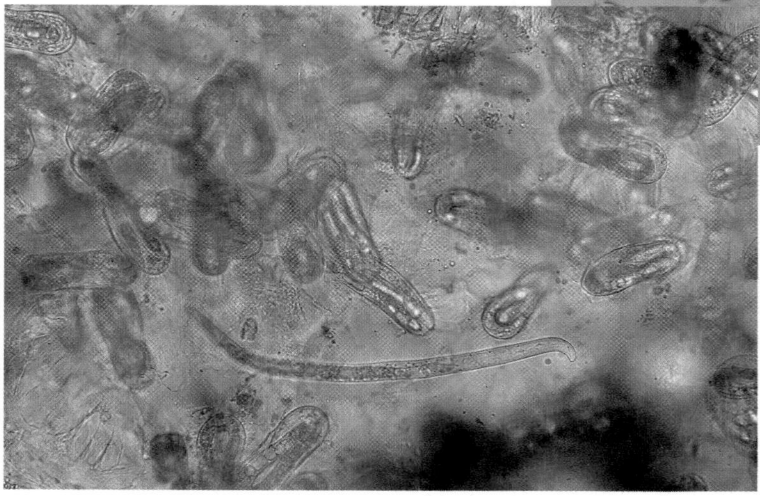

**416** Around the roots we can distinguish mature eggs containing a young larva with a clearly visible stylet. ***Meloidogyne* sp.**

417
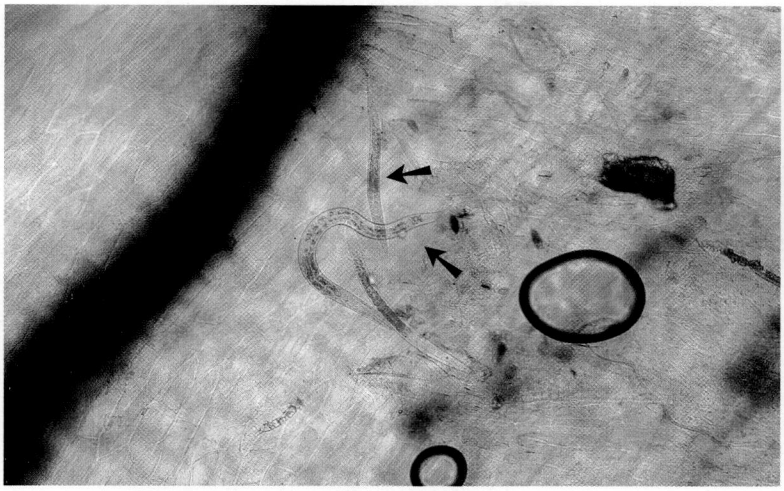

418
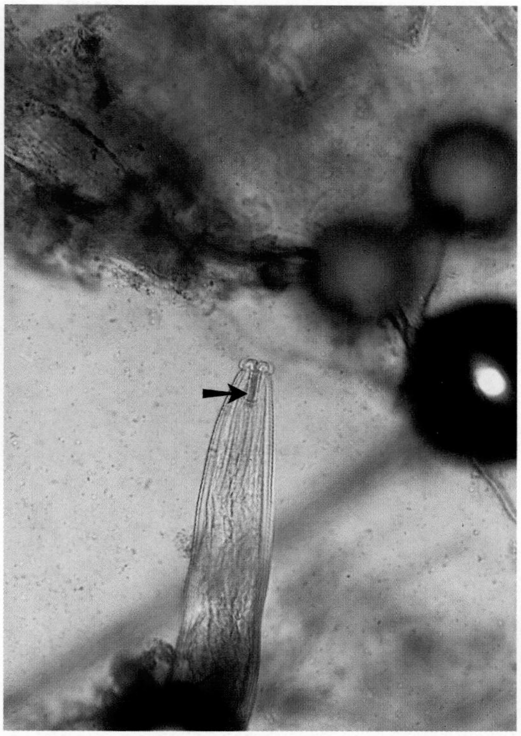

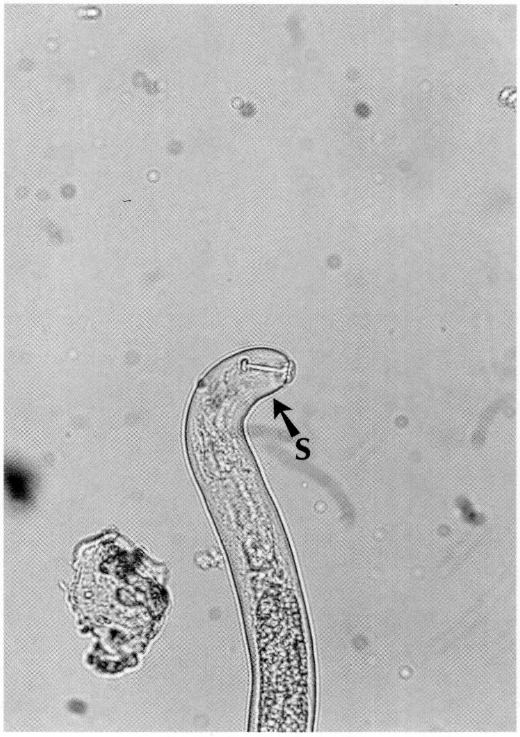

419

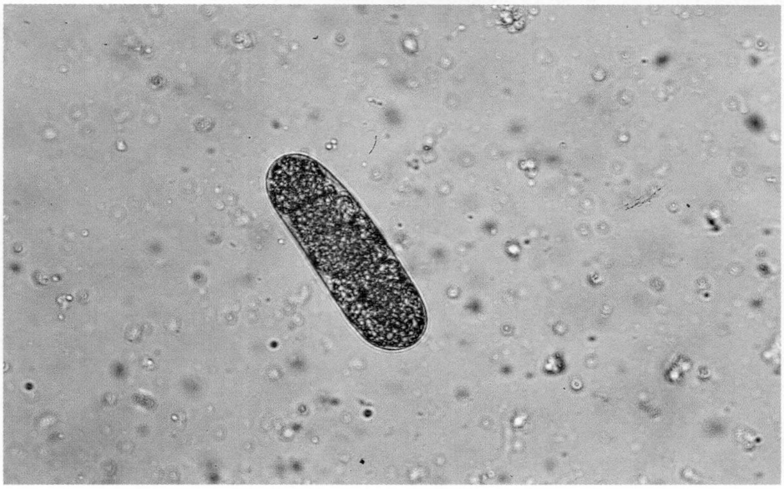
420

## Principal structures which can confirm the presence of phytophagous nematodes in lettuce roots or in their proximity

**417** The presence of nematodes in the roots or close by is confirmed by observing **minuscule worms** using a stereoscopic or ordinary microscope.

**418 Saprophytic nematodes** do not have any stylet.

**419 Phytophagous nematodes** have a hollow buccal **stylet** (S) which allows them to pierce the cells in order to absorb their content.

**420 Eggs** which guarantee the conservation of nematodes are often visible in or close to damaged tissue.

## Table 24: Some characteristics of the principal nematodes affecting l.s.v. roots (also see page 195)

| **Names of nematodes recorded on l.s.v.** | *Meloidogyne incognita, M. javanica, M. hapla, M. arenaria,* and so on.'Root-knot nematode'. | *Pratylenchus penetrans* (Cobb) Filipjev & Schuurmans Stekhoven. 'Lesion nematode'. | *Longidorus africanus* Merny. 'Needle nematode'. | *Rotylenchus robustus* (de Man) Filipjev. 'Spiral nematode'. |
|---|---|---|---|---|
| **Behaviour** | Root knot nematodes with endoparasitic females. | Migratory endoparasitic nemaode. | Ectoparasitic nematode. | Ectoparasitic or semi-endoparasitic nematode. |
| **Damage to lettuce varieties** | Swelling roots also have root knots. Branching of roots and proliferation of rootlets associated with the effects of *M. hapla.* Slowed growth of plants which are late in forming a head. | Numerous reddish brown lesions on the roots, rotting of root system. Plants are stunted; they may wilt and become yellow. | Inhibition of root elongation, proliferations and swellings located at the extremity of young roots, presence of necrosis. Some roots may be forked and shorter. Affected plants often grow poorly. | There are numerous areas of root necrosis located at the nutritional sites of this nematode. |
| **Distribution and effect on l.s.v.** | Marked. | Marked. Recorded in several production areas (Canada, USA, France, and so on). | Minor. Recorded in Africa, Israel and USA (in California). | Minor, serious on occasion. |
| **Parasitic process** | Penetration of roots and secretion of salivary juices causing hypertrophy of root cells. This results in the formation of the characteristic root knots and blisters. | Penetration and destruction of root cortex cells. | Perforates the root cells with its stylet and punctures the cytoplasmic content. Normally it does not penetrate the roots. *L. africanus* attacks only the shoot of roots, thus preventing them from growing longer. | Larvae penetrate young roots, then devour the cells. Females become hypertrophic and remain attached to the roots by their head. |
| **Other characteristics** | See fact file 36. | See fact file 37. | This migratory ectoparasite completes several cycles a year, essentially in the soil. Damage is particularly serious in damp soil. | Infests numerous hosts, particularly in sandy soils. Mainly attacks ornamental ligneous plants. Adults measure 0.6–1.9 mm in length and have a curved body. |

## Table 25: Principal nematodes recorded on lettuce, endive and chicory

| | ***Lactuca sativa*** (lettuce) | ***Cichorium endivia*** (endive) | ***Cichorium intybus* var. *sativum*** (chicory) | ***Cichorium intibus* var. *foliosum*** (witloof) |
|---|---|---|---|---|
| **Ectoparasitic nematodes** | | | | |
| *Hemicycliophora similis* | + | | | |
| *Longidorus africanus* | + | | | |
| *Longidorus maximus* | + | | | |
| *Paratrichodorus christiei* | + | | | |
| *Paratrichodorus minor* | + | + | | |
| *Paratylenchus projectus* | + | | | |
| *Rotylenchulus reniformis* | + | | | |
| *Rotylenchulus robustus* | + | | | |
| *Tetylenchus joctus* | + | | | |
| *Tylenchorynchus clarus* | + | | | |
| *Tylenchorynchus claytoni* | + | | | |
| **Migratory endoparasitic nematodes** | | | | |
| *Pratylenchus crenatus* | + | | + | + |
| *Pratylenchus penetrans* | + | + | + | + |
| *Aphelenchoides ritzemabosi* | + | | | |
| *Ditylenchus dipsaci* | | | + | + |
| **Sedentary endoparasitic nematodes** | | | | |
| *Heterodera schachti* | | | + | + |
| *Meloidogyne arenaria* | + | + | + | + |
| *Meloidogyne hapla* | + | + | + | + |
| *Meloidogyne incognita* | + | | | |
| *Meloidogyne javanica* | + | + | + | + |
| *Nacobuccus aberrans* | + | | | |
| *Nacobuccus batatiformis* | + | | | |
| *Nacobuccus serendipiticus* | + | | | |

*Belonolaimus gracilis* and *Trichodorus primitivus* have also been recorded on lettuce. It is advisable to remember that ectoparasitic nematodes from the *Xiphinema*, *Paratrichodorus* and *Trichodorus* genera can be virus vectors: TRSV in the case of the first genus and TRS for the other two.

## Figure 20: Location and behaviour of principal nematodes on plant roots

1 Ectoparasitic nematodes
2 Migratory endoparasitic nematodes
3 Nematodes with cysts
4 Nematodes with root knots, sedentary endoparasites

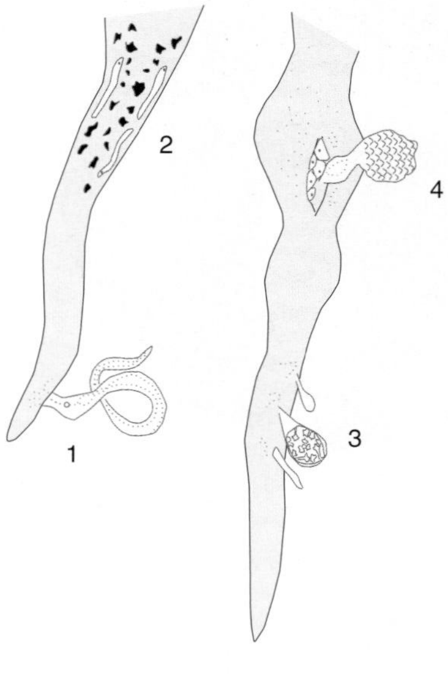

**421** Numerous yellowish-white aphids are present all along the roots; their presence gives the latter a cottony-white appearance.
***Pemphigus bursarius***

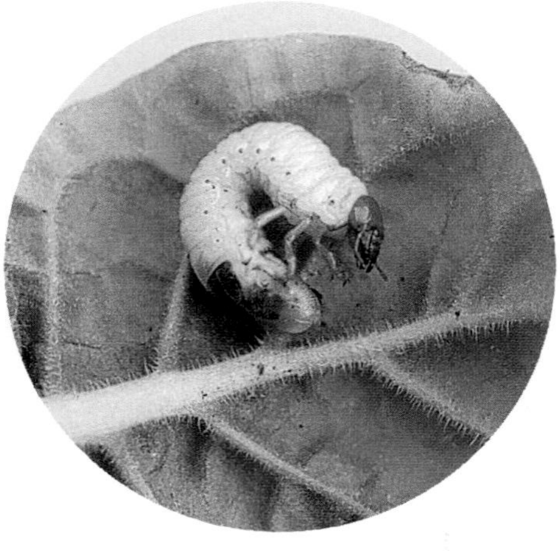

**422 White worm** larva.

**423 Yellow worm** larva.

**424** Adult **tipula**.

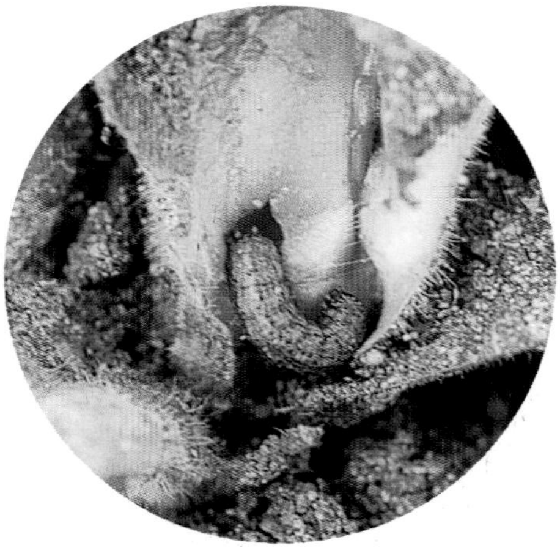

**425 Grey worm** caterpillar.

- *Pemphigus bursarius* (root aphid)

This aphid, which has a distinctive biology, colonizes the roots of lettuce, endive, and other members of the Asteraceae family such as the sowthistle. The wingless insects prosper under these conditions. Affected l.s.v. are often rather weak and chlorotic. Marked attacks may lead to complete wilting of plants. We can also see, all along the roots, numerous yellowish-white insects bearing tufts of white wax which can be seen fairly easily (**421**).

Attacks of this aphid are seen above all in the open field close to poplar trees. In fact, this species winters on this tree in the form of eggs, then in the form of apterous females in the spring. These insects feed on the petioles and cause the formation of root knots. Attacks occur less frequently under protection.

- **Other soil-based pests**

Several pests are likely to gnaw and devour the roots and crown of l.s.v. Adult insects, caterpillars, and larvae attack both seedlings and adult plants.

In many cases the guilty party or parties will not be far away. If you look carefully at affected plants and their surroundings, you will find the pest or pests responsible. The principal pests which may be involved in this type of damage are listed in *Table 26*.

**Table 26 : Principal pests responsible for root damage on l.s.v.**

| Organs attacked and type of damage | Name | Description of the pest |
|---|---|---|
| Roots devoured. | ***Melonontha melolontha*** (white worm, common maybug). | Larvae are 3–3.5 cm in length, white in colour. Head and feet are shiny brown (**422**). |
| Roots devoured and perforation of the crown. | ***Agriotes* spp.** (yellow worm, wire worm, click beetles). | Cylindrical larvae, 2.2–3 cm in length, shiny yellowish brown incolour, cylindrical to sub-cylindrical (**423**). |
| Crown gnawed. | ***Hepialus* spp.** (hepiali). | Caterpillars are 3.5 cm in length, white, shiny and transparent. Light brown head and prothoracic plate. |
| Crown and roots gnawed. | ***Tipula* spp.** (tipula). | Larvae are 3.5–4.5 cm in length, dull greyish colour. Small black head, difficult to see (**424**). |
| Crown gnawed and leaves close to the soil nibbled away. | ***Bourletiella hortensis*** (garden springtail). | Adults are black to dark green in colour, 1.5 mm in length. Large head with long antennae. Prominent black eyes, encircled with yellow |
| Crown gnawed and leaves torn. | ***Agrotis* spp.** (grey worm, terricolous moths). | Caterpillars are 3.5 cm in length, fleshy, vary in colour from greyish to greenish, surmounted by darker coloured spots or stripes (**425**). |

426

427

428

429

## Examples of damage occurring on and in the taproot and stem of l.s.v.

**426** The internal tissues of the taproot and stem of a healthy plant do not normally reveal any particular damage.

**427** However, the taproot of this lettuce is narrower in its distal part and is also a darker brown.
***Pythium tracheiphilum***

**428** In some cases the taproot can swell and have hardly any roots.
***Rhizomonas suberifaciens***

**429** The vessels of this lettuce, affected by ***P. tracheiphilum***, have become brown.

# Internal and/or external damage and abnormalities of the taproot and stem

## Possible causes

- *Fusarium oxysporum* f. sp. *lactucum* (fact file 13)
- *Phytophthora porri* (fact file 8)
- *Pythium tracheiphilum* (fact file 11)
- *Sclerotinia minor* (underground attack) (fact file 6)
- *Verticillium dahliae* (fact file 12)
- *Erwinia carotovora* subsp. *carotovora* (fact file 18)
- *Rhizomonas suberifaciens* (parasitic corky root) (fact file 19)
- Hollow heart
- Non-parasitic corky root (see page 195)
- Vitrescence of the taproot

## Arguments in support of the diagnosis

Several parasitic and non-parasitic diseases are likely to cause various symptoms on and in the taproot and stem of l.s.v. These symptoms are accompanied by other signs, sometimes more characteristic, which occur on different organs. Amongst these diseases, a number are caused by a few vascular micro-organisms, i.e. organisms located and moving in the vessels. The location and multiplication of these micro-organisms in the vessels, the reactions of the plants (numerous 'gummy' substances are secreted to form obstructions), also cause other equally characteristic symptoms, resulting in wilting, yellowing, and drying out of leaves, often in sectors.

Generally speaking, if vessels are colonized, they become yellow and brown, to varying degrees. Adjacent tissues, such as the cortex or pith, are sometimes affected.

In order to identify these vascular diseases, plants must be observed very carefully in order to identify the symptoms listed on the following pages. You will see that these diseases often cause fairly comparable symptoms on l.s.v., so you will be well advised to have your diagnosis confirmed by a specialist laboratory which will carry out vascular isolation on an artificial medium.

430 Some plants are smaller in size.
**Corky root - R. *subferifaciens***

431 When the plant is pulled out, its root system has numerous rather corky lesions.
**Corky root - R. *suberifaciens***

432 In addition, numerous roots have gone yellow and brown; in places corky striations and bands can be observed.
**Corky root - R. *suberifaciens***

433 The taproot may be hypertrophied, extremely corky, sometimes with broad corky striations.
**Corky root - R. *suberifaciens***

## Parasitic corky root (*Rhizomonas suberifaciens*)

• **Non-parasitic corky root**
***Rhizomonas suberifaciens***

Extensive swelling of taproots and lateral roots can sometimes be seen on some l.s.v. plants. These symptoms can be caused by parasitic and non-parasitic diseases, the effects of chemicals or excessively nitrogenous fertilizer, and the influence of a phytopathogenic bacterium (*R. suberifaciens*) (also see page 197).

This bacterium causes root lesions which change from yellow to brown. We also see surface corkiness and longitudinal splitting on the roots and taproot (**431** and **432**). The taproot may be particularly badly affected; in this case it becomes excessively hypertrophied. Affected plants grow slowly and have small heads (**430**). The extremely corky tissues of the taproot are brittle. In particularly critical situations only a small portion of the taproot and a few adventitious roots survive (**433**).

• ***Sclerotinia minor*** **('Drop')**

As we pointed out in the section entitled 'Damage to leaves, in contact with the soil, and the crown', this fungus mainly causes damp rotting. This affects the parts of the plants in contact with the soil, particularly the senescent lower leaves, fairly quickly taking over the crown and all the leaf tissues, which decompose and collapse.

Attacks deeper into the taproot have also been recorded. In fact, under certain conditions *S. minor* can directly colonize the tissues of the taproot located 3–10 cm below the surface of the soil (**fig. 21**). Invaded tissues go brown and progressively necrotic. If the taproot is entirely girdled it may break. Affected plants grow more slowly and look slightly wilted.

**Figure 21: Fairly atypical location of damage from *Sclerotinia minor* on the taproot of a lettuce.**

*Verticillium dahliae*

*Fusarium oxysporum* f. sp. *lactucum*

434

435

*Pythium tracheiphilum*

Various types of virus

436

*Erwinia carotovora* subsp. *carotovora*

Hypothetical diagnoses made in the presence of these symptoms

## • *Pythium tracheiphilum*

This aquatic fungus is capable of colonizing the vascular system of the taproot of some Compositae, particularly artichoke, groundsel, and lettuce. On lettuce plants we see that the taproot is sometimes deformed and that the secondary roots are fewer in number (**440**). Vessels show browning, to a greater or lesser degree, on the roots, taproot, and stem (**437–438**). In some cases, adjacent tissue is also affected. Growth of diseased plants is slowed down, or even halted. They are very much smaller than the apparently healthy surrounding plants (**439**). In addition to being stunted, or even becoming dwarfed, some plants have a fairly significant proportion of leaves becoming yellow and wilting. Initially this wilting occurs at the hottest times of the day and plants may recover during the night. Subsequently, this damage may sometimes become irreversible, leading to the drying out and death of some plants scattered around the plot.

Necrotic lesions can also be observed on leaves in the USA; this occurs in heavily infected plots in the presence of sustained humidity. *P. tracheiphilum* establishes itself on areas of damage on young leaves with succulent, tender tissues. Soil particles, dispersed after rainfall or sprinkler irrigation, harbour the sporangia and oospores of the fungus which cause the contamination. These leaf symptoms have never been observed in France.

Recorded in a few countries, such as the Netherlands and the USA (in Wisconsin), *P. tracheiphilum* is now fairly widespread in various French production areas. Although not considered to be a major lettuce pathogen, this fungus is becoming progressively worrying in some plantings where it causes chronic and significant damage.

We should point out that rotting of the lettuce stem has been observed in Australia, without any damage to the roots. Affected plants wilt and collapse within 2–4 weeks. Within the damaged tissues we can see the oogonia and sporangia of a *Phytophthora*. ***Phytophthora porri*** has been identified as being responsible; this mainly attacks the lettuce stem and is believed to be incapable of living as a parasite on leeks and cabbage.

## • *Erwinia carotovora* subsp. *carotovora* ('Soft rot')

Fairly abrupt collapse of l.s.v. is sometimes seen after rainy periods or in plots which have been over irrigated, close to harvest-time. This is caused by the colonization of the internal tissues of the lettuce by extremely polyphagous bacteria belonging to the species *Erwinia* and, more particularly to the species *E. carotovora* subsp. *carotovora*.

Invaded by this bacterium, the vessels of the stem and taproot first of all become pink, then brown (**441**). The pith is soon affected. It becomes glassy and takes on a greenish hue (**442–444**). A nauseating odour accompanies these symptoms. The pectinolytic enzymes produced by *E. carotovora* subsp. *carotovora* progressively liquefy the tissues. Damp, dark brown to black rotting first of all destroys a few leaves then the whole head (**443** and **445**). Ultimately, affected lettuce literally liquefies and collapses.

## • Vitrescence of the taproot

A glassy symptom has been recorded in published works. This takes the form of browning and liquefaction of the taproot pith. This symptom is rather reminiscent of the damage caused by *E. carotovora* subsp. *carotovora*.

On lettuce this physiological disorder appears mainly in summer, close to harvest-time, and in plots where nitrogenous fertilizing is excessive.

437

438

439

440

***Pythium tracheiphilum* ('Pythium wilt')**

437 The longitudinal section of this lettuce clearly reveals vascular browning associated with the presence of ***P. tracheiphilum***.

438 The vessels and adjacent tissues are darker, bordering on black, in colour.
***P. tracheiphilum***

439 Generally speaking, affected plants grow very slowly, ultimately becoming stunted in comparison with the surrounding healthy plants.
***P. tracheiphilum***

440 As well as displaying vascular browning, diseased plants have a fairly small root system.
***P. tracheiphilum***

441

442

443

444

445

## *Erwinia carotovora* subsp. *carotovora* ('Soft rot')

**441** After cutting a cross section of the stem, we can see that the vessels are pinkish brown in colour. Subsequently, the bacterium colonizes the pith which becomes glassy and takes on a greenish hue.
***E. carotovora* subsp. *carotovora***

**442** The pith and vascular system progressively liquefy; completely rotted tissues have a fairly characteristic black colour.
***E. carotovora* subsp. *carotovora***

**443** The bacterium sometimes attacks the main vein of lower leaves. Localized or extensive damage can clearly be seen. This is damp, and brown to black in colour.
***E. carotovora* subsp. *carotovora***

**444** Ultimately the whole stem is affected; now rotting is spreading to the leaves in the head.
***E. carotovora* subsp. *carotovora***

**445** In certain situations the vein, as well as the surrounding lamina, can decompose entirely.
***E. carotovora* subsp. *carotovora***

**446a** These plants, which have grown poorly, have several leaves which are rather wilted and/or dried out.
***Fusarium oxysporum* f. sp. *lactucum***

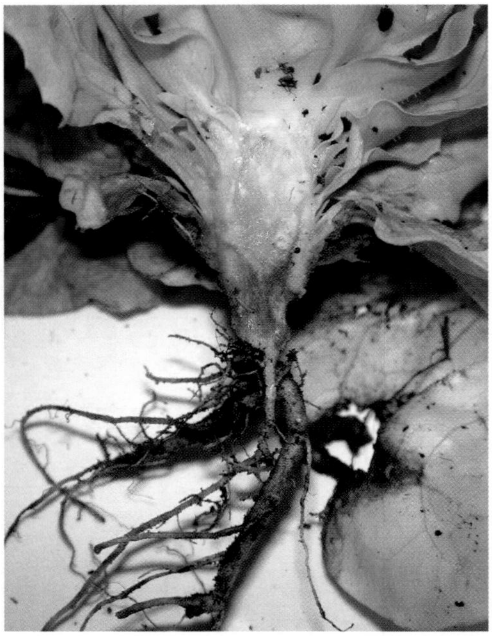

**446b** The vessels are rusty brown in colour.
***Fusarium oxysporum* f. sp. *lactucum***

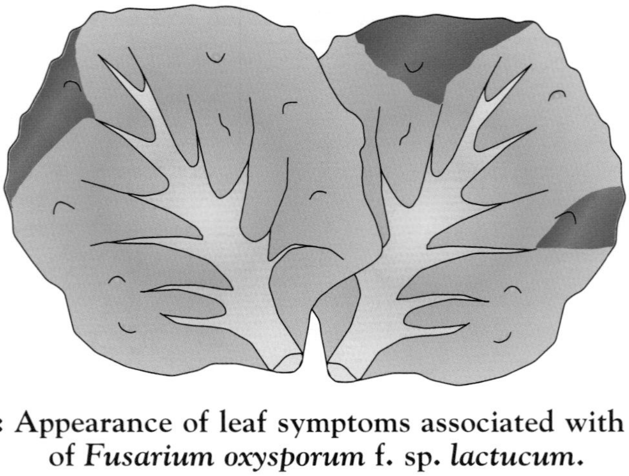

**Figure 22: Appearance of leaf symptoms associated with parasitism of *Fusarium oxysporum* f. sp. *lactucum*.**

• *Fusarium oxysporum* f. sp. *lactucum* ('Fusarium wilt')

This vascular fungus, essentially reported in several Californian and Arizonan counties, Japan, and a few European countries, causes vessels to go brown at the root, taproot, and stem level. Some leaves located on one side of the lettuces become yellow and/or display necrosis around the periphery of the lamina. They also display vein browning (**446a** and **446b**). Sometimes plants are dwarfed or fail to form a head. Symptoms reported in Germany are fairly comparable. Wilting and drying out of one side of the lamina, in a V-shape (**fig. 22**), are characteristic of this vascular disease. Some parts of the vessels are rusty brown in colour.

Signs of the disease have also been reported on seedlings in the rosette stage. In this case, reddish brown lesions can also be seen in the vascular system of the roots and taproot, while the seedlings wilt and die.

• *Verticillium dahliae* ('Verticillium wilt')

Although present in numerous countries and affecting numerous plant species, both cultivated and uncultivated, this vascular fungus has never been reported on lettuce except in California. Symptoms observed are comparable to those caused by Fusarium. We see discolouration of the vascular system which takes on a fairly dark greenish black hue. Some lower leaves wilt, become partially yellow at first, and then dry out. Subsequently they adhere very strongly to the head.

In France, *V. dahliae* affects several market garden crops; to our knowledge it has never been reported on lettuce.

• **Hollow heart**

This non-parasitic disease of endive, which for a long time was attributed to inappropriate nitrogenous fertilization, probably starts up very early in nurseries exposed to high temperatures, particularly in the autumn. Recent observation has allowed us to propose the hypothesis that floral initiation probably begins under these conditions, but plants cannot go to seed, because the days are too short. Apical dominance is probably inhibited. Axillary buds probably develop with the symptom of hollow heart appearing.

Other factors responsible for the destruction of the terminal bud, such as certain particularly 'aggressive' pesticides, or burning after water has been retained on the shoot, may lead to the development of hollow heart.

In certain cases, secondary micro-organisms can multiply and lead to surface browning of the tissues (**447**) (also see the chapter entitled 'Abnormalities in the growth of l.s.v. and/or in the shape of their leaves' and page 33).

447

**'Hollow heart'**

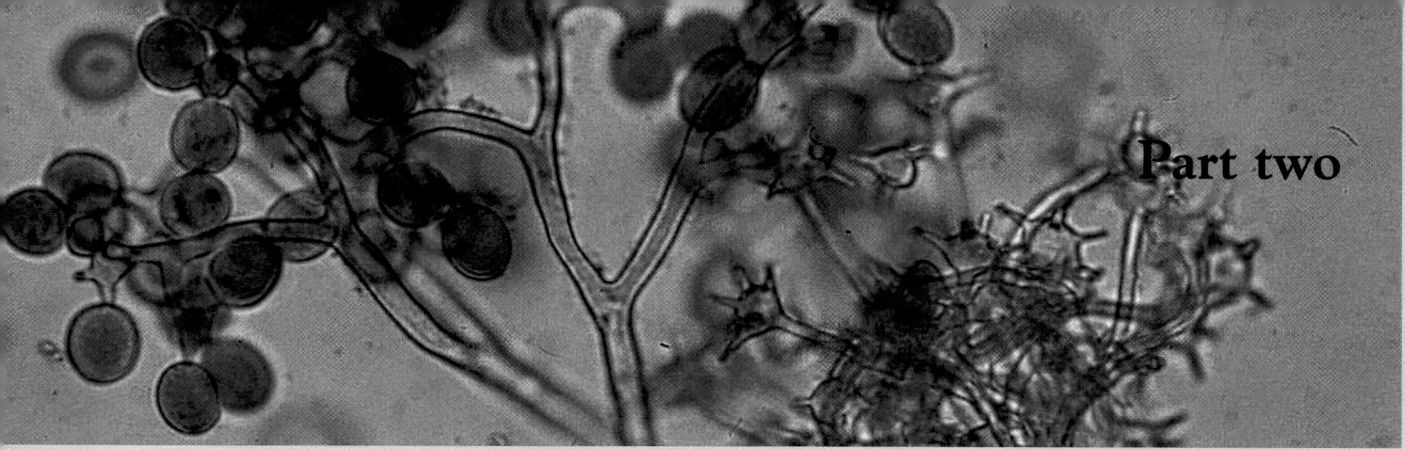

# Part two

# Principal characteristics of pathogenic agents and methods of protection

## Advice to the reader

Numerous fungicides, recorded in published works as being effective against fungi which are pathogenic to l.s.v., are referred to in several fact files in this part of the book. We would like to draw your attention to the fact that a number of these fungicides may no longer be authorized or available by the time the book is published.

This situation is due to several European directives which, over recent years, have been making progressive and drastic changes to the pesticide market, with favourable effects on their use in Europe. In particular, directive No. 91/414/CE of the European Union Council has, in the initial instance, harmonized the conditions for obtaining marketing authorization for plant health products in member states. It also initiated the re-examination programme, to last a period of several years, for old pesticides and miscellaneous substances (already authorized in at least one of the member states before 25 July 1993). This is why numerous pesticides, considered not to meet the requirements of farming policies that respect both the environment and consumers, have been progressively removed from the market in Europe.

Under these conditions, it was not possible for us to supply a list of pesticides suitable for use on lettuce varieties in step with legislation under continual development. Consequently it is quite clear that the information provided in the fact files, in particular that relating to pesticides, will have to be updated to reflect the consequences of applying this directive and the legislation in force in the individual reader's country.

# Fungi

## General information

Phytopathogenic **fungi** are organisms which are generally more complex than bacteria, small in size, sometimes visible to the naked eye but more usually under the microscope. They are characterized by the formation of filaments (hyphae), either free or intertwined, the whole of which is given the name of **mycelium**. This may be **coenocytic** (non-compartmentalized) (*Pythium* spp., *Bremia lactucae**, *Phytophthora* spp.*, and so on) or compartmentalized (*Botrytis cinerea*, *Sclerotinia* spp., *Rhizoctonia solani*, and so on). We should point out that some aquatic fungi do not have any mycelium, for example *Olpidium brassicae*, a virus vector. Very often they form amiboid or plasmodic structures, sporangia, as well as uni- or biflagellate **zoospores**.

On l.s.v. they also produce special structures, very resistant to unfavourable elements and perfectly adapted for their survival. The most common are the **chlamydospores**, spores with a thick wall, formed by, for example, *Thielaviopsis basicola*, *Fusarium oxysporum* f. sp. *lactucum*, and *Pythium* spp. *Mycocentrospora acerina* produces **toruloid mycelium** whose conservation properties are comparable to those of chlamydospores. *Pythium* spp., like *Bremia lactucae*, can also survive in the form of **eggs** (oospores) deriving from sexual reproduction. More visible, with variable dimensions, and equally effective in ensuring the lasting presence of a fungus, the **sclerotia**, dense mycelium masses, characterize the presence of certain cryptogams on and/or in affected plant tissues fairly clearly: *Sclerotinia sclerotiorum*, *Sclerotinia minor*, *Sclerotium rolfsii*, *Botrytis cinerea*, *Rhizoctonia solani*, and so on.

During the **contamination** phase, fungi come into contact with l.s.v. plants. Fairly quickly the propagules, spores, germinate and differentiate an attachment structure known as **appressorium**, which also enables them to penetrate their host. Certain fungi prefer young tissues, others senescent organs. They also have a certain specificity towards the organs attacked. It is easy to distinguish fungi which basically predominate on the roots and/or crown of lettuce (*Olpidium brassicae*, *Pythium* spp., *Pyrenochaeta lycopersici*, *Thielaviopsis basicola*) or the vessels (*Fusarium oxysporum* f. sp. *lactucum* and *Verticillium dahliae*), whereas others are limited solely or mainly to aerial organs (leaves, stems, fruit, flowers) (*Erysiphe cichoracearum*, *Bremia lactucae*, *Septoria lactucae*, *Cercospora longissima*, *Puccinia opizii*).

Penetration of the plant can occur in different ways:

- directly across the cuticle and plant cell wall (*Erysiphe cichoracearum*, *Microdochium panattonianum*, and so on);
- through the natural openings present on plants, such as the stomata (*Bremia lactucae* and partly, *Cercospora* longissima, and so on);
- through damaged areas or senescent tissues present on plant organs for example, old leaves (*Botrytis cinerea*, *Sclerotinia sclerotiorum*, *Sclerotinia minor*, and so on), breaks in epidermal hairs, sites from which roots emerge, damage from nematodes (*Fusarium oxysporum* f. sp. *lactucum*, *Thielaviopsis basicola*) and so on.

Once they have penetrated their host, they put in place various parasitic processes which allow them to establish themselves and invade the host; this corresponds to the **infection** phase. Up to this stage, the fungi have lived off their reserves; now they will start to take various substances from their host. These withdrawals can be made in a way which causes minimum trauma to the plant cells or can very quickly result in their destruction. In the first case, the fungi are completely dependent on living cells, and this is why they spare them; these are described as obligate parasitic fungi (*Erysiphe cichoracearum*, *Bremia lactucae*, and so on). For example, in l.s.v. mildew, host invasion is very limited. Its mycelium remains outside the cells, but has special structures, **haustoria** (a type of sucker), which allow it to take its nutrients from the cells without destroying them, initially at least. In the second case the fungi have considerable saprophytic potential, and host survival is not very important. These **facultative parasites** rapidly destroy the plant cells, mainly by means of

* For many years the oomycetes, which contain almost 550 species, have been classified as Fungi. Their ultrastructure, biochemistry and molecular sequences indicate that they belong to the Chromista, which includes algae (green and brown), diatoms, and so on.

pectinolytic enzymes (*Botrytis cinerea*, *Sclerotinia sclerotiorum*, *Sclerotium rolfsii*, and so on) and sometimes very necrotic toxins. In both situations, after having colonized their host's tissues with varying degrees of rapidity (**invasion phase**) and if climatic conditions are favourable, the fungi form specialist structures which allow them to produce numerous spores (**sporulation**). These structures have different names depending on their origin: asexual plant multiplication (anamorph form) (sporocystospores, conidiophores, pycnidia, acervuli, and so on) or sexual reproduction (telomorph form) (oospores, zygospores, cleistothecia, perithecia, apothecia, and so on). They are at the basis of the botanical classification of fungi and allow them to be identified. They are frequently visible inside or on the surface of damaged tissues and are very varied in appearance. Some of them help the fungi to survive, but their essential role is to ensure their **dissemination** by producing spores (zoospores, sporangia, conidia, **basidiospores**, ascospores, and so on).

The **incubation period** corresponds to the period of time between the parasitic fungus penetrating the plant and the appearance of the first symptoms of disease. The **latency period** also begins at the time of initial contamination but ends when sporulation occurs.

Spores are then disseminated in various ways:

- by the **wind or air currents** which carry them directly, or in the dust from contaminated soil. Liberation usually occurs passively, depending on climatic conditions and, more importantly, surrounding humidity (*Bremia lactucae*, *Erysiphe cichoracearum*, *Botrytis cinerea*, and so on);
- by **water** which transports the fungal propagules either passively as it flows over and into the soil (*Pythium* spp., *Phytophthora* spp., and so on), or more actively as a result of leaching, splashing, or splattering occurring during rainfall or sprinkler irrigation (*Microdochium panattonianum*, and so on). Some poorly evolved fungi are more particularly adapted to aquatic life. They have spores, known as zoospores, which are equipped with flagella which allow them to move easily in a damp environment (*Olpidium brassicae*, numerous species of *Pythium* and *Phytophthora*).

These two methods of transport frequently act together. For example, frequently fine droplets of water fix several conidia of a phytopathogenic fungus present on the leaves of a diseased plant and are picked up and carried away by the wind before being deposited a few metres away on one or several healthy plants.

Secondly, growers and their tools in the course of tending growing plants, specialists in the course of their successive inspections, can easily spread the various known propagules of fungi. They may also be disseminated by various vectors for example, certain aerial insects or seeds (*Septoria lactucae*). The latter method of dissemination is fairly unusual in pathogenic fungi of l.s.v.

Fungi development is influenced by climatic conditions. Temperature is not the most important factor. However, just like plants, they require minimum temperatures in order to develop. In the winter, as temperatures are often very low, these micro-organisms reduce or cease their activities. These resume as soon as temperatures rise and take place across a very broad band of temperatures, from a few degrees to over 30 °C. Temperature can intervene indirectly by making plants fragile (freezing of certain organs, and so on) or by exacerbating certain symptoms (more intense wilting in the case of root or vascular diseases).

The predominating factor is certainly **humidity** which takes various forms: rain, irrigation, dew, fog, and so on. It is essential at practically all stages of development of aerial fungi, particularly for germination of spores and penetration of the germination tube in the host. Germination of spores frequently requires relative humidity higher than 95%; in certain cases the presence of free water on the leaves is essential (*Bremia lactucae*, *Cercospora longissima*, and so on). The formation of spores, their longevity as well as their dissemination, are also significantly affected by humidity.

The great majority of fungi appreciate damp environments, whether within the plant cover or around the roots. It is now clearly accepted that soil soaked in water is also very conducive to attacks of soil-based cryptogams, such as numerous species of *Pythium* and *Phytophthora*, *Thielaviopsis basicola*, and so on. In addition, the roots of plants placed in these asphyxiating conditions are much more susceptible.

A few rare fungi prefer a drier environments, at least at certain periods of their cycle.

We have already seen how the **wind** helps to liberate and disseminate spores. It can sometimes have a negative effect on the development of a mycosis by reducing environmental humidity. Effects of **light** are fairly insignificant. If this is reduced, plants become etiolated, which makes them more susceptible to certain fungi such as *Botrytis cinerea* in indoor nurseries.

Soil **pH** and **plant nutrition** also play varying roles in the development of certain fungi. It is not always easy to show their effects clearly.

448

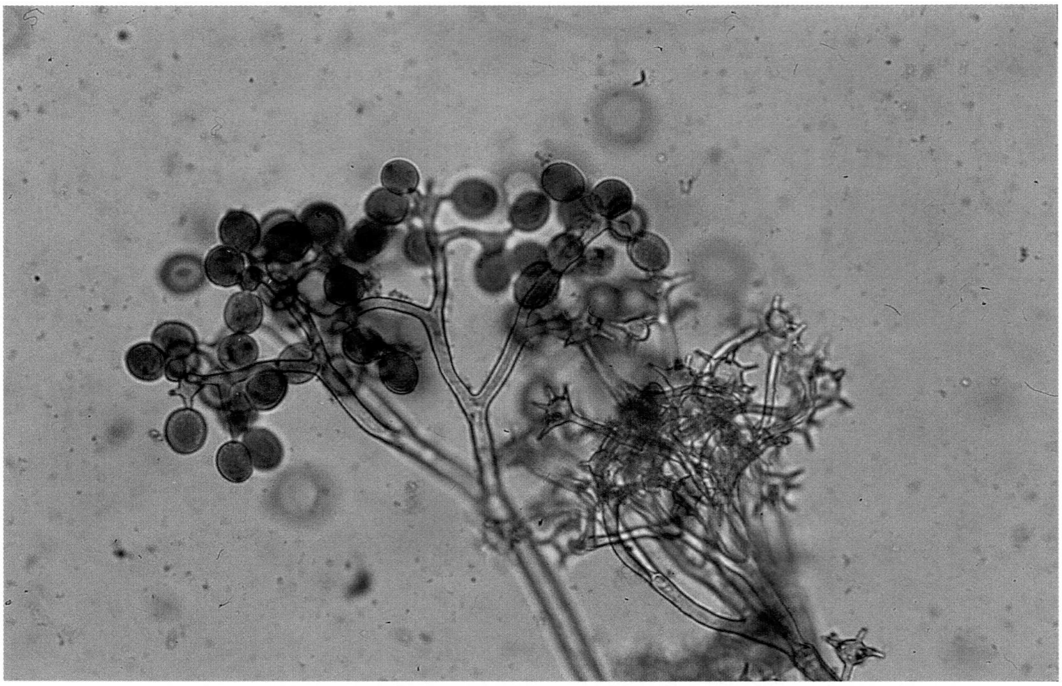

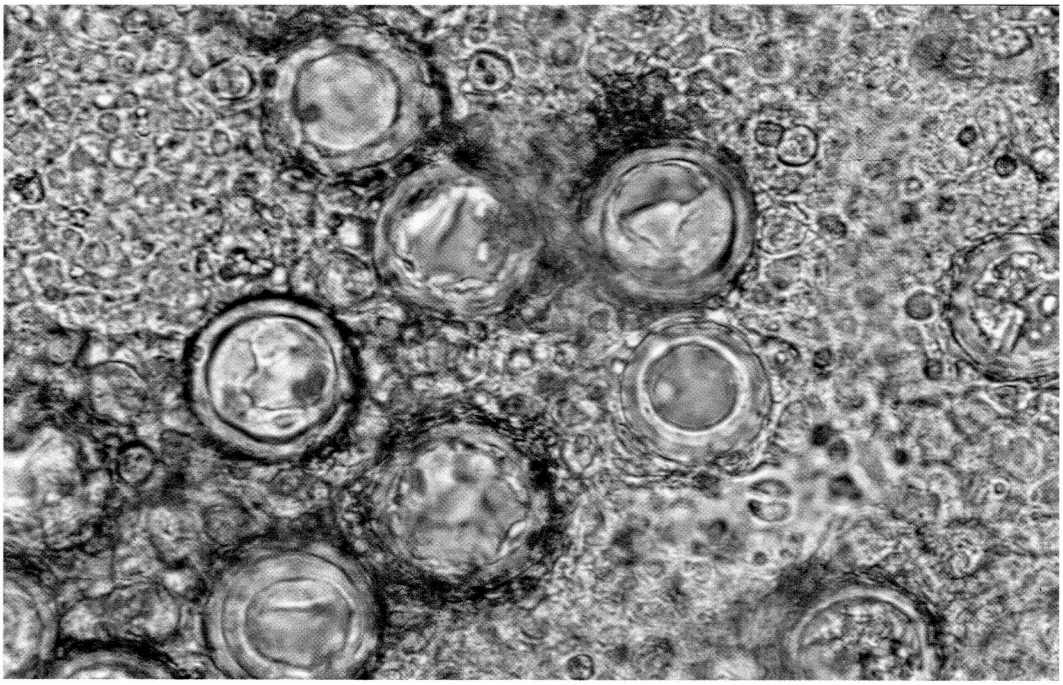

449

**448** ***Bremia lactucae*** produces bushy dichotomous sporangiophores that carry, at the extremity of their sterigmata, fairly circular sporangia, the diameter of which varies between 12–30 µm.

**449** In damaged tissues, we can sometimes see brown oospores measuring between 27–30 µm in diameter. They ensure the sexual reproduction of this heterothallic fungus.

# Fungi which mainly attack leaves

## *Bremia lactucae* Regel — Downy mildew

(Eukaryotes, Pseudofungi, Chromists*, Oomycota, Oomycetes, Peronosporas, Peronosporaceae)

## Principal characteristics

*Bremia lactucae* is a parasite on approximately 230 plants from the Asteraceae family (e.g. Compositae). In fact, there are probably numerous special forms adapted to a fairly extensive range of hosts.

### • Frequency and extent of damage

Mildew is one of the oldest, most frequent and most feared diseases affecting cultivated lettuce, both in the open field and under protection. It mainly attacks in production zones where there are prolonged periods of humidity (rainfall, sprinkler irrigation, fog, dew, and so on) and in cool weather, particularly in the USA and Europe. It is responsible for considerable losses and, if climatic conditions are particularly favourable, may destroy entire crops in a few days. *B. lactucae*, which mainly affects lettuce, has been recorded in most lettuce-producing countries; it often constitutes a constant threat and requires preventative fungicide treatment.

Wherever the production areas, controlling mildew is not easy, for two main reasons:

- firstly, numerous physiological strains, capable of circumventing the majority of resistant genes introduced into lettuce varieties, have been recorded in several countries, rendering any locally implemented genetic control useless for the time being;
- secondly, strains resistant to certain fungicides from the phenylamide family (metalaxyl, oxadyxil, and so on) have been described, calling into question their use.

In countries with temperate climates, *B. lactucae* is very frequent and feared by all lettuce growers, especially during winter production periods. These growers implement a set of methods for circumventing this fungus at a very early stage, right from the nursery, in order to control it in the best possible way.

### • Principal symptoms

*B. lactucae* can attack lettuce throughout its growing cycle.

Young plants are particularly susceptible to mildew. *B. lactucae* develops very quickly on cotyledons which it covers with its numerous white sporangiophores. It invades the leaf tissues which become chlorotic; the lamina of some young leaves is rolled around the edge. It also causes stunting of seedlings and, ultimately, their death.

On older plants, it develops initially on the leaves in the crown. It produces broad pale green to yellow spots, delimited by the veins and therefore fairly angular. Eventually the spots become necrotic and light brown in colour. *B. lactucae* sporulates fairly abundantly, particularly on the lower face of leaves before or after chlorotic spots become visible on the lamina. Bushy sporangiophores, carrying the sporangia, emerge from the stomata and constitute white matting which varies in density; this lies at the origins of the previously used common name of 'mildew' (**448**). Subsequently spots develop on the more internal leaves and on those of the heart. Badly affected leaves, on which spots have converged, become entirely necrotic and die.

Systemic infections sometimes occur. Browning of internal stem tissues and the base of leaves is then visible.

* For many years the oomycetes, which contain almost 550 species, have been classified as Fungi. Their ultrastructure, biochemistry and molecular sequences indicate that they belong to the Chromista, which includes algae (green and brown), diatoms, and so on.

We should point out that damaged tissues constitute nutrient bases for secondary bacterial (*Pseudomonas* spp., *Erwinia carotovora* subsp. *carotovora*, and so on) or fungal (*Botrytis cinerea*, and so on) invaders which, if conditions are damp, cause damp soft rotting in the field and sometimes even during storage.

(See **6**, **147**, **195**, **199–215**, **274**, **281**, and **282**)

## • Biology, epidemiology

**Survival, sources of inoculum:** *B. lactucae* seems to be able to survive in different forms depending on the production zones:

- via its oospores (from its sexual reproduction; this fungus is heterothallic) (**449**) and its mycelium, both of which form in the necrotic tissues and are subsequently found in the soil, alongside plant debris;
- on surrounding lettuce, which may be wild (*Lactuca serriola*, and so on) or cultivated, on which the fungus sporulates to varying degrees;
- the seeds sometimes harbour this fungus; however, they have never been shown to lie at the origins of primary contamination.

**Penetration and invasion:** the oospores seem to be able to germinate and contaminate young seedlings directly in the nursery. The sporangia, which have a lifetime of 6 days, are responsible for the large majority of contamination which takes place within a few hours. They germinate in the presence of free water and emit a germination tube on the surface of the leaves which directly penetrates the cuticle and epidermal cells. Contamination via the stomata is possible. It is probably fairly rare for zoospores to form. Infection can take place within 3 hours. Subsequently, the mycelium progressively invades the cells of the mesophyll. Its spread may be inter- and intracellular. If climatic conditions are favourable, yellow spots appear between 4 and 7 days after the initial contamination. *B. lactucae* may be systemic and completely overrun plants.

**Sporulation and dissemination:** the appearance of sporangiophores and therefore sporangia requires the presence of high humidity. It takes place during the night. Sporocystospores emerge through the stomata. Sporulation can take place between 5–24°C.

Once formed at the extremity of sterigmata, numerous sporangia are liberated thanks to the joint action of a rise in temperature and a reduction in humidity. The spores are mechanically ejected, especially during the morning. They are immediately carried away by the wind and air currents which transport them to neighbouring plants or to those further away in other plots; they are responsible for secondary contamination. Splashes of water can also help to disseminate the fungus.

The disease is sometimes disseminated via contaminated plants. In some countries, growers have specialized in producing large quantities of plants which they market in different production zones. Plant contamination sometimes passes unnoticed and they are then sent out in a diseased state to distant growers, thus contributing to early development and proliferation of the disease in the crop.

**Conditions favourable to its development:** this obligate parasitic fungus is greatly influenced by climatic conditions. It appreciates prolonged periods of cool, damp, cloudy weather (with a relative humidity close to 100%). Long periods in the morning, when leaves become damp, particularly favour infection. Sprinkler irrigation encourages mildew more than other watering methods.

The range of temperature conducive to germination of its sporangia is between 10–15°C. Infection can take place within 2–3 hours for a range of temperature of between 2–20°C. It sporulation is intense if nocturnal temperatures are in the order of 5–10°C with daytime temperatures varying between 12–20°C. On the other hand, as soon as the weather becomes more clement, with the temperature rising above 20°C and a fall in humidity, sporulation is greatly reduced. Beyond 25°C, the mildew has an increasingly reduced activity up to 30°C.

Lettuce which has undergone stress while growing, such as temperatures which are too low, poor availability of light, lack of transient water, seem more susceptible to this cryptogam. If climatic conditions are particularly favourable, *B. lactucae* can complete a full cycle in less than 5 days.

## Protection*

### • During cultivation

As soon as the very first spots are noticed, **fungicide** treatments must be carried out immediately. You should be aware that curative treatments, whose efficacy is relative, more easily generate the appearance of fungicide-resistant strains. Preventative treatments are preferable. Several active materials are used throughout the world to control *B. lactucae*.

**Contact fungicides**, the Dithiocarbamates (maneb, mancozeb, zineb**, proponeb, metiram-zinc, and so on), are traditionally used. These products do not produce good results against this mildew. To be relatively effective they must be applied every week. However, they do have the advantage of being fairly versatile and are not involved in phenomena of resistance. The use of these fungicides has been limited and/or very closely regulated as they leave residues in lettuce.

Copper is used on lettuce for combating mildew in the USA. However, it must be used with care as repeated applications can result in yellowing and necrosis of the lamina.

**Penetrating and/or systemic fungicides** such as cymoxanil, propamocarb HCl are used throughout the world, mostly in combination amongst themselves and/or with the preceding or following fungicides. In order to reduce the risk of the appearance of resistant strains, the phenylamides, in particular oxadyxil and metalaxyl, are used in several countries in combination with maneb, mancozeb, or zineb. In spite of these precautions, their use involves a few risks associated with the possible adaptation of *B. lactucae* to the fungicides of this chemical family. This situation has already been encountered in several countries: the USA, Australia, South Africa, the Netherlands, Great Britain, France, and so on.

Metalaxyl was, or is still, used in a fairly specific way in certain countries. It is actually added to the substrate in the nursery or incorporated in the soil at pre-planting stage. Given the risks of resistance, this strategy is very dangerous. Very rapidly it leads to the selection of resistant strains. The combination with a contact fungicide can no longer have an anti-resistant role since only metalaxyl migrates into the plants, with the contact fungicide remaining in the soil. This practice must be abandoned.

We recommend that you consult table 26; it provides the most up-to-date recommendations for lettuce treatments proposed by specialists for controlling mildew.

In the nursery, attempts can be made to restrict the initial outbreak(s) by treating them with a more concentrated dose of fungicide and eliminating the affected seedlings. Plants produced under these conditions can only be marketed if the attack of mildew has been brought under complete control and the purchasing growers are warned of the risks. In addition to fungicide treatments, a number of prophylactic measures must be applied.

**In the nursery**, shelters must be ventilated as much as possible in order to reduce humidity. Try to avoid sprinkler irrigation in late evening and especially the morning, as contamination is much more likely to take place during the morning. If this is not possible, it must be carried out while the weather is hot and sufficiently early in the afternoon to allow the plants time to dry off before nightfall.

**During cultivation** the same irrigation recommendations as those suggested for the nursery must be adopted. In the greenhouse it may be necessary to provide heating in order to reduce surrounding humidity. You must do your utmost to prevent a film of water forming on the plants.

When the crop has finished growing, as much plant debris as possible must be eliminated rapidly; the remaining residues must be buried deeply in order to encourage their decomposition.

### • Subsequent crop

The next **nursery** must be set up where it is sunny, and certainly not damp and shady. If it is set up in the same shelter as in the previous year, you are advised to apply the hygiene and disinfecting measures recommended on page 357.

During cultivation, you must avoid planting in poorly drained soil or where there is surface water retention, and in soil which does not have sufficient organic matter. Crop rotations lasting at least 3 years are recommended. New planting must not be carried out close to crops already affected. Any fertilizer applied must be balanced, and under no circumstances excessive. In countries or regions where this is possible, planting density can be reduced in order to produce better ventilated plots, in which humidity within plant cover is lower. If possible sowing blocks must be laid out in the direction of the prevailing winds in order to encourage maximum plant ventilation. Species of wild

* In order to give the propsed control methods a 'universal' nature, we have produced a fairly comprehensive list of the methods and fungicides used in the various grower countries.
**It is clear that these proposals must be modified according to the country concerned and the pesticide legislation in force.**

** **Zineb is now prohibited in France.**

**Table 26: Examples of recommended treatments for lettuce most frequently proposed by specialists in France**

| Stages of application | Associated active or unusable materials |
|---|---|
| **Nursery** | 3 powderings/week of zineb or 2 sprinkler applications/week of mancozeb, as soon as germination takes place. |
| **Just before planting, after the last watering** | |
| **During cultivation** | |
| As plants take up | oxadyxil + cymoxanil + mancozeb or phosetyl-Al + mancozeb |
| 7–9 leaf stage | oxadyxil + cymoxanil + mancozeb or phosetyl-Al + mancozeb |
| 11–13 leaf stage | oxadyxil + cymoxanil + mancozeb or phosetyl-Al + mancozeb |
| 16–18 leaf stage | |

lettuce must be eliminated from plots and their surroundings.

Given the rapid progression of this mildew and the risks it poses for the crop, **preventive fungicide treatments** are essential while the crop is growing, just as they are in the nursery. The choice of products and the rates at which they are used must be established with your expert in accordance with local growing practices. Bear in mind that the period between two treatments should not exceed 12 days. You must be particularly vigilant when mixing pesticides or when applying treatments under extreme climatic conditions as there are risks of incompatibility between products and phytotoxicity. The Dithiocarbamates, posing problems with residues, can basically be used at the very start of the growing season: in the nursery and just after planting. Be sure to alternate fungicides that have different methods of action in order to limit selection of resistant strains. No more than 2 applications of oxadyxil-based products should be made per campaign and nor should they be used once attacks have become established. These remarks apply equally to other phenylamides used in other countries apart from France.

**Resistant varieties** are available in several countries. They possess several resistance genes in order to control the numerous strains present in the field. A list of genes used and strains described can be seen in the appendix on page 353. In some countries there is no information about which strains are present. Experiments must be carried out locally in order to define whether the chosen resistant varieties will be capable of controlling the indigenous strains. In fact, although they do represent a real advantage, these varieties must be used in combination with complementary chemical controls.

# *Erysiphe cichoracearum* DC.

(Eukaryotes, Fungi, Ascomycota, Erysiphales, Erysiphaceae)

# Powdery mildew, White mould

## Principal characteristics

### • Frequency and extent of damage

Powdery mildew is considered to be a secon-dary disease affecting l.s.v. (lettuce, endive - frisée and escarole) in virtually all the countries in which it attacks. It has been reported in several states of the USA, Canada, Korea, Japan, in various European countries (Switzerland, Great Britain, Germany, the Netherlands, Italy, Greece, and so on), and in the Mediterranean basin (Israel). Its damage is rarely significant and may occur in the open field and under shelter.

Powdery mildew is often sporadic and brings down the quality of l.s.v., but only occasionally causes serious damage. It mainly attacks endive, chicory, and lettuce. Its attacks on lettuce seed-plants pose a few problems with regard to seed production.

### • Principal symptoms

Plants affected by *E. cichorearum*, whatever the age of the plants, display very characteristic symptoms. A greyish, powdery white mycelium appears first of all in the form of spots on the upper faces of old leaves. These gradually extend and converge to cover a fairly significant portion of the lamina. Spots are sometimes visible under the lamina. Affected tissues are often chlorotic with irregular brown lesions which are due to the death of plant cells.

The presence of black minuscule, spherical masses, the cleistothecia of the fungus, has sometimes been reported on colonized tissues. Fairly old plants and old leaves are the most affected.

(See **273**, **277–280**)

### • Biology, epidemiology

The epidemiology of this obligate parasitic fungus, with regard to l.s.v., is not well known.

**Survival, sources of inoculum:** *E. cichoracearum* can persist from one year to the next in several ways:

- thanks to its globular cleistothecia (perithecia without ostioles) which ensure its sexual repro-duction; they sometimes form on leaves affected by downy mildew. They are not found very commonly, but they do nevertheless constitute a form of survival in winter. Generally, the asci present in the cleistothecia form in autumn. It is only the following spring when the ascospores will be projected:
- via other hosts. The species *E. cichoracearum* is very polyphagous. It is capable of attacking a large number of hosts, over 200 species from 25 different families. In fact, this species probably contains relatively specialized forms infecting different plants. The parasitic specialization of strains predominating on l.s.v. is not very well understood; they must be able to survive and multiply on alternative hosts, cultivated plants, or wild ones. *Lactucae serriola* and other types of wild lettuce are capable of harbouring it.

**Penetration and invasion:** The host is infected by a primary inoculum consisting of either ascospores or conidia. Once these contact the host, they germinate rapidly within 2 hours if surrounding conditions are conducive (18–25°C, 95–98% relative humidity). They form an appressorium, then directly penetrate the epidermal cells, developing haustoria. The latter act as suckers which are able to extract the elements needed for the fungal mycelium to grow.

450

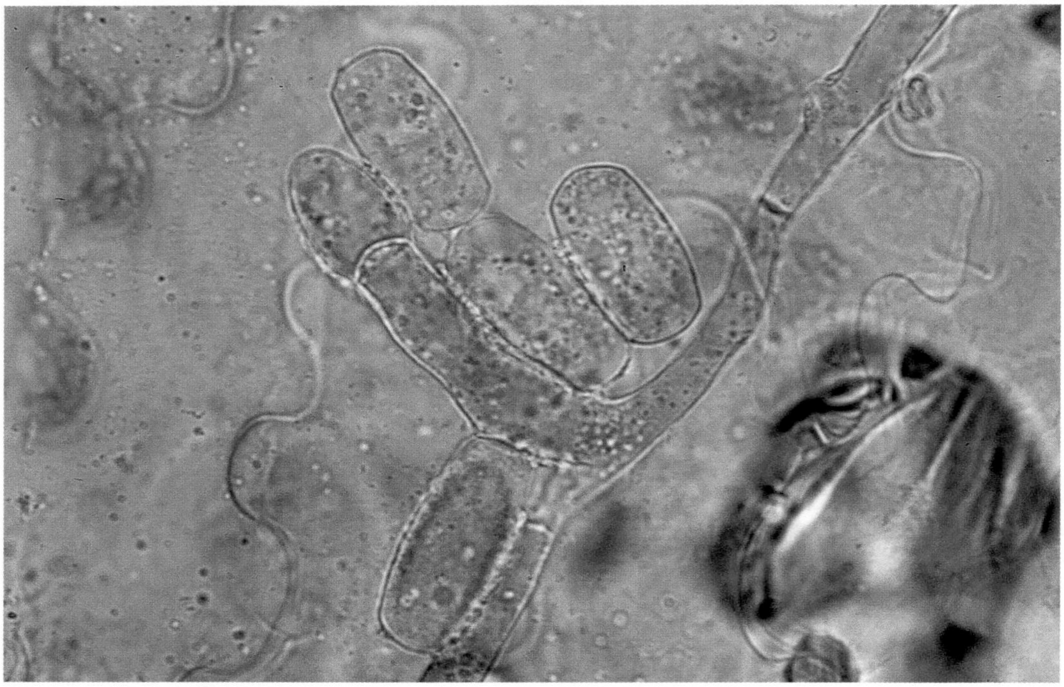

451

**450** A superficial mycelial network covers the tissues affected by powdery mildew. Young conidiophores can be distinguished

**451** *Erysiphe cichoracearum* frequently forms chains of hyalin, cylindrical to doliform conidia (25–45 × 14–26 µm), positioned at the end of short conidiophores (anamorph form: powdery mildew).

**Sporulation and dissemination:** Within the 4–6 days following contamination, short conidiophores appear on the secondary hyphae formed on the surface of the lamina and produce conidia (**450** and **451**). Sporulation is very abundant. The conidia are quite fragile and only live for a few hours, or at most a few days if conditions are favourable. They are very light and are therefore easily transported and disseminated by the wind (up to 200 km according to certain authors) and, also by rain or sprinkler irrigation, when splashing occurs. Dissemination reaches its maximum at night, between midnight and 4 a.m.

**Conditions favourable to its development:** *E. cichoracearum* seems to be capable of multiplying at temperatures between 4–32°C. Its thermal optimum is around 18–25°C. The presence of free water on the leaves does not seem conducive to its development.

Light conditions influence its growth. It prefers diffuse light, whereas direct sunlight impedes its spread. This is partly why it is found more fre-quently on lower leaves within the plant cover, where the surrounding microclimate is parti-cularly favourable. Young plants are probably less susceptible. *E. cichoracearum* can complete a full cycle in 4 days if surrounding conditions are favourable.

## Protection*

### • During cultivation

When confronted with an attack of powdery mildew, it is very often too late to intervene. It is actually very difficult to control a declared epidemic of powdery mildew, especially if it takes place at an advanced stage of crop growth.

Very few **products** are **approved** for use on l.s.v., apart from sulphur on endive. On this type of l.s.v., this product may be applied up to the 18-leaf stage. We should point out that treatment is only effective if applied at the right time, at the correct dose, with sufficient volumes of drench, and equipment appropriate for application to l.s.v. Some of these treatments are likely to be toxic to plants if applied on hot dry days. They should be carried out during cooler cloudy periods.

A few **prophylactic measures** can improve control of this mycosis. These include:

- eliminating plant debris from plots and their environment, otherwise quickly bury it deeply;
- destroying, in the plot and its surroundings, weeds which can act as relay plants for the parasitic fungus (in particular *Lactuca serriola*).

### • Subsequent crop

When the next crop is planted, a certain number of additional **preventative measures** may be put in place in order to limit the risks of powdery mildew appearing.

Consequently, crop rotation should be carried out. This does not necessarily need to take long since the cleistothecia of powedery mildew cannot be conserved for long periods in soil. The position of the future plot must be carefully chosen to ensure that it is situated in a place which is sufficiently sunny and ventilated. Balanced fertilization must be provided, and crops already harbouring *E. cichoracearum* must not be planted nearby. Untimely watering must be avoided and all stagnant water in the plot should be removed.

In areas where powdery mildew is more constant from one year to the next, preventive treatments can be carried out using sulphur. In France, the recommendation is to treat frisée endive at 12–14 leaf stage in particular, with an additional treatment at the 18-leaf stage.

**Differences in sensitivity** between species of *Lactuca* and between cultivars within the same species have been recorded. Behaviour recorded in published works sometimes varies from one author to another. This situation may be explained by the existence of specialized forms of *E. cichoracearum*; some authors have even referred to strains.

* In order to give the proposed control methods a 'universal' nature, we have produced a fairly comprehensive list of the methods and fungicides reported in the various grower countries. **It is clear that these proposals must be modified according to the country concerned and the pesticide legislation in force.**

452

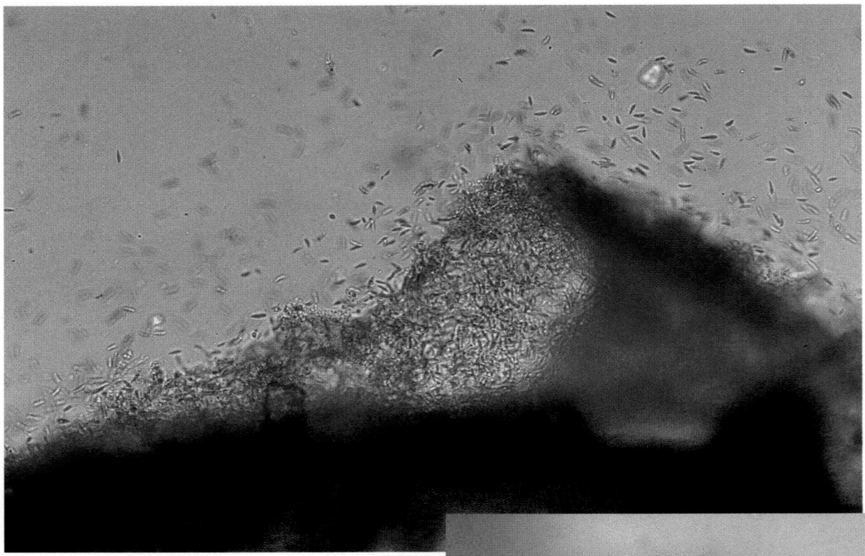

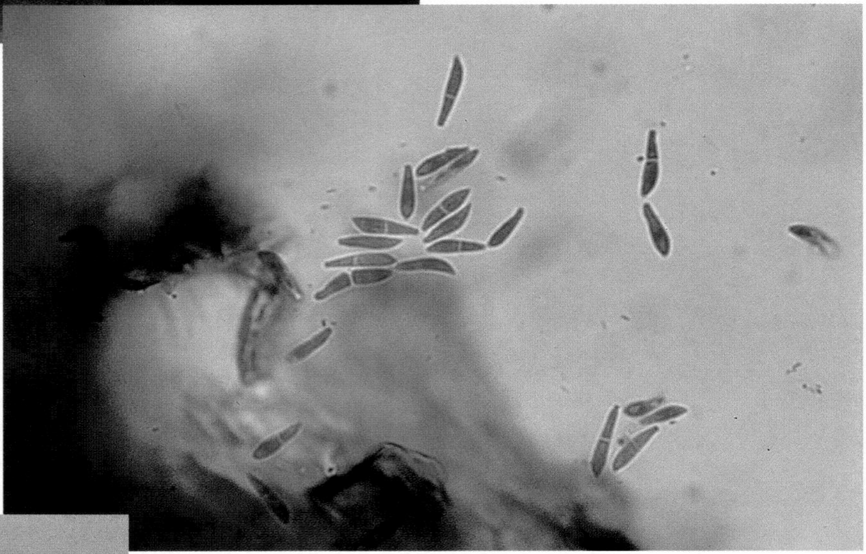

454

453

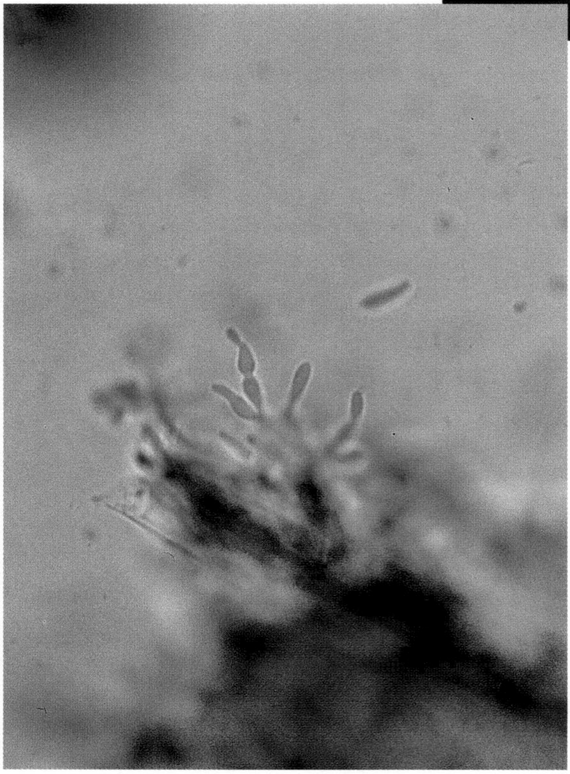

**452** Acervuli (or sporodochia) of *Microdochium panattonianum* can be seen in necrotic leaf tissues.

**453** The conidiophores which can be observed in the acervula are smooth and hyalin, sometimes compartmentalized; they show shrinking in the region where the conidia are formed.

**454** Numerous spindle-shaped and hyalin conidia are produced at the extremity of the conidiophores. They are generally bicellular when mature, slightly curved inwards, and have a constriction in the partition. Length varies between 5–17 µm.

# *Microdochium panattonianum* (Berl.) Sutton *et al.* in Galea *et al.* ex. *Marssonina panattoniana* (Berl.) Magnus

(Eukaryotes, Fungi, Mitosporic fungi)

# Anthracnose, Shot-hole, Ring spot, Rust

## Principal characteristics

• ***Frequency and extent of damage***

This fungus has been recorded in numerous countries over several continents (USA, Australia, New Zealand, and so on), and in Europe (Germany, Great Britain, Denmark, Netherlands, Switzerland, Italy, France, Hungary, Serbia, and so on). It is found in both temperate and subtropical production zones. Under certain particularly favourable conditions, it may cause considerable damage both while the crop is growing and after harvesting, during storage and transport. It affects cultivated and wild lettuce, endive, and chicory.

It has been known for many years, but its effects have often been relatively limited. Anthracnose seems to have been on the increase since the early 1990s, especially on lettuce, and more randomly on endive. At the moment it is a disease encountered more frequently on crops growing in the open field in the autumn. Now it can often be seen in damp shelters during the winter.

• **Principal symptoms**

M. *panattonianum* may attack young plants, which are particularly susceptible. When attacks are severe, and they are covered with damp spots, seedlings become chlorotic and their growth is halted.

This fungus is responsible, above all, for spots on all parts of leaves close to the ground. On the lamina, these spots are initially small and damp. Subsequently, they spread and become more circular. Their delimitation by the veins may give them an angular appearance. Damaged tissues take on an orange to brown colouration. They quickly become lighter as they dry out, split, and fall. Leaves then have a riddled appearance.

Spots are more elongated on the veins, slightly depressed and usually convergent. They are also an orange colour, which varies in intensity.

Around the periphery of the spots, unobtrusive masses of whitish to pink spores form; these are the acervuli (or sporodochia) of M. *panattonianum* which are responsible for its asexual reproduction (anamorph form).

If contamination takes place at a late stage, lesions only appear during storage and transport, lowering the quality of l.s.v. ready to be marketed.

(**See 38, 39, 196, 216–224, 248, 254, 378**)

• **Biology, epidemiology**

**Survival, sources of inoculum:** Essentially M. *panattonianum* survives from one year to the next on the debris present in the soil where the crop is growing or in its surroundings. Some authors report that it can persist in the soil for 4 years. In certain production regions where susceptible wild lettuce (*Lactuca serriola*) survives, this may contribute to its lasting presence. It may also be harboured by other weeds, particularly *Sonchus aster*. Short term survival of this fungus on seeds has been reported, but this now seems to be called into question.

In fact, the origins of initial contamination mainly lies on the 'microsclerotia' (35–65 μm in diameter) which form in the cells of damaged tissues. These microsclerotia present on the soil may be projected onto the lower leaves of lettuce during rainfall or sprinkler irrigation. Subsequently they germinate on the damp lamina, producing a short appressorium. Conidiophores may also form on the sclerotia present on the soil surface. The conidia produced are also responsible for primary contamination. They are deposited on the leaves by the wind and splashes of water.

**Penetration and invasion:** The presence of free water on the lamina is essential for contamination which takes place directly through the cuticle or, more rarely, through the stomata. M. *panattonianum* colonizes the tissues progressively (intra- and intercellular colonization) and these quickly become necrotic and fill with micro-sclerotia. These spread to the soil at the same time as the decomposed tissues. Duration of incubation lasts on average 4–12 days depending on climatic conditions.

**Sporulation and dissemination:** High humidity allows the fungus to sporulate abundantly on the spots. The presence of free water is necessary initially; subsequently relative humidity of 100% is sufficient. The numerous conidia formed under the cuticle (**452–454**) lie at the origins of secondary contamination. These conidia are encouraged by rainfall, sprinkler irrigation and the wind, which carries micro-droplets over distances which vary depending on its strength. Market garden owners tending their plants, and the tools they use, may also help to spread the disease.

**Conditions favourable to its development:** M. *panattonianum* is a fungus which is particularly likely to attack in damp environments following periods of fog, dew, and especially rainfall or sprinkler irrigation. At the same time it appreciates fairly low temperatures; its thermal optimum is around 17–19°C. Symptoms may even occur at around 2°C. The conidia can no longer germinate beyond 28°C. The fungus is probably destroyed after 10 minutes' exposure at 40°C.

## Protection*

### • During cultivation

As soon as the first symptoms appear, the main advice is to **reduce the humidity** of the shelter by ventilating it as much as possible and avoding sprinkler irrigation, especially in the evening. It is preferable for workers to tackle affected plots when the plants are dry.

**Fungicide treatments** can be carried out, particularly if favourable damp conditions persist. In some countries, chemical means of combating this disease have been made difficult by the fact that there is no fungicide approved for this 'use'. The Dithiocarbamates (mancozeb, maneb) used alone or in combination with penetrating or systemic active materials for combating mildew, help to limit the spread of this disease. In addition, chlorothalonil and benomyl are probably very effective. In published works several other fungicides have been reported as being effective: prochloraze, flutriafol, propiconazole (phytotoxic to l.s.v.). Some authors have stated that spraying carried out with significant volumes of drench are more effective. As soon as climatic conditions become drier, the activity of the fungus is greatly reduced and treatments can be spaced out.

It is essential to **eliminate plant debris** during and at the end of the growing season, given its fundamental role in the survival of this organism. It is also important to dig over the soil to a significant depth in order to bury any remaining debris effectively and encourage its decomposition.

### • Subsequent crop

It is essential to envisage **crop rotation** lasting for at least one year; several years are preferable.

If the crop is being grown under protection, **surface disinfecting** of the walls, with a disinfectant, may be considered. The ground should be treated in a similar way. In order to do this, a fumigant (methyl bromide** is effective) or steam may be used, or a fungicide may be spread over the ground before planting. Poorly drained plots must be dug over in order to eradicate this fault. Wild lettuce (*L. serriola*) must be removed from the area surrounding crops.

**Fungicide treatments** will be effective only if they are carried out as a preventative measure prior to rainfall or extensive irrigation.

Plants must not be handled, and the soil must not be dug over in crops while the lettuce plants are wet. The initial infected plants must be eliminated and destroyed quickly, as well as old leaves displaying spots.

If **lettuce seed is produced** in hot, dry zones, unfavourable to the fungus, seeds which are free from the fungus can be obtained.

The wild species *L. saligna* is resistant to *M. panattonianum*. No varieties possess this resistance.

* In order to give the proposed control methods a 'universal' nature, we have produced a fairly comprehensive list of the methods and fungicides reported in the various grower countries. **It is clear that these proposals must be modified according to the country concerned and the pesticide legislation in force.**

** **Methyl bromide will be prohibited after 2005.**

Nevertheless we find differences in sensitivity between cultivars. For example, romaine and head lettuce seem to be less susceptible. Australian studies report that *L. serriola*, endive, and chicory seem likely to be immune to *M. panattonianum* under their experimental conditions. In published works there seem to be differences in results between those carrying out experiments. This may be due to the plant material tested or to pathotypes differing from one country to another. In fact, five pathotypes have been described in California.

455

456

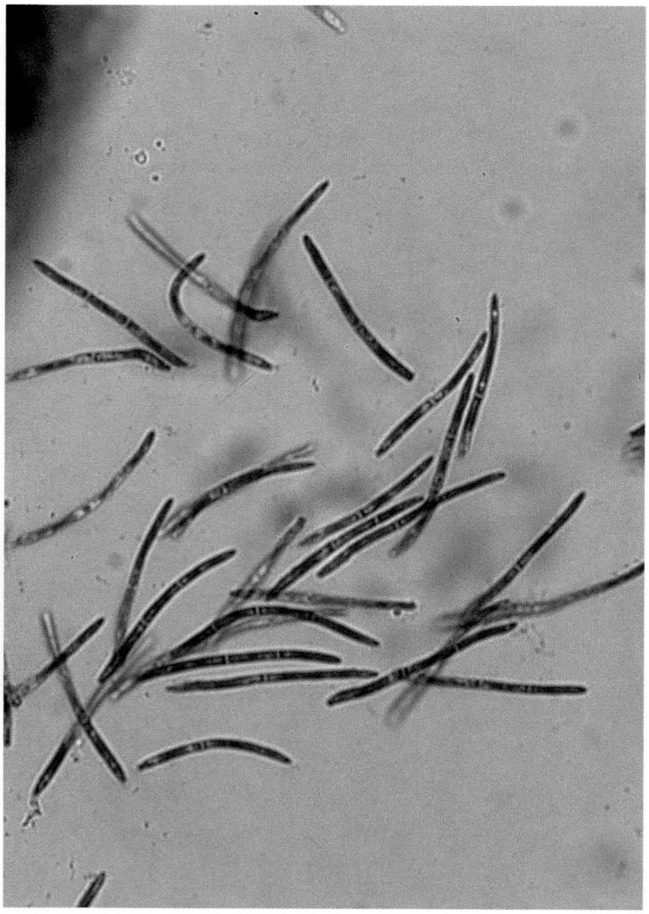

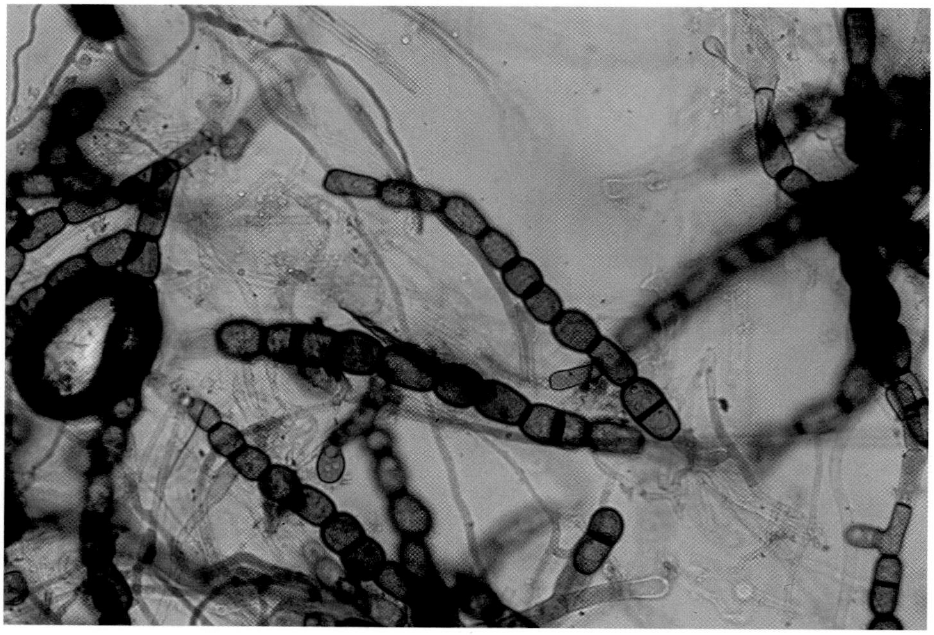

457

**455** Pycnidia of *Septoria lactucae* partly immersed in the leaf tissues (100–200 μm in diameter). A cluster of spores has been liberated at ostiole level.

**456** The conidia produced are hyalin and spindly; they have 1–3 partitions. Their dimensions are variable: 25–40 × 1.5–2 μm.

**457** In addition to its multicellular conidia, *Mycocentrospora acerina* produces groups of cells (17–30 μm) brown and swollen (chlamydospores) which help to conserve it.

# Other principal parasitic fungi of leaves

| | ***Septoria lactucae* Pass.*** **(Septoriosis, 'Septoria leaf spot')** (Eukaryotes, Fungi, Mitosporic fungi) | ***Mycocentrospora acerina* (Hart.) Deighton** **('Mycocentrospora leaf spot')** (Eukaryotes, Fungi, Mitosporic fungi) |
|---|---|---|
| **Distribution and damage** | This fungus is found worldwide. It is not considered to be a serious parasite on the various types of lettuce. Observed from time to time in France during the 1960s, it seems to have disappeared from intensive lettuce crops. It has reappeared recently in the Parisian region in summer crops. | This fungus, which is fairly widespread worldwide, has been recorded on rare occasions on lettuce, essentially in England. In France, it has been encountered sporadically on this plant during February or March in the Midi, on lettuce grown in the open field. |
| **Symptoms** | Irregular, chlorotic leaf spots, delimited by the veins, which are brown and sometimes very broad when they converge. A yellow halo surrounds them. They are scattered with numerous pycnidia containing spores which are generally spindly, hyalin, with 1–3 partitions. The centre of the necrotic tissues pulls away and leaves are consequently covered in holes. These spots mainly develop on old leaves. (See **111, 150, 169–173**) | Small initially transparent spots, rather rounded in shape, rapidly extend and take on a mid- to dark brown colour. Necrotic tissues dry out, becoming paper-like, and fall. Ultimately, numerous holes are scattered on the lamina. Black, deliquescent rotting may start up and progressively spread to the head. The taproot may be affected, via the insertion point of one or several rotten leaves. In this case, the plant suddenly wilts in the same way as during attacks to the crown caused by *Botrytis cinerea* or *Sclerotinia* spp. (See **174** and **175**) |
| **Biological information** | *S. lactucae* is preserved on plant debris, seeds and wild lettuce (*Lactuca serriola*). Some weeds harbour it. Its conidia, formed inside pycnidia (**455** and **456**) are responsible for con-tamination and dissemination of the disease following splashing. Seeds can also spread the fungus over long distances. Working tools and animals also make their contribution. The complete cycle of this fungus may take place within 5–10 days. Damp and relatively hot periods of weather encourage the spread of this fungus. Its thermal optimum is between 20–24°C. Insufficiently fertilized or etiolated plants are probably more susceptible. | *M. acerina* survives on plant debris thanks to its dense brown mycelium, with thick walls like those of the chlamydospores (toruloid shape) (**457**). Its highly polyphagous nature allows it to attack a large number of vegetable plants (parsley, parsnip, celery, carrot, and so on) and flowers (primrose, cyclamen, pansy, petunia, and so on). These flowers, as well as various susceptible wild plants (shepherd's purse, nettle, and stitchwort) ensure that this fungus can multiply and play the role of reservoirs. Contamination takes place during cold damp periods, following splashing. Once in the tissues, the fungus colonizes them rapidly and symptoms are visible after 5–6 days. It sporulates on damaged tissues forming characteristic elongated hyalin conidia. These have 5–11 partitions as well as a thin, tapering lateral appendage (see **fig. 6**). These spores are scattered after rainfall and ensure secondary contamination. In certain plants, seeds can also spread the fungus over long distances.<br>Periods of damp and relatively cold weather (5–15°C) encourage the development and spread of this fungus. |

458

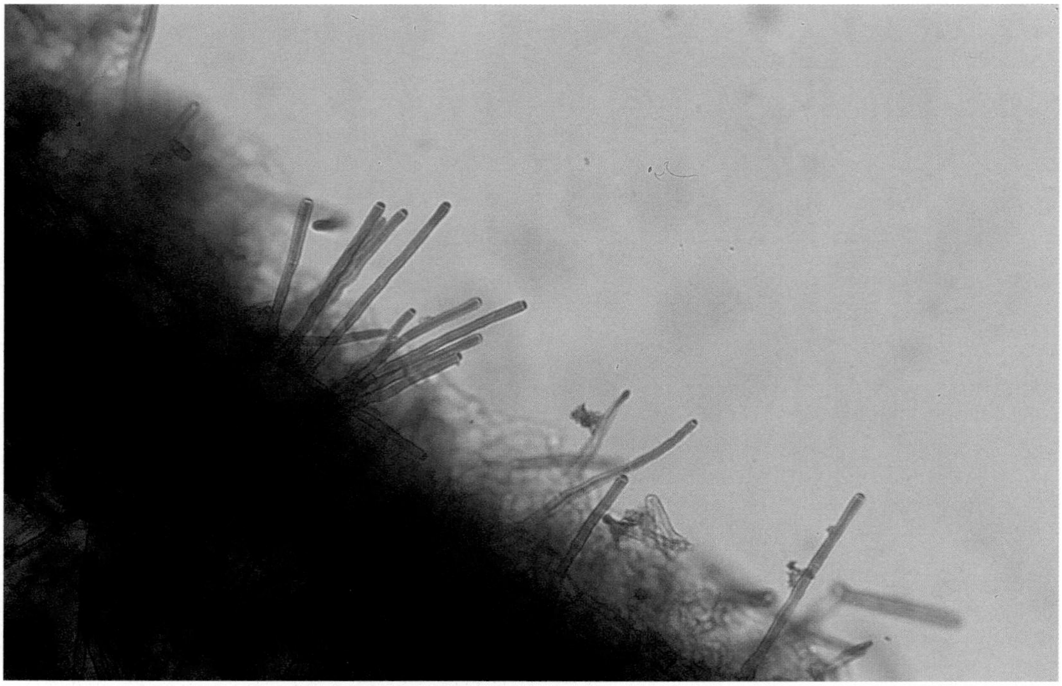

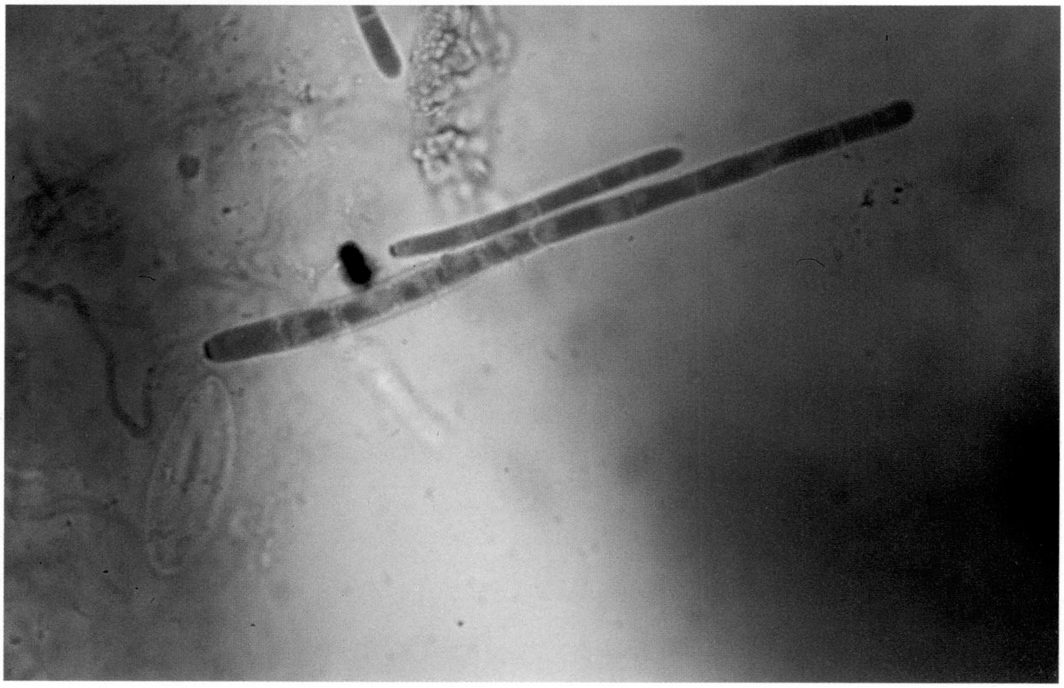

459

**458** Several non-branched and upright conidiophores of *Cercospora longissima* are visible on the epidermis of this leaf. Initially hyalin, they end up by gradually becoming brown.

**459** The conidia formed at the extremity of the conidiophores are hyalin and cylindrical. They have 1–18 partitions and measure 11–170 µm long.

| | ***Septoria lactucae* Pass*** **(Septoriosis, 'Septoria leaf spot')** (Eukaryotes, Fungi, Mitosporic fungi) | ***Mycocentrospora acerina* (Hart.) Deighton ('Mycocentrospora leaf spot')** (Eukaryotes, Fungi, Mitosporic fungi) |
|---|---|---|
| **Protection**** | During major attacks, treatments using fungicides based on Benzimidazoles may be carried out. More recent products, such as IBS, ought to be effective. Take care when using benomyl, as the selection of resistant strains is possible. Old diseased leaves must be eliminated when the plants are harvested, and must not be abandoned on the ground. In the open field, crop rotation lasting more than 4 years should be tried. The ground should be well drained. If you are in any doubt about the health quality of seeds, they can be disinfected for 30 minutes in water at approximately 48°C. New planting should not be carried out close to infected crops. The soil in the nursery, as well as the structure of shelters, must be disinfected. Seeds can be produced in relatively dry zones, free from the fungus. No variety of resistant lettuce is currently available. | Several measures must be taken in those rare plantings affected by this fungus:<br>• carry out fairly long crop rotation not involving susceptible crops;<br>• disinfect the soil with a fumigant; dazomet produces good results;<br>• coat seeds with thirame or captane.<br>Preventive fungicide treatments must be carried out in cultivations which are particularly badly affected. Anti-botrytis treatments based on iprodione are normally effective.<br>Diseased old leaves must be eliminated when the crop is harvested; under no circumstances must they be abandoned on the soil.<br>New planting should not be carried out close to affected crops. The soil in the nursery, as well as the structure of shelters, can be disinfected. No resistant variety is currently available. |

* Several other species of Septoria have been described as affecting *Lactuca*: ***S. ludiviciana***, ***S. fernandezii***, ***S. schebelli***, ***S. unicolor***, ***S. sikangensis*** and ***S. lactucina***. More recently ***S. intybi*** has been reported in Italy on *Cichorium endivia*.

** In order to give the propsed control methods a 'universal' nature, we have produced a fairly comprehensive list of the methods and fungicides reported in the various grower countries.
**It is clear that these proposals must be modified according to the country concerned and the pesticide legislation in force.**

| | ***Stemphylium botryosum f. lactucum* Wallr.** (anamorph form: Eukaryotes, Fungi, Mitosporic fungi)<br>**(Syn.: *Stemphylium herbarum Simmons)***<br><br>***Pleospora herbarum f. lactucum* (Pers.) Rabenh.** (telomorph form: Eukaryotes, Fungi, Ascomycetes, Dothideales, Pleosporaceae)<br><br>**(Stemphyliosis, 'Stemphylium leaf spot')** | ***Cercospora longissima Cugini ex.* Traverso non Cooke & Ellis & Everhart (Cercosporiosis, 'Cercospora leaf spot')**<br><br>(Eukaryotes, Fungi, Mitosporic fungi) |
|---|---|---|
| **Distribution and damage** | Described several years ago in several countries, particularly those in Europe (Great Britain, Italy, Netherlands, Spain and Portugal), in Israel, South Africa, and the USA, this fungus is not currently considered to be a major pathogenic agent of lettuce, except perhaps in Israel. *S. herbarum* has recently been reported on chicory in Italy. | This fungus causes mild attacks in numerous countries worldwide (USA, India, Iran, and so on); it is particularly damaging in tropical zones, especially in some African countries (Somalia), Asia (Japan, Hong Kong, and so on) and in Central America. It has never to our knowledge been reported in France over recent decades. On the other hand, it seems to be present in Italy. |
| **Symptoms** | Spots which are initially small and damp, progressively taking on variable shapes: circular to oval, sometimes angular if restricted by the veins. Concentric patterns, which may be fairly dark, can be seen on these spots. Dark brown to black conidial matting covers them during damp periods. Once the central tissues have become necrotic and dried out, they may fall; subsequently leaves have a number of perforations. Damage is observed mainly on old leaves (see **186**). | Small brown, damp spots, surrounded by a pale green halo. They spread and form areas of brown damage circumscribed by the veins; the tissues become progressively necrotic. They may converge; under these conditions, portions of leaves are completely necrotic. Sometimes they are covered with a greyish to beige cover made up of conidiophores and hyalin, cylindrical conidia with 1–18 partitions (**458** and **459**). These spots can mainly be found on old leaves (see **151, 176–180**). |
| **Biological information** | This fungus has significant saprophytic potential. It is capable of surviving on plant debris, in its mycelial form, via its conidia or its perithecia (ensuring its sexual reproduction). In addition it can probably survive on seeds. Contamination mainly takes place during cold, damp climatic periods, via the stomata. Subsequently, the fungus rapidly colonizes the tissues. If conditions are favourable, the symptoms can already be seen a few days after the initial contamination. It correlates abundantly on damaged tissue, producing large muriform spores. These are dispersed by the wind and splashes of water following rainfall or irrigation. Morning dew also encourages its development, as well as damp, relatively cold climatic periods. It develops between 4–34°C. Its thermal optimum is between 22–28°C. Symptoms probably still develop after harvesting, during storage. | This fungus survives in soil on plant debris. In addition it can probably survive on seeds and on a certain number of hosts, particularly wild lettuce: *Lactuca serriola*. It develops during damp periods, when there is free water on the leaves. The conidia present on the surface of the lamina germinate and penetrate the stomata. It appreciates temperatures close to 25°C. Under these conditions it rapidly invades the tissues: symptoms are already visible after 3 days and it sporulates 5 days after contamination. The spores formed are liberated as soon as humidity drops. They are also spread by splashing, following rainfall or irrigation. |
| **Protection*** | Crop rotation should be considered. This fungus can also be controlled by using healthy seeds. Fungicide based treatments can be carried out. Maneb and chlorothalonil seem to be effective against *Stemphylium*. This fungus survives easily on plant debris. It will be essential to eliminate this debris during or after harvesting. Sprinkler irrigation must be avoided or must be carried out during the day in order to allow plants to dry rapidly. Sources of resistance have been identified in *Lactuca saligna*: resistance controlled by a dominant allele (Sm) and a recessive allele (sm). | In tropical and subtropical countries which are particularly badly affected, treatments which prevent or stop it can be carried out using fungicides suchas benomyl, chlorothalonil, and mancozeb. Certain IBS, used as an anti-mildew, are also effective against *Cercospora* spp. Take care to eliminate plant debris and to carry out crop rotation. Future crops must be established in well-drained plots free from disease, with limited plant density. Healthy seeds must be used. There are no resistant varieties. |

***Alternaria cichorii* Nattrass (anamorph form)**

(Eukaryotes, Fungi,
Mitosporic fungi)
**(described under several synonyms, see page 106) (Alternariosis, 'Alternaria leaf spot')**

| | |
|---|---|
| **Distribution and damage** | Currently this fungus is not considered to be a serious pathogenic agent on l.s.v., except perhaps on chicory. It has been reported in the USA, Japan, Argentina, Slovenia, Germany, Italy, and so on. In France, although previously recorded on escarole in the Eastern Pyrenees, it no longer seems to attack lettuce and endive. |
| **Symptoms** | Spots, which are small and damp at first, progressively become circular to oval in shape. Concentric patterns are visible on the spots, leading to confusion with Stemphyliosis. Once the central tissues have become necrotic, they lighten. |
| **Biological information** | This fungus has significant saprophytic potential. It is capable of surviving on plant debris in its mycelial form or via its conidia. It probably also survives on seeds. Contamination takes place above all during damp climatic periods accompanied by temperatures of between 20–25°C. Spores germinate on the surface of the lamina, penetrating it directly across the cuticle, occasionally via the stomata. Subsequently, the fungus rapidly colonizes the tissues. It sporulates on damaged tissues, producing large isolated conidia at the extremity of the club-shaped conidiophores. These are dispersed by the wind and splashing following rainfall or irrigation. Morning dew also encourages its development, as well as damp climatic periods. |
| **Protection** | This fungus can be controlled by using healthy seeds. It survives on plant debris, so eliminating this is essential after harvesting. Certain authors recommend weeding and rapidly crushing this debris as well as carrying out rotation with green cover crops. Sprinkler irrigation must be avoided or carried out during the daytime in order to allow the vegetation to dry out rapidly. Certain anti-botrytis fungicides are effective on this fungus. |

* In order to give the propsed control methods a 'universal' nature, we have produced a fairly comprehensive list of the methods and fungicides reported in the various grower countries.
**It is clear that these proposals must be modified according to the country concerned and the pesticide legislation in force.**

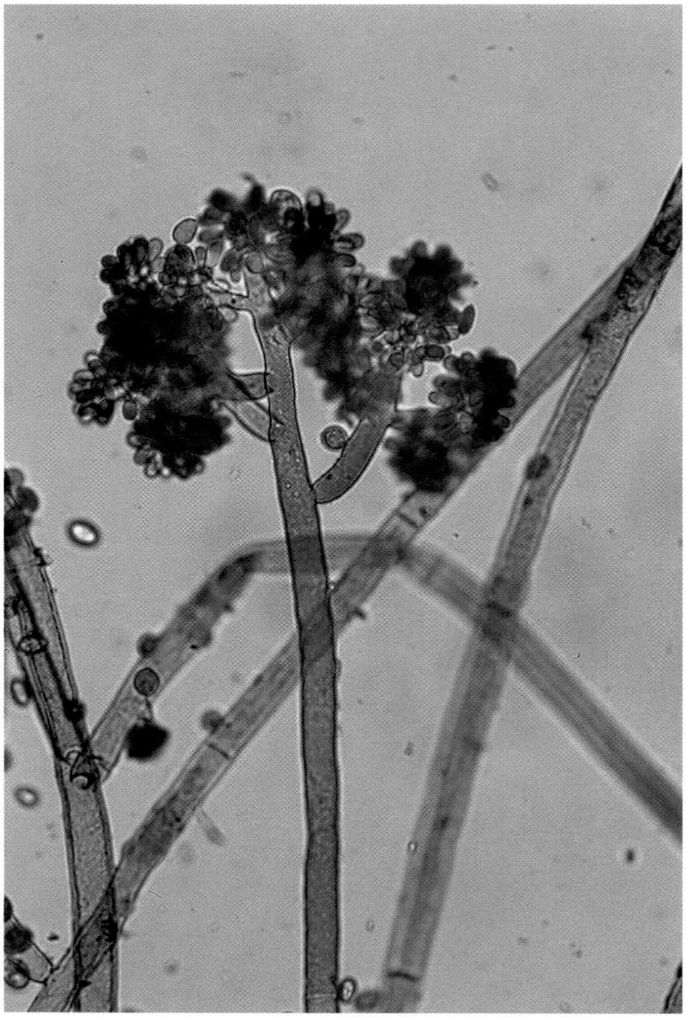

**460** *Botrytis cinerea* projects long robust conidiophores, irregularly branched, progressively becoming black from the base.

**461** Conidia form at the extremity of the conidiophores where we can see a blister shaped bulge. They are carried by a sterigmata, ovoid to elliptical in shape and are hyaline to slightly pigmented in colour (6–18 × 4–11 μm).

# Fungi which mainly attack the lower leaves, in contact with the soil, and the crown



## *Botrytis cinerea* Pers. — Grey mould

(anamorph form: Eukaryotes, Fungi, Mitosporic fungi)

## *Botryotinia fuckeliana* (de Bary) Whetzel

(telomorph form: Eukaryotes, Fungi, Ascomycetes, Leotiales, Sclerotiniaceae)

### Principal characteristics

**• Frequency and extent of damage**

This ubiquitous fungus is very polyphagous and is observed practically wherever lettuce is grown all over the world (China, Korea, Japan, India, USA, Europe: Germany, Italy, Netherlands, Spain, France, and so on). It may attack alone and cause significant damage, particularly under shelter. It is usually reported as forming part of the parasitic complex of the lower leaves and crown of lettuce, particularly in combination with *Sclerotinia sclerotiorum* and *S. minor* (Italy, China, Canada, Germany). This complex may vary from one country or plot to another. Given its biological characteristics, it mainly affects autumn and winter crops.

*B. cinerea* can be found both in the open field and under shelter, where attacks are often more severe. This situation is partly due to the fact that the plant tissues produced are more tender and succulent. In addition, in shelters, plants grow in an environment which is often damper, thus encouraging the development of *B. cinerea*. Although observed from time to time in nurseries, it is after planting, and especially once the head has formed, that this fungus develops and becomes more difficult to control. It is a key stage in the cycle of l.s.v. In fact, it is at this point that plant growth will stop, since the plants are now well developed and have spread out to completely cover the soil. Old leaves, covered by younger ones, will receive much less light and be more or less crammed down against the soil. Fairly quickly they will become chlorotic and senescent, remaining permanently damp because of the lack of ventilation of the plant crown at this stage of the crop. You must remember that *B. cinerea* is an opportunist fungus which colonizes weakened tissues more easily. Old leaves (senescent floral parts as well) therefore constitute nutrient bases which strongly favour its establishment on l.s.v.

We sometimes note damage after harvesting, and during storage; some affected plants establish sources of rotting in boxes.

The biological variability of this fungus is not well understood. Strains resistant to several fungicides belonging to different chemical families are known (Dicarboximides, cyclic imides, and so on). Recent research in molecular biology has revealed significant genetic variability within the species *B. cinerea*.

**• Principal symptoms**

In the nursery, damping-off is rarely due to *B. cinerea*. However, this fungus, sometimes present on seeds which have ripened under rainy conditions and in the same environment as seedlings, is responsible for failed emergence and mortality of seedlings. Once these have collapsed they are sometimes covered by a sterile white mycelium web, known as 'web disease'. This is what happens when sowing is too dense; sometimes the fungus attacks seedlings establishing itself on senescent cotyledons or lower leaves in contact with damp soil. The damage formed is reddish-brown in colour and soft in consistency. Generally speaking, damping-off is more likely to be caused by several species of *Pythium* spp. (fact file 8) and by *Rhizoctonia solani* (see fact file 7).

While the crop is growing and as harvest time approaches, fungal attacks may be indicated by abrupt wilting of plants, isolated or in groups, after the vascular system has been damaged.

As we stated earlier, it affects old senescent leaves which are located under plant cover and in contact with soil after the head has formed. It invades them rapidly, producing damp, mid-brown to dark brown rotting. This rotting spreads to other adjacent healthy leaves and the crown, which is damaged to varying degrees. Vessels are therefore affected and plants may wilt and/or perish gradually or abruptly depending on climatic conditions.

Aerial attacks on young leaves of the head also occur. Damp brown spots develop mainly around the edge of the lamina. Fairly rapidly they cause rotting of the same nature which spreads to the head. This may occur on lettuce which is harvested too late. In China, damage to the lettuce stem has also been noted.

Significant attacks may take place during the period when the seed plant flowers. Senescent floral parts constitute nutrient bases which greatly favour the establishment of *B. cinerea* on inflorescences, causing rotting and contamination of a few harvestable seeds.

Whatever the parts of the plants involved, affected tissues are covered with very characteristic grey mould consisting of the con-idiophores and conidia of the fungus. Black sclerotia (2–5 mm in diameter) are sometimes visible on the same tissues. Sometimes we see the formation of a dirty white downy mycelium which may be confused with that of *Sclerotinia* spp.

(See **2, 148, 181, 182, 274, 275, 283–287, 290, 296, 302, 323, 327, 332, 335–343, 366,** and **368**)

## • Biology, epidemiology

**Survival, sources of inoculum:** This fungus is sometimes found on seeds. It can survive in the soil on an extremely wide variety of plant debris in several forms: conidia, mycelium, sclerotia. The latter can persist for several years in the soil. If they are close to leaves in contact with the soil, they germinate and spread to these leaves from the mycelium. The saprophytic potential of *B. cinerea* allows it to survive easily on organic matter. It is polyphagous and capable of attacking and colonizing several hundred cultivated or wild plants which help it to survive and constitute potential sources of inoculum.

This is the case with the majority of market garden plants. It sporulates abundantly on these plants and on its sclerotia. Contamination is therefore often aerial; in this case it involves conidia which are very easily transported by the wind. These spores germinate on leaves in the presence of water.

**Penetration and invasion:** Subsequently, the germination tube of the conidia or the sclerotial mycelium penetrate the tissues, particularly the leaf parenchyma, destroying the cell walls and their content. Penetration takes place either directly across the cuticle, or through various types of damage. *B. cinerea* also invades all necrotic and/or dead tissues such as those observed following frost damage, water stress resulting in 'tip burn', sunburn, and so on. It also colonizes tissues which have already been damaged by various pathogenic agents such as *Sclerotinia minor* or *sclerotiorum*, *Rhizoctonia solani*, *Bremia lactucae*, and so on, or by pests. It rapidly spreads to the tissues where it can cause rotting in a few days, via hydrolysis of peptic substances. Other secondary micro-organisms, particularly bacteria, spread to these tissues, and also help to decompose them.

**Sporulation and dissemination:** On all its hosts, as on plant debris, it produces mycelium and numerous long, branched conidiophores. Ovoid to spherical conidia are produced at their extremities, ensuring the dissemination of *B. cinerea* (**460** and **461**). The wind and air currents are mainly responsible for this, assisted, to a lesser degree, by rain and splashing. Workmen also spread it as they are tending the plants. The mycelium lies at the origins of contamination caused by diseased tissues coming into contact with healthy tissues. *B. cinerea* also forms on the damaged tissues small flat sclerotia which ensure its survival.

**Conditions favourable to its development:** As with numerous aerial fungi, it is particularly fond of damp environments. Relative humidity of around 95% and temperatures of 17–23°C are conditions which are very conducive to its attacks. These occur under shelter but also in the open field, during rainy periods or following sprinkler irrigation. Over-vigorous or etiolated plants are particularly vulnerable. Certain sub-stances, liberated as seedlings become senescent, probably encourage its development. Screen or covers are sometimes used to protect plants from insects. *B. cinerea* causes more damage under these covers, because of the resulting higher humidity.

Under plastic tunnels, the quality of the covering probably influences the dev-elopment of certain diseases, in particular, *B. cinerea*. In fact, less significant damage has been noted under EVA, compared with PVC and polythene. EVA seems to provide better light transmission, lower humidity, and reduced formation of droplets of water on the walls of the shelters.

## Protection*

- **During cultivation**

To date *B. cinerea* has been very **difficult** to control. There are several explanations for this situation:

- particularly receptive plants (succulent and tender leaves) often grow in a climatic context favourable to *B. cinerea,* particularly under shelter;
- an obligation to halt fungicide treatments a relatively long time before harvesting, particularly when l.s.v. are at their most vulnerable;
- the particular ability of this fungus to adapt rapidly to the fungicides used to combat it;
- insufficient number of fungicides approved for use with *Botrytis* on l.s.v. (essentially contact fungicides, including cyclic imides).

If you see symptoms of *B. cinerea* appearing in your crop and a preventative treatment programme has not been set up, we recommend that you apply one of the approved fungicides: iprodione, procymidone, vinchlozoline, thirame, pyrimethanil and its combination with maneb, cyprodinil + fludioxonil, and so on. From this point onwards you will need to follow the schedule of treatment recommended below and comply with the cut-off times before harvesting, which vary from one fungicide to another.

In addition it is advisable to implement several **prophylactic measures** to complement the chemical measures, both in the nursery and in the open field. Shelters must be ventilated as much as possible, in order to reduce ambient humidity and, in particular, to avoid the presence of free water on the plants. Setting up a thermal screen on the l.s.v., such as agrotextiles (unwoven mesh, woven fabric), helps to increase humidity and reduce the amount of light available. In the case of severe attacks, it is preferable to remove these covers. Irrigation should be carried out during the morning and in the early afternoon, but never in the evening. This system will allow the plants to dry out as rapidly as possible. In some cases, shelters may need to be heated in order to reduce humidity, and in particular to eliminate dew present on the leaves.

**Plant debris** must be eliminated very quickly while the crop is growing, particularly affected plants on which *B. cinerea* sporulates abundantly and sometimes forms sclerotia. Any stress to plants resulting in growth spurts must be avoided.

Nitrogenous fertilizing must be controlled. It must not be excessive (causing very receptive succulent tissues), or insufficient (sources of chlorotic leaves constituting nutrient bases which are ideal for *B. cinerea*).

**Once the crop has finished growing,** plant debris must be rapidly eliminated from plots in order to prevent this debris being buried in the soil at a later date, thus allowing the fungus to survive. Digging the soil to a good depth will facilitate decomposition of any remaining residues.

- **Subsequent crop**

If the **nursery** is established every year in the same place and/or under the same shelter, it will be essential to implement the measures recommended on page 357.

The efficacy of **crop rotation** is disappointing; this situation is certainly due to the polyphagous nature of *B. cinerea* and to the fact that the inoculum mainly comes from the crop environment.

The soil of **future plots** of l.s.v. must be well prepared and drained in order to avoid the formation of puddles of water conducive to late attacks after the head has formed. In the open field, planting rows must be arranged if possible in the direction of the prevailing winds so that the plant crown and plant cover are well ventilated.

* In order to give the propsed control methods a 'universal' nature, we have produced a fairly comprehensive list of the methods and fungicides reported in the various grower countries.
**It is clear that these proposals must be modified according to the country concerned and the pesticide legislation in force.**

Cultivating lettuce on sowing blocks will help to ventilate their crown and avoid water stagnating close to them. Excessively dense planting must be avoided as this may damage plants.

Preventative **fungicide** treatments will be essential. Currently the schedules proposed in other countries are fairly comparable. They involve 3–5 treatments in addition to those carried out in the nursery (often taking place after planting, at the 7–9, 11–13, and 16–18 leaf stages). Their number fluctuates according to the period of the year, the type of l.s.v., the type of crop (open field or under shelter), and production zones.

Currently growers have more fungicides available than previously:

- multi-purpose **traditional products** to which *B. cinerea* has not developed any tolerance (thirame, dichlofluanide, and tolclofosmethyl, and so on);
- more specific products against which resistant strains have been detected on other crops, and also on l.s.v. (the 3-dicarboximides, or cyclic imides – iprodione, procymidone, vinchlozoline). The dicarboximides, which initially performed very well, rapidly led to the selection of (non-persistant) resistant strains that are now very frequent in numerous crops. The benzimidazoles, used at one time on *B. cinerea*, have also been affected by resistance. Currently we can still find strains tolerant to fungicides from this chemical group in the field (benomyl, carbendazime, and so on);
- **new products** (pyrimethanil, cyprodinil + fludioxonil, and so on), not yet involved in phenomena of resistance on l.s.v., allow an anti-resistance chemical control (based on alternating active materials with different modes of action) to be set up.

The other **diseases** and **predators** must be **controlled** as they cause areas of damage and tissue necrosis conducive to the establishment of *B. cinerea*. The appearance of symptoms of 'tip burn' should also be avoided for the same reasons. As soon as there is significant plant growth, you must be vigilant, especially during cloudy weather and as l.s.v. begin to form heads.

Although no **varietal resistance** has been found in l.s.v., as with numerous other hosts of this fungus, slight differences in susceptibility have been observed between cultivars. These are probably associated with the relatively upright appearance of certain types (romaine, Latin) or with a thicker leaf cuticle. In addition, endive seems to be less susceptible than lettuce.

A number of **original methods** have been or are being experimented with in several countries. Thus, compost extracts have been used in the UK; sprayed on plants, they reduce the damage due to *B. cinerea* and can increase yield. Anti-oxidants have also been used in Israel to limit the development of this fungus.

Several **antagonistic** fungi and bacteria have been evaluated *in vitro* or *in vivo* for controlling *B. cinerea*. Among these we can list: *Streptomyces* spp., such as *S. griseoviridis*, and so on, *Gliocladium virens*, *Trichoderma harzianum*, *Ulocladium atrum*, and so on.

# *Sclerotinia sclerotiorum* (Lib.) de Bary *Sclerotinia minor* Jagger

(Eukaryotes, Fungi, Ascomycetes, Leotiales, Sclerotiniaceae)

# Sclerotiniosis ('Sclerotinia drop', 'Watery soft rot')

## Principal characteristics

### • Frequency and extent of damage

These two fairly polyphagous ascomycetes are widespread and their damage has been reported in all areas of production worldwide: Australia, New Zealand, Japan, China, USA, Canada, Europe (Belgium, France, Netherlands, Germany, Spain, Italy, UK, Portugal, and so on), Mediterranean basin (Israel, Turkey, and so on). The financial losses registered in some countries have at times been significant. In the USA, for example, damage may fluctuate from one plot to another, from under 1% of diseased plants to up to 75%.

These two fungi attack with varying degrees of seriousness in numerous plots, alone or in combination with other fungi such as *Botrytis cinerea*.

They are capable of attacking l.s.v., both in nurseries and in the open field. In numerous countries they are considered to be serious pathogenic agents for all types of l.s.v. They may cohabit in the same plot, but generally one or the other predominates in a given site.

### • Principal symptoms

These two fungi cause very similar symptoms on l.s.v., after planting and especially when their heads are forming, and as harvest time approaches.

They are responsible for damp, light brown damage affecting the parts of plants in contact with soil, particularly senescent leaves. This damage develops very quickly into rotting which spreads to the layers of leaves close to the soil. The petioles and principal veins as well as the crown are invaded, resulting in chlorosis and wilting of external leaves, then the whole plant, sometimes in less than 2 days. Subsequently rotting spreads to all the leaf tissues which decompose and collapse. When plants are pulled up they offer no resistance.

Initially spots, then areas of damage comparable to those described previously, may start to appear higher up on leaves exposed to the ascospores of *S. sclerotiorum*. The spots spread rapidly and inevitably result in rotting of several leaves and the head.

In the past, slow and late wilting of l.s.v. has also been recorded following attacks of *S. minor* deeper in the soil. This fungus causes brown damage on the taproot, 5–10 cm below the surface of the soil.

Whatever the location of the attack, a fairly downy white mycelium forms on certain parts of affected tissues. Here we can see structures which allow us to differentiate between the two *Sclerotinia* spp.:

- a few large black sclerotia, fairly elongated, 2–20 mm in length by 3–7 mm wide, for *S. sclerotiorum* (**463**);
- a conglomeration of small black sclerotia, irregular, rather circular, from 0.5–2 mm in diameter, for *S. minor* (**462**).

The telomorph (perfect) form of these fungi is sometimes visible on the surface of the soil, this is essentially in the case for *S. sclerotiorum*. It takes the form of small 'trumpets', the apothecia (**464**), especially on the largest sclerotia. These apothecia produce ascospores, which lead to aerial contamination.

(See: *S. minor*: **328, 333, 344–347, 369**; *S. sclerotiorum*: **288–290, 329, 334, 348–352, 364, 367**)

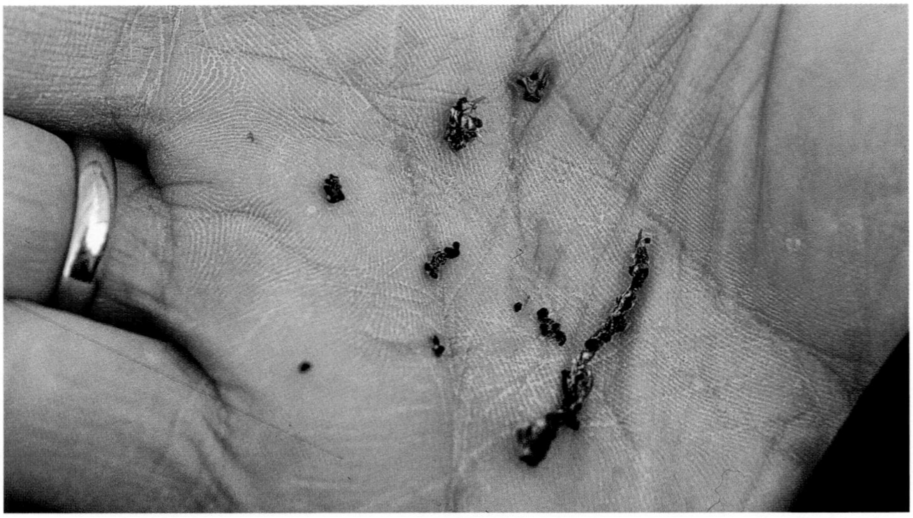

462

463

464

**462** *Sclerotinia minor* forms small black sclerotia which often form a mass, measuring 0.5–2 mm in diameter.

**463** The sclerotia of *Sclerotinia sclerotiorum* are also black but much bigger (2.5–20 mm) and irregular in shape.

**464** The telomorph shape of *S. sclerotiorum* materializes in the production of apothecia on the sclerotia. Their colour is variable (yellowish white, light brown to slightly dark brown).

- **Biology, epidemiology**

**Survival, sources of inoculum:** *S. sclerotiorum* and *S. minor* have considerable saprophytic potential. They can survive in the soil for 8–10 years thanks to the sclerotia produced on affected organs and/or the mycelium present in the plant debris abandoned on plots. In addition, these are very polyphagous fungi which can be found on numerous host plants.

*S. sclerotiorum* is more polyphagous and has been recorded on over 400 different plants, both cultivated and wild. It infects numerous vegetable crops used in rotation with l.s.v., such as bean, tomato, pepper, several types of Cucurbitaceae, and so on.

*S. minor* attacks fewer hosts; nevertheless it has been reported on over 90 plant species. The level of attacks of this fungus is closely correlated with the number of sclerotia present in the soil.

These numerous hosts are capable of multiplying the fungi and acting as sources of inoculum when, after harvesting, they are incorporated in the soil with the sclerotia of these two fungi.

Contamination from *S. minor* takes place essentially through the sclerotial mycelium found close to the lower leaves of l.s.v. These sclerotia need to dry out for a certain time before they can germinate.

Contamination of lettuce varieties by *S. sclerotiorum* may take place in the same way. On the other hand, this fungus forms apothecia on its sclerotia. These organs ensure its sexual reproduction and produce numerous asci containing the ascospores. Thus, millions of ascospores are liberated into the air over a period of 2–3 weeks; they initiate aerial contamination. They can germinate on the leaves only in the presence of free water, from rainfall, sprinkler irrigation, or dew.

**Penetration and invasion**: Whatever the nature of the inoculum (mycelium, ascospores), these two fungi easily penetrate and rapidly invade senescent or dead leaves of l.s.v. They progress towards the healthy tissues, which they rot by means of numerous lytic enzymes. For example *S. sclerotiorum* produces endo- and exopectinases, hemicellulases, and proteases. It also synthesizes oxalic acid which influences both the expression of its pathogenic power and the receptivity of its host.

If ambient humidity allows it, these two *Sclerotinia* spp. produce a white mycelium, varying in density, as well as sclerotia on damaged tissues. *S. minor* has been observed to produce up to 12287 sclerotia per plant, whereas *S. sclerotiorum* only formed 63 under the same conditions. If crop residues are ploughed in and therefore incorporated in soil, 70% is found in the top 8 cm.

**Sporulation and dissemination:** The sclerotia are sometimes able to transmit these fungi to other plots, for example, they can be transported via the soil present on ploughing tools or seedlings. As we pointed out previously, unlike *S. minor* (heterothallic species), *S. sclerotiorum* (homothallic species) generates apothecia more easily, especially if temperatures are not very high, between 8–16°C. The ascospores produced, several billion per apothecia, initiate dissemination of the disease by wind, sometimes over several hundred metres.

**Conditions favourable to its development:** These two *Sclerotinia* spp. are capable of developing at temperatures of 4–30°C. Their thermal optima are located slightly below 20°C. They are encouraged by damp rainy periods and particularly affect lettuce which has reached a stage of advanced development.

Light soils rich in humus are more conducive to the development of *S. sclerotiorum*. This fungus is sensitive to $CO_2$ and it is this property which explains its location in the very top centimetres of soil. Temperature and humidity of the soil also influence the survival of the sclerotia of these fungi. Apothecia also form following rainfall, storms, and irrigation which increase soil humidity.

## Protection*

- **During cultivation**

Preventative control of the two *Sclerotinia* spp. has also posed problems associated with lower efficacy of the cyclic imides, particularly in France when the soil is treated. In fact, premature degradation of iprodione and vinchlozoline has been recorded in certain soils in the Eastern Pyrenees where these fungicides were used repeatedly. This degradation is associated with both the pH and the microbial activity of the soil. Moreover, strains resistant to the benzimidazoles (benomyl, carbendazime, and so on) and to quintozene have been reported in published works.

If you observe symptoms due to these two *Sclerotinia* spp. in your crop, and if a preventative treatment programme has not been put in place, we recommend that you apply one of the **fungicides** approved for use against rotting of the lettuce crown: thirame, iprodione, procymidone, vinchlozoline and its combination with maneb, pyrimethanil, cyprodinil + fludioxonil, and so on. From this point onwards you will need to follow the treatment schedule recommended below, and comply with the cut-off times before harvesting (which vary from one fungicide to another).

In addition, several **prophylactic measures** should be put in place to supplement chemical control. In order to reduce ambient humidity and avoid the presence of free water on plants, shelters must be ventilated as much as possible. Irrigation must be carried out preferably during the morning and in the early afternoon, but never in the evening. This system will allow the plants to dry out as rapidly as possible. In some cases, shelters may need to be heated in order to reduce humidity, particularly by eliminating the dew present on the leaves.

**Plant debris**, especially affected plants on which these two fungi produce numerous sclerotia, must be eliminated very quickly while the crop is growing..

**Nitrogenous fertilization** must also be controlled. It must not be excessive (producing very receptive succulent tissues), or insufficient (sources of chlorotic leaves constitute nutrient bases ideal for the fungi).

**Once the crop has finished growing**, plant debris must be rapidly removed from plots and destroyed in order to avoid it being subsequently buried in soil where the fungi can survive. If no crop is being grown, flooding of infested plots may reduce the number of viable sclerotia present in the soil.

## • Subsequent crop

If **nurseries** are established every year in the same shelter, it is essential to implement the hygiene measures recommended on page 357.

The efficacy of **crop rotation** is rather disappointing. This situation is certainly due to the polyphagous nature of these two *Sclerotinia* spp. Rotations of at least 5 years must be carried out in heavily contaminated soils. It will be extremely advantageous to alternate crops of l.s.v. with cereals. However, we must point out that alternating crops of l.s.v. with broccoli and burying the residues of this plant have succeeded in reducing the number of sclerotia present in soil in the USA and the damage caused by *S. minor*. In France, maize, cereals, onions, and spinach are also believed to discourage the development of these two *Sclerotinia*. Green manure susceptible to these fungi must not be grown either, although certain organic fertilizers reduce the damage caused by *S. sclerotiorum*.

Repeated crops of plants susceptible to one or both *Sclerotinia* spp. on the same plot will inexorably lead to an increase in the inoculum of the soil. In this case, several methods of control can be envisaged as preventative measures for controlling these fungi. The aim of some will be to reduce the level of surface inoculum of the soil. This is particularly so with regard to **methods for disinfecting the soil**. Several fumigants can be used (Methyl bromide**, dazomet, metam-sodium, and so on); in the case of methyl bromide, bromine residues are highly undesirable and its use must be managed with this concern in mind. You should be aware that the efficacy of these fumigants varies and some of them pose problems associated with equipment and have a certain number of disadvantages:

- destruction of natural antagonistic micro-organisms of certain pathogenic agents;
- increased receptivity of disinfected ground to parasites;
- appearance of toxicity phenomena (excess of exchangeable manganese, excess of ammonium after more or less completely halting nitrification, and so on). In the USA metam-sodium is applied to combat *S. minor* by means of spraying or using a localized irrigation system.

In sunny regions, energy from the sun is also commonly used. **Solar disinfecting** of the soil (solarization or pasteurization) is increasingly implemented in numerous countries and is now being used in France. It consists of covering the soil to be disinfected, which must have been very well prepared and humidified in advance, with a polythene film 35–50 μm thick. This is kept in place for at least 1 month, at a very sunny period of the year. This is an effective, economical method, which can control the fungi which colonize the surface of this soil. In addition it can reduce the phenomenon of degradation of cyclic imides noted in certain soils. There has been renewed interest in using steam disinfection of soil to replace methyl bromide which will soon be

* In order to give the proposed control methods a 'universal' nature, we have produced a fairly comprehensive list of the methods and fungicides reported in the various grower countries.
**It is clear that these proposals must be modified according to the country concerned and the pesticide legislation in force.**
**** Methyl bromide will be prohibited after 2005. Its use, which is no longer considered to be good plant health practice, involves risks with residues; abundant rinsing of the soil in water is therefore essential.**

prohibited. Now machines are available which are well adapted to market garden crops; automatic equipment exists if large service areas need to be disinfected.

Several **fungicides** can be applied **to the soil** before planting (quintozene, cyclic imides). We should remind you that iprodione and vinchlozoline can be prematurely degraded in certain soils. This phenomenon, which only begins after repeated use, does not persist if they are not used for at least 3 years.

The **soil** in plots where l.s.v. are to be grown in the future must be **well prepared** and drained in order to avoid the formation of puddles of water conducive to late attacks after head formation and to the formation of apothecia responsible for aerial contamination by *S. sclerotiorum*. **Digging over the soil to a significant depth** can bury the sclerotia deeply and they will be destroyed more rapidly by antagonistic soil-based micro-organisms. In the open field, planting rows must be arranged in the direction of prevailing winds so that the plant crown and plant cover are well ventilated. **Planting** l.s.v. on **sowing blocks** will also encourage ventilation of their crown and will prevent water from stagnating nearby. The use of plastic mulching can partly isolate old leaves from the soil and therefore help to reduce contamination, at least from the two *Sclerotinia* spp and *Rhizoctonia solani*.

**Preventative fungicide treatments** can be applied after planting. The same schedule of treatments can be used as that recommended for *Botrytis cinerea*. Thus, 3–5 applications will be planned just after planting (especially for *S. minor*) and at the 7–9, 11–13, and 16–18 leaf stage, in order to reach the old leaves before the plants cover the soil. Their number will fluctuate according to the period of the year, the type of l.s.v., the type of crop (open field or under shelter), and production zones. The products referred to above can be used.

Other **diseases** and **predators must be controlled** as they lie at the origins of damage and tissue necrosis favourable to the establishment of these two fungi. As soon as plant growth becomes significant, you must be vigilant, especially during cloudy weather and as the period approaches when lettuces form their head.

**Irrigation** must be carried out preferably during the morning and at the beginning of the afternoon, and never in the evening, in order to allow plants to dry out as quickly as possible. It must never be excessive and must never leave surface water permanently on the ground. Spraying carried out close to harvest time must be avoided as conditions (presence of water, greater receptivity of plants, and so on) are very conducive to these mycoses. Experimenters report that drip irrigation associated with light tillage of the soil probably reduces damage due to *S. minor*, compared with furrow irrigation carried out after ploughing the soil in the traditional way.

Although no **varietal resistance** has been found in l.s.v., differences in susceptibility have been observed between cultivars.

Several antagonistic fungi and bacteria have been evaluated *in vitro* or *in vivo* for controlling *Sclerotinia* spp. Among these we can list: *Streptomyces griseoviridis*, *Coniothyrium minutans* (which recently received marketing authorization), *Gliocladium virens*, *Sporodesmium sclerotivorum*, *Talaromyces flavus*, and *Trichoderma harzianum*.

465

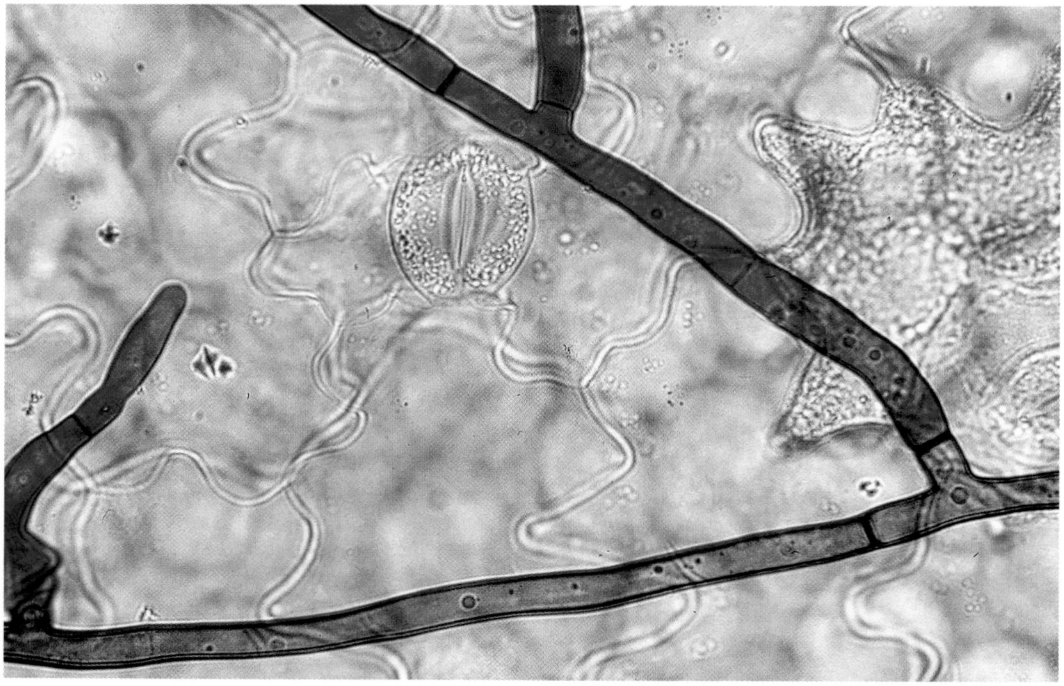

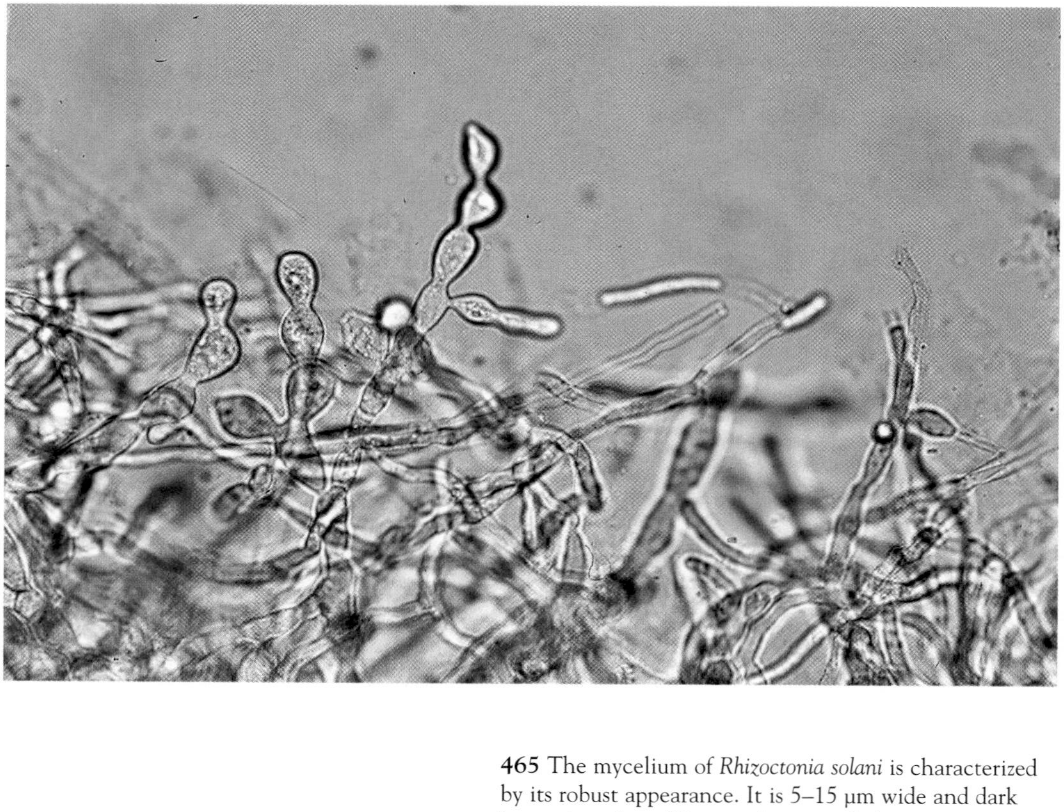

466

**465** The mycelium of *Rhizoctonia solani* is characterized by its robust appearance. It is 5–15 μm wide and dark brown in colour. We can also see a slight constriction in the lateral branchings as well as the presence of partitions.

**466** Keg-shaped formations can be observed on the mycelium; they are believed to be incipient sclerotia.

# *Rhizoctonia solani* Kühn

(anamorph form: Eukaryotes, Fungi, Mitosporic fungi)

# Damping-off
# Bottom rot

# *Thanatephorus cucumeris* (A. B. Frank) Donk

(telomorph form: Eukaryotes, Fungi, Basidiomycetes, Ceratobasidiales, Ceratobasidiceae)

## Principal characteristics

Strains of *R. solani* have been the subject of numerous studies aimed at characterizing them. Among the differentiation criteria studied, their anastomosis affinities allowed several groups (AG) and sub-groups to be established. We still have a limited knowledge of the strains which attack l.s.v. Several strains belonging to groups with different anastomoses are capable of causing damping-off and bottom rot on lettuce. This is the case with strains belonging to the AG-1 (AG-1-1B and AG-1-1C), AG-2-2 and AG 4 groups. The AG-1 group incorporates numerous very polyphagous strains attacking the dicotyledons, but also the leaf sheaths of Gramineae. The AG-2 group, subdivided into two sub-groups, and possibly into three very shortly, comprises strains relatively specialized to the Cruciferae (AG-2-1) or to maize and sugar beet (AG-2-2). Strains belonging to AG-4 have *Thanatephorus praticola* as their telomorph form.

Investigations carried out in France into l.s.v. have enabled us to detect strains belonging to anastomosis groups AG1 and AG5, with the latter being less aggressive.

### • Frequency and economic importance

*R. solani* is found worldwide and has frequently been recorded in all areas where l.s.v. are grown, particularly in the USA and in Europe (Netherlands, England, and so on). It affects lettuce as well as endive. It is capable of infecting numerous (over 100) different hosts. On l.s.v., this soil fungus is known, above all, for causing damping-off in the nursery as well as extremely damaging bottom rot on plants when they approach maturity, both in the open field and under shelter.

In France, it causes some damping-off damage mainly in rather extensive nurseries. However, it is most feared as harvest time approaches. Growers are forced to trim plants whose leaves have rotted, sometimes extensively, reducing the plant's market weight.

It is considered to be a biological marker of 'tired' market garden soils, which have grown several crops of l.s.v. and/or other susceptible vegetable crops.

### • Description of symptoms

*R. solani*, just like the *Pythium* spp., causes mortality of seedlings associated with girdling of the stem at ground level. This appears rather brown and fairly dry in consistency.

This fungus is feared more for the symptoms it causes on adult plants. The first symptoms are visible as harvest-time approaches, on leaves in contact with the soil. They may appear in slightly different ways depending on the type of salad variety. We may see:

- numerous reddish to brown areas of damage, dry in consistency, both on the petioles, the principal vein, and the lamina. They seem to develop more quickly on the lamina where the tissues give the impression of 'melting';
- browner lesions located on the lamina temporarily following the principal vein.

These spots spread rapidly and tissue rotting is established. Its dampness depends on climatic conditions. Initially it affects the lower leaves which may wilt and become yellow. Subsequently it spreads to the leaves of the head and sometimes even the stem. Ultimately the crown may be completely encircled.

On or close to damaged tissue, we can sometimes see unobtrusive whitish to brown filaments moving along the stem, the veins, and from one leaf to another. Poorly defined masses, brown in colour, are sometimes arranged all along the veins.

We should point out that *Botrytis cinerea* and bacteria (*Erwinia carotovora* subsp. *carotovora*, *Pseudomonas marginalis*) may colonize tissues which have already been damaged by *R. solani* and help to accelerate rotting of tissues.

(See **253**, **330**, **353–363**, **365**, **388**, and **395**)

- **Biology, epidemiology**

**Survival, sources of inoculum**: *R. solani* is frequently found in numerous soils which have grown a succession of vegetable crops. It has saprophytic potential which allows it to survive in the soil in the absence of any susceptible hosts. It is found as mycelium and pseudo-sclerotia, often in organic matter and in an extremely wide variety of plant debris which it easily colonizes. It develops easily in the soil, especially if the latter has been disinfected and cleared of potential antagonistic micro-organisms. It is a very polyphagous parasitic fungus which can attack and survive on an extremely wide variety of hosts and on their debris. It may be present in certain substrates and composts, sometimes in certain types of peat or on a few seedlings which have been bought in. Quite often it pollutes equipment which has been used in the nursery without having been disinfected.

**Penetration, invasion**: Contamination takes place via the mycelium present in the soil or deriving from the sclerotia. It colonizes the surface of leaves in contact with the soil. Subsequently it penetrates the lamina directly via the cuticule, via the stomata, or via various areas of damage. Its inter- and intracellular development is often very rapid, especially if it encounters favourable climatic conditions. This parasitic process lies at the origin of damping-off and bottom rot on l.s.v.

**Sporulation, dissemination**: The fungus forms mycelium from the damaged tissues (**465**) and this moves through the tissues and over the soil, spreading to other healthy l.s.v. These sclerotia, mixed with particles of soil dirtying various pieces of equipment, also help to disseminate the fungus. It also has a sexual form which is responsible for its aerial dissemination. This has been reported on other plants and involves the basidiospores. These spores are formed on the basidia present on the surface of a hymenium which is sometimes found on the ground or on the leaves. These spores can be disseminated by the wind and air currents, over quite significant distances. Their role in the epidemiology of *R. solani* on l.s.v. has not, to our knowledge, ever been clearly described.

**Conditions favourable to its development**: *R. solani* can develop in both damp, heavy soils; and lighter, drier soils, at acid or alkaline pH, and at temperatures between 5–36°C. Soils which are too dry or too wet seem to inhibit it. It mainly attacks l.s.v. as harvest time approaches, since at this stage of its development conditions are very favourable. In fact the plants then cover the soil completely and consequently the micro-climate under the plant cover has changed. The lack of ventilation at the foot of the plants leads to an increase in humidity; this then creates a real 'damp chamber'. Free water on the leaves is not essential. Unlike *Botrytis cinerea* and the *Sclerotonia* spp., attacks of *R. solani* on lettuce take place mainly when temperatures are clement, at around 23–27°C, in the presence of humidity. Some authors have reported 'cold' strains which develop at low temperatures. The minimum temperature required for infection is 9°C. At this temperature, the incubation time is 11–15 days whereas at 20°C it only lasts for 3 days. Given the diversity of strains present in the field, it is difficult to specify the optimum conditions for the development of this fungus.

## Protection*

- **During cultivation**

If you are confronted with attacks of *R. solani* in the nursery, we advise you to eliminate diseased plants, as well as those in the nearby area. If you are in any doubt, you can carry out **leaf treatments** using the following fungicides: iprodione, pencicuron, mepronil, and so on. Subsequently, especially after planting out, two to three applications can be carried out, with the final treatment taking place at the 11–13 leaf stage at the very latest.

* In order to give the proposed control methods a 'universal' nature, we have produced a fairly comprehensive list of the methods and fungicides reported in the various grower countries. **It is clear that these proposals must be modified according to the country concerned and the legislation in force on the subject of pesticides.**

In the greenhouse, the climate **must be controlled as much as possible** in order to avoid excessive humidity and temperature. For this reason shelters must be well ventilated. The ground should be well dried after rainfall and sprinkler irrigation. **Balanced fertilization** must be maintained and plant debris and diseased plants must be carefully eliminated once the crop has finished growing.

- **Subsequent crop**

In order to avoid introducing this parasitic fungus into your planting, it is essential to use a healthy substrate and good quality seedlings. Seedlings produced in mini sowing blocks must not be placed on the soil, especially if it has not been disinfected; a plastic film must be used for insulation. If you produce your own seedlings we advise you to apply the measures recommended on page 357.

In contaminated soils, one of the following options can be chosen, depending on the grower's technical and commercial requirements:

- in the case of a crop grown in the open field, you can combine **soil treatment** prior to planting with a fungicide (quintozene, pencicuron, mepronil, iprodione, tetrathiocarbonate, and so on) mulching this product with plastic film. After planting, leaf treatments with some of these fungicides can be carried out in order to complete the efficacy of these measures, using the same methods as those specified above;
- under shelter, an approach comparable to the one described above can be envisaged, excluding perhaps plastic mulching of the soil.

**Soil disinfection** using a fumigant may be considered. The ones traditionally used (methyl bromide**, metam-sodium, dazomet) are effective against *R. solani*. The same is true for steam. You must be particularly vigilant in order to avoid re-infecting soil which has been disinfected.

In growing areas where this is applicable, **solar disinfecting** of the soil (solarization) can be carried out; fairly spectacular results have been recorded, particularly in certain Mediterranean countries. This consists of covering the soil, which has been previously very well prepared and moistened, with a polythene film 35–50 μm thick.

This will be kept in place for at least 1 month during a very sunny period of the year. It is an economical and effective method capable of eliminating this fungus which colonizes the surface of the soil.

Obviously the **agricultural measures** recommended while the crop is growing must be implemented. In addition, damp, heavy soils must be drained, and the soil plowed before planting in order to destroy some of the sclerotia which are destroyed more rapidly in deep soil. Crop rotation can be carried out with cereals, sweet corn, fodder crops, and onions. Lettuce should be planted on sowing bricks. This will prevent lettuce leaves coming in contact with water and will encourage better ventilation of the plant crown. Fertilization must be balanced, and under no circumstances be too low in nitrogen.

There is no **resistant variety**. Some types of l.s.v. with a more erect appearance, such as romaine lettuce for example, appear to be less badly affected.

Experiments have been carried out with **antagonistic** micro-organisms to control *R. solani* (*Trichoderma harzianum*, *Trichoderma koningii*, *Burkholderia cepacia*, and so on). Although sometimes promising, the use of these micro-organisms is not yet sufficiently reliable to be able to recommend them in the field.

** Methyl bromide will be prohibited after 2005. Its use, which is no longer considered to be good plant health practice in Europe, involves risks of residues; abundant rinsing of the soil in water is therefore essential

# Fungi which mainly attack the roots, crown, and stem

## Various types of Pythiaceae (Pythium spp., Phytophthora spp.)

## Damping-off, Pythium diseases

(Prokaryotes, Pseudo-fungi, Chromists*, Oomycota, Oomycetes, Pythiales, Pythiaceae)

## Principal characteristics

Numerous Pythium spp. and some *Phytophthora* spp. (***Phytophthora cryptogea*** **Pethybr. & Lafferty,** ***Phytophthora porri*** **Foster)** are likely to attack, or have been recorded on, l.s.v. The types of *Pythium* are listed in *Table 27*.

**Table 27: Some characteristics of *Pythium* spp. encountered on l.s.v.**

| **Species of *Pythium*** | **Nature of principal fruiting bodies** | |
|---|---|---|
| | **Sporangia** | **Oogones** |
| *Pythium megalacanthum* de Bary Globular | Globular | Sculpted |
| *Pythium irregulare* Buisman (syn. *P. debaryanum* Hesse) | Globular | Poorly sculpted |
| *Pythium polymastum* Drechsler | Globular | Sculpted |
| *Pythium uncinulatum* Plaats-Niterink & Blok | Globular | Sculpted |
| *Pythium spinosum* Sawada | Absent | Sculpted |
| *Pythium dissotocum* Drechsler | Filamentous | Smooth |
| *Pythium aphanidermatum* (Edson) Fitzp. | Filamentous | Smooth |
| *Pythium catenulatum* Matthews | Filamentous and swollen | Smooth |
| *Pythium myriotylum* Drechsler | Filamentous and swollen | Smooth |
| *Pythium tracheiphilum* Matta | Filamentous and swollen | Smooth |
| *Pythium rostratum* Butler | Globular | Smooth |
| *Pythium ultimum* Trow | Globular | Smooth |
| *Pythium sylvaticum* Campbell & Hendrix | Absent | Smooth |
| *Pythium violae* Chesters & Hickman | Absent | Smooth |

We have covered all the Pythiceae (excluding *Bremia lactucae* and *Pythium tracheiphilum*) in the same fact file for at least three reasons:

- they cause comparable symptoms;
- they are often influenced by the same conditions;
- the methods implemented to control them are identical.

* For many years the oomycetes which contain almost 550 species, have been classified as Fungi. Their ultrastructure, biochemistry, and molecular sequences indicate that they belong to the Chromists which includes, in particular, algae (green and brown), diatoms, and so on.

- **Frequency and extent of damage**

Pythiaceae are present worldwide, attacking in all areas where lettuce is grown. They cause damage in nurseries, which sometimes means that sowing has to be repeated. Adult plants, particularly their roots and basal leaves, may also be attacked. As crops are increasingly grown in soil-less culture, these aquatic fungi have found a perfect breeding ground; in some market gardens they constitute a production-limiting factor.

The species most damaging to crops is certainly *Pythium tracheiphilum*, responsible for lettuce dwarfism. We have associated it with the vascular fungi (see fact file 11).

Rare types of *Phytophthora* attack lettuce. *Phytophthora porri* has been recorded in the soil in southern Australia and *Phytophthora cryptogea* in crops being grown in soil-less culture in California and in hydroponic crops grown in Belgium.

- **Principal symptoms**

In addition to causing seed rotting (failed emergence), *Pythium* spp., just like *Rhizoctonia solani*, are responsible for the soft, damp damage which starts on seedling stems on the soil surface. In this case, the crown looks as though it has been pinched. Affected tissues gradually become brown. Sometimes the roots are affected; they become light brown in colour and their cortex breaks up to varying extents. Seedlings soon collapse (*Pythium aphanidermatum*, *Pythium ultimum*, *Pythium irregulare*, *Pythium dissotocum*, *Pythium uncinulatum*, *Pythium violae* and so on) (**Damping-off**).

Some species cause root browning which is sometimes located at root extremities. We also see feeder roots disappearing and rotting of the root cortex and taproot. These symptoms sometimes appear in the soil, but more frequently in soil-less culture. They are accompanied by leaves becoming yellow and wilting, which may be reversible. In some cases, lettuces grow poorly and their size is reduced (*Pythium aphanidermatum*, *Pythium uncinulatum*, *Pythium myriotylum*, *Pythium irregulare*, *Pythium dissotocum*, *Pythium polymastum*, and so on) (**Pythium root rot** and **Wilt**).

Some *Pythium* spp. are responsible for leaf damage. Fairly extensive damp brown spots and bottom rot are the signs of the effects of these fungi (*Pythium aphanidermatum*, *Pythium uncinulatum*, and so on) (**Pythium bottom rot** and **Leaf blight**).

Whatever the symptoms noted, if we use a light microscope we are often able to see the presence of oospores and/or chlamydospores in the damaged tissues.

*Phytophthora cryptogea* causes browning and root rot on lettuce grown in soil-less culture. *Phytophthora* mainly attacks the stem causing hard, dark rotting starting in the soil.

(See **374**, **375**, **387**, and **394**)

- **Biology, epidemiology**

**Survival, sources of inoculum:** *Pythium* spp. are capable of living in a saprophytic state at the expense of the organic matter. Their low parasitic specificity allows them to attack a large number of hosts, which also ensures their multiplication and survival. *Phytophthora cryptogea* attacks numerous vegetable crops such as cucumber, tomato, cabbage, celery, asparagus, pepper, and so on. *Phytophthora porri* has a more limited spectrum of hosts, restricted to the alliums, white cabbage and carrots. It can survive in the soil thanks to structures of resistance, such as oospores, chlamydospores and, to a lesser degree, the sporangia (**367–370**). In certain species, the oospores can survive for 2–12 years. The survival forms of these fungi are stimulated by the exudates emitted by the seeds or the roots. They produce either a germination tube or zoospores.

**Penetration, invasion:** These fungi directly penetrate the epidermal tissues or pass through damaged areas. They rapidly invade the tissues, thanks to the joint action of various pectinolytic and cellulytic enzymes, moving between and into the cells. Sporangia and oospores form inside the tissues or on their surface.

**Dissemination:** Pythiaceae are perfectly well adapted to living in the aqueous phase of soil and in the nutrient solution of crops grown in soil-less culture. Water is largely responsible for their dissemination, thanks to their numerous flagellate zoospores produced by the sporangia. It is also favoured by certain substrates and by seedlings. In nurseries where there is a high density of seedlings, *Pythium* spp. is transmitted from plant to plant as it progresses within the soil of the mycelium. Aerial dissemination is sometimes possible if splashing occurs during sprinkler irrigation or heavy rainfall.

**Conditions favourable to their development:** It is worth pointing out that not all Pythiaceae have the same pathogenic power. In addition they require special conditions in order to infect seedlings:

- the presence of water is almost always essential. High soil humidity, and reduced gaseous exchanges, constitute an ecological advantage for these fungi, at the expense of other fungi and micro-organisms which sometimes compete for the organic material in the soil;
- temperature influences the behaviour of these fungi in different ways. There are species which appreciate cold soil, at temperatures of around 15°C, such as *Pythium ultimum* (temperatures: optimum = 15–20°C, minimum 2°C, maximum 42°C), others have higher thermal optima. This is particularly the case with *Pythium aphanidermatum* (temperatures: optimum = 26–30°C, minimum 5°C, maximum 41°C), which are found mainly in hot tropical zones and in crops grown out of the soil;
- host receptiveness is not constant throughout its life. Young seedlings and succulent tissues are very susceptible. Subsequently adult plants may become susceptible, essentially when they suffer various types of climatic or growth stress. It is mainly the nutrient rootlets which are affected. This is sometimes the case in crops grown in soil-less culture.

Certain interactions, unfavourable to l.s.v., between *Pythium* spp. and phytophagous nematodes (*Meloidogyne hapla*) have been observed.

## Protection*

### • During cultivation

**In the nursery**, irrigation must be limited if it has been excessive and ventilation of shelters must be encouraged as much as possible. It is also essential to saturate the whole substrate with an anti-Pythiaceae fungicide solution. Diseased seedlings and those nearby must be eliminated (also see page 357).

**While the crop is growing** and following attacks on the roots, chemical applications may differ depending on the context:

- in the soil, generally only a few plants are affected and no treatment is needed;
- in soil-less culture, fungicides may be added to the nutrient solution. Lower doses must be selected in order to avoid chemical injury.

Leaf attacks are normally well controlled by means of anti-mildew leaf treatments.

Several **fungicides** are currently used worldwide in order to combat Pythiaceae: metalaxyl, furalaxyl, propamocarb HCl, etridiazole. It is advantageous to alternate products from different chemical families with different modes of action. In fact, there is a risk of selecting strains resistant to some of these fungicides. Strains of *Pythium* spp. (in particular *Pythium aphanidermatum*) tolerant to metalaxyl and furalaxyl, have already been reported.

You also need to control the **climate within the greenhouse** in order to avoid excessive humidity. To achieve this, shelters must be well ventilated. The soil must be well drained and dried following sprinkler irrigation and rainfall in the open field. Balanced **fertilization** must be maintained and you must avoid stressing the plants. Diseased plants and plant debris must be carefully eliminated, particularly the root systems.

### • Subsequent crop

It is advisable to use a healthy or disinfected substrate in order to produce seedlings. In countries where growers produce this substrate themselves (for example from sand, recovered soil, or various composts) there are risks of contamination. The same is true for growers who mix the substrate they have bought with the ingredients referred to above. Sowing blocks must not be placed on the soil, especially if it has not been disinfected. It would be preferable to place them on a plastic film or on shelves. In contaminated soil which has not been disinfected, as a preventative measure the seed bed could be soaked in an anti-Pythiaceae fungicide solution. We recommend incorporating propamocarb HCl into the soil or using it to soak the sowing blocks. This option can involve cryptonol. The **hygiene measures** recommended for the nursery on page 357 can be implemented.

In certain particularly badly affected soils, disinfecting with a **fumigant** must be undertaken. Those which are traditionally used (Methyl bromide**, metam-sodium, dazomet) are effective against *Pythium spp*. The same is true for **steam**.

* In order to give the proposed control methods a 'universal' nature, we have produced a fairly comprehensive list of the methods and fungicides reported in the various grower countries. **It is clear that these proposals must be modified according to the country concerned and the pesticides legislation in force.**

**** Methyl bromide will be prohibited after 2005. Its use, which is no longer considered to be good plant health practice, involves risks with residues; abundant rinsing of the soil in water is therefore essential.**

467

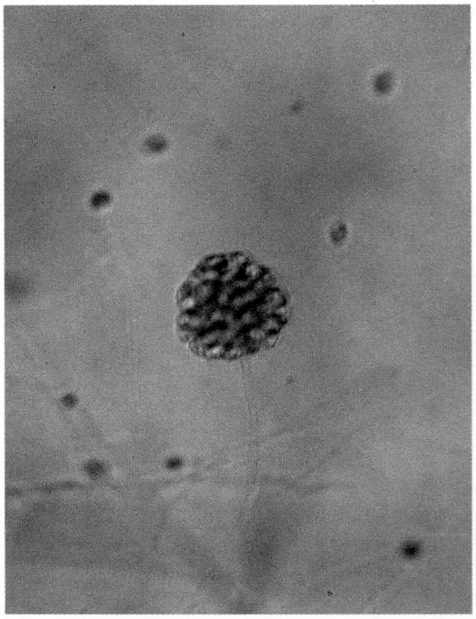

468

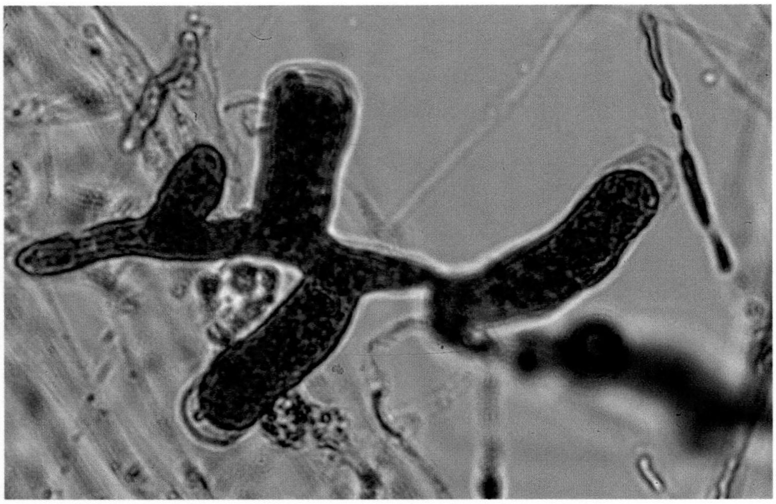

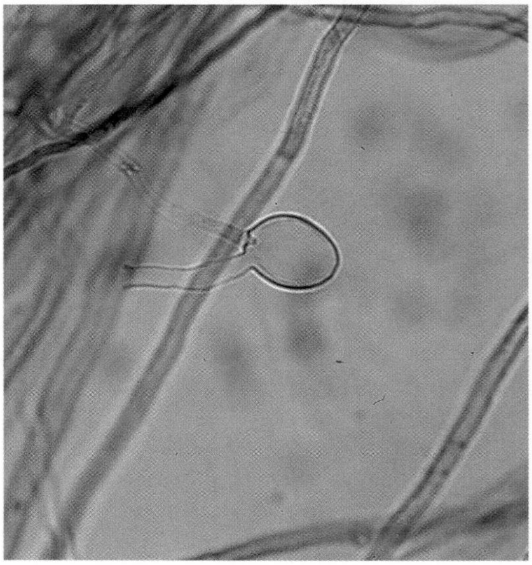

469

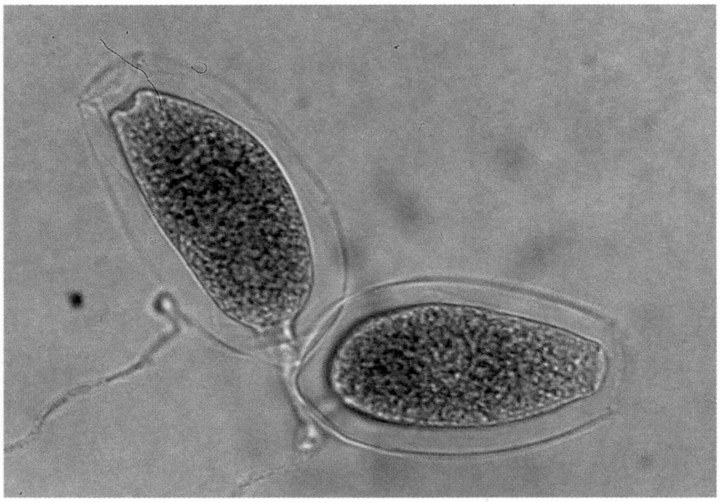

470

## Principal structures observed among the Pythiaceae

**467** Filamentous sporangia (*Pythium* sp.).

**468** Swollen filamentous sporangia (*Pythium aphanidernatum*).

**469** Globular sporangia (*Pythium* sp.).

**470** Sporangia of *Phytophthora crypotegea*.

**471** Oogones (O) fertilized by an antheridia (A), intercalary hyphal swelling (HS) (*Pythium* sp.).

**472** Sculpted oogones (*Pythium* sp.).

**473** Appressorium of *Pythium myriotylum*.

In production zones where this is practicable, **solar disinfecting** of the soil (solarization) can be carried out. Fairly spectacular results have been recorded, particularly in certain Mediterranean countries. It consists of covering the soil to be disinfected, which must have been very well prepared and humidified in advance, with a polythene film 35–50 μm thick. This is kept in place for at least one month, at a very sunny period of the year. It is an effective, economical, broad spectrum method, capable of controlling Pythiaceae.

Heavy damp **soils** must be drained. **Crop rotation** must be carried out using cereals and fodder crops. L.s.v. must be planted on raised beds. This will prevent water being retained at the base of the plants. Fertilization must be balanced.

In some countries seeds are coated in an anti-Pythiaceae fungicide.

When planting takes place, seedlings must not be placed in soil which is too wet or too cold; irrigation carried out at this stage of the crop must not be excessive.

In crops grown in soil-less culture, you must be vigilant about checking the **healthy quality of irrigation** water coming from channels or pools which may have become contaminated. Disinfecting the nutrient solution could be envisaged. Low pressure UV units have proved to be effective against *Pythium* spp. *Olpidium brassicae* (see fact file 14) in various countries, particularly Belgium, reducing the level of damage caused by these micro-organisms. Other methods are currently being implemented in order to treat nutrient solutions recycled from numerous crops: biofiltration, chloration, and so on. Their spectrum of efficacy is broad and incorporates several fungi, particularly the Pythiaceae.

There are no **resistant varieties.**

Experiments have been conducted with some **antagonistic** micro-organisms in order to control some species of *Pythium* (*Coniothyrium minutans*, *Trichoderma harizianum*, *Trichoderma koningii*, *Gliocladium virens*, *Bacillus subtilis*, *Enterobacter cloacae*, *Burkholderia cepacia*, and so on). Their use is already effective in certain countries.

471

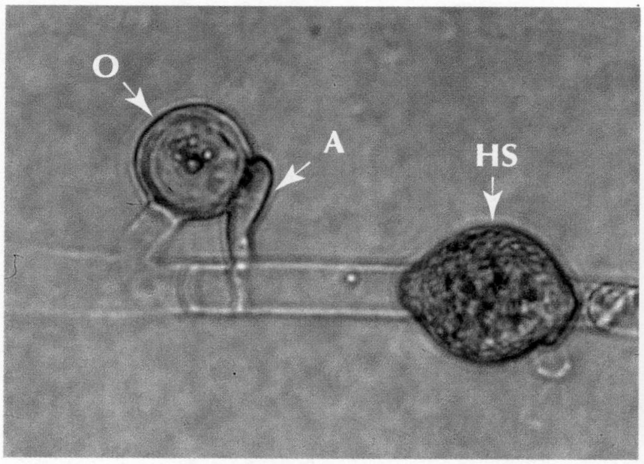


473

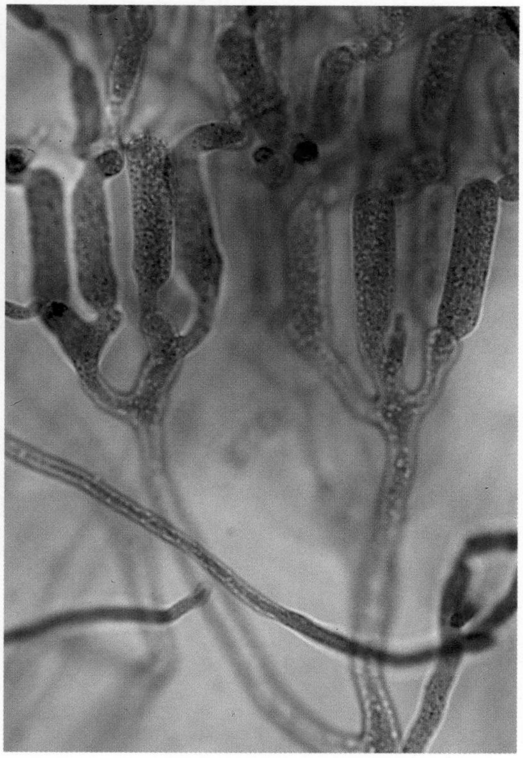

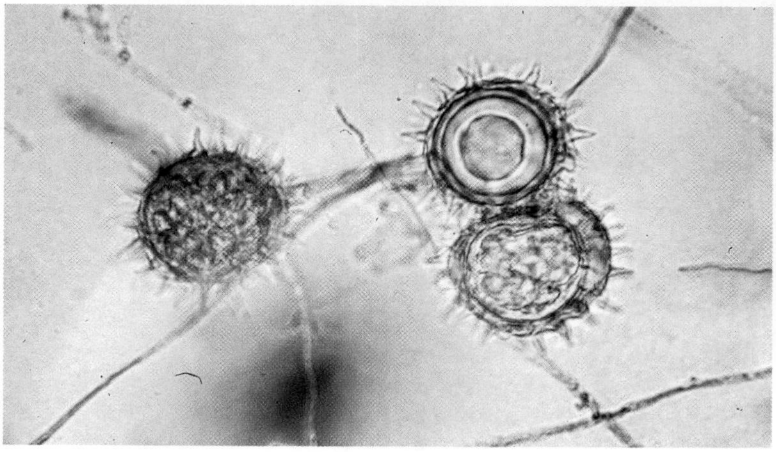

472

474

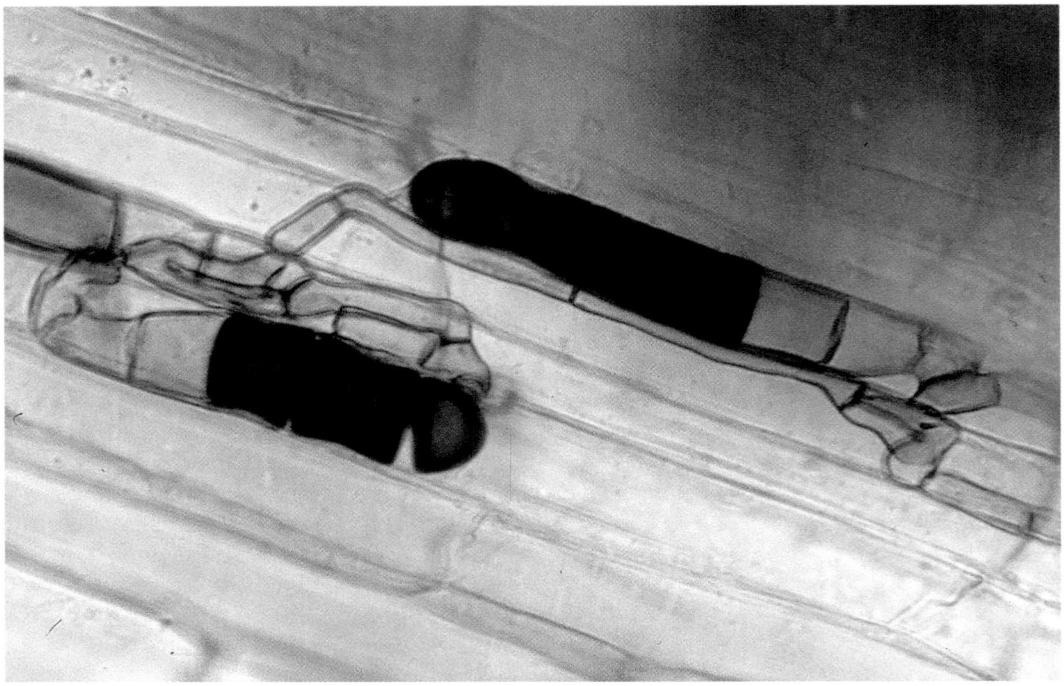

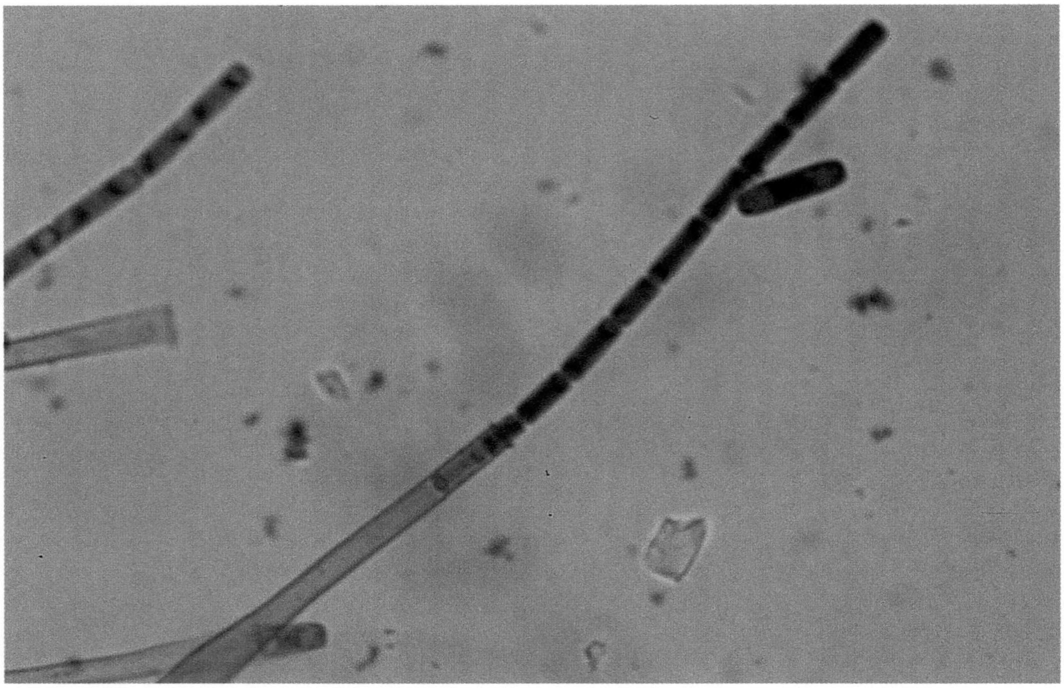

475

**474** In the root cortex cells, we can distinguish chlamydospores in the shape of stacked-up kegs (25–60 × 10–12 μm) progressively becoming black (*Thielaviopsis basicola*).

**475** A very elongated phialid produces endoconidia (10–23 × 3–5 μm) which are cylindrical and hyalin (*Chalara elegans*).

# *Thielaviopsis basicola* (Berk & Broome) Ferraris

# Black root rot

(Chlamydospore form: Eukaryotes, Fungi, Mitosporic fungi)

# *Chalara elegans* Nag. Raj. & Kendrick

(Conidian form: Eukaryotes, Fungi, Mitosporic fungi)

## Principal characteristics

### • Frequency and extent of damage

This fungus, present in numerous countries worldwide, mainly attacks chicory and has been recorded on very rare occasions on lettuce, essentially in the USA and Australia. In the USA, it has not been considered as a fungus which is pathogenic to lettuce. In Australia it was recorded in the early 1990s in Queensland. In this state it also attacks tobacco, *Lathyrus odoratus* and *Lupinus angustifolius*. Several hectares of lettuce had to be turned under because of poor plant growth.

In France, we have only seen damage from *Chalara elegans* once, on endive (frisée) grown in the Eastern Pyrenees during the autumn of 2000.

### • Principal symptoms

This soil fungus may be responsible for significant damage on the underground part of lettuce. It progressively colonizes the roots, initially causing unobtrusive, minor damage, light brown in colour. Subsequently these spots extend, becoming dark brown to black over their entire length. A large number of the nutritive rootlets are destroyed. Ultimately a significant proportion of the root system presents fairly characteristic black rot. The large roots, slightly less susceptible, have brown to black corky bands on their surface.

Arthroconidia (chlamydospores), typical of *T. basicola*, can be seen in the root tissues; they allow the fungus to be identified with certainty (**474**).

Affected lettuce is often poorly developed and chlorotic. It sometimes wilts on hot days.

(See **13, 389–393,** and **396**)

### • Biology, epidemiology

**Survival, sources of inoculum:** *T. basicola* can survive for a very long time in the soil, thanks to its chlamydospores. It is capable of colonizing the organic matter and infecting numerous host plants, both cultivated and uncultivated, which help to multiply and preserve it. In fact, over 120 species belonging to at least 15 botanical families have been recorded. Among vegetable crops, we can list bean, pea, cucumber, melon, watermelon, carrot, eggplant, tomato, and so on Not all these hosts show the same degree of susceptibility to this fungus, and allow it to multiply to different extents. It is well known that strains of *T. basicola* have different levels of pathogenicity.

In nurseries, dust from contaminated soil constitutes a significant source of inoculum. It is also conserved on the equipment used for producing seedlings. In Australia it has been found, prior to sowing, in certain peat-based substrates.

**Penetration and invasion:** The chlamydospores, and, to a lesser degree, the endoconidia, germinate close to the roots and penetrate them either directly across the epidermis, or through areas of damage. The fungus rapidly colonizes the tissues in the cortex and vessels which it then rots. It produces numerous chlamydospores in the damaged tissues. It also forms these on the root surface, alongside a multitude of endoconidia (**475**).

**Dissemination:** the chlamydospores and endoconidia are easily disseminated by water and soil dust. It is likely that the soil, present on tools used to work the soil, helps to propagate them. The same is true for contaminated lettuce seedlings.

**Conditions favourable to its development:** *T. basicola* is known above all for having serious effects on various plants when growing conditions are difficult. This is the case during cold, wet springs. Under these conditions, the root development of plants is reduced, or even halted, and the fungus takes advantage of this, causing a serious attack. It therefore appreciates wet and especially cold soil. Its thermal optimum is normally located at around 17–23°C. Australian strains that have been recorded probably have a higher thermal optimum, of around 23–26°C, as a result of having to adapt to the subtropical conditions of the affected region. Soil pH influences the behaviour of this fungus; at an acid pH (at around 5.6) it is normally less active. Adding calcium to the soil can increase its development.

## Protection*

### • During cultivation

Unfortunately, there is no way of 'reestablishing' plants which have suffered attacks from *T. basicola* in the nursery. Given the risk of these plants spreading the disease, we suggest that you eliminate them rapidly. This warning is particularly important if you grow seedlings.

The use of **fungicides** from the Benzimidazole family (benomyl, thiophanate-methyl) or certain inhibitors of sterol biosynthesis (propiconazole, and so on) have a certain degree of efficacy against this fungus. They are used in advance to soak the substrate or soil carrying the lettuce seedlings. Strains of *T. basicola* resistant to benomyl have been recorded in published works.

The situation is virtually identical when the crop is growing. In order to avoid enriching the soil with plant debris and chlamydospores, it is essential for the **root systems** of l.s.v. to be lifted from the plot and destroyed.

### • The subsequent crop

In the nursery it is important to use healthy substrates (see page 357). Seedlings produced in mini sowing blocks should not be placed on the soil, especially if the latter has not been disinfected; a plastic film must be used to insulate them. Preventative treatments are sometimes carried out in the nursery using the products referred to above. We also advise you to apply the methods of hygiene recommended in the chapter on *Sclerotinia* spp. (fact file 6).

**Crop rotation** must be implemented as an essential requirement if an increase in the inoculum level of the soil is to be avoided. To be effective rotation periods must be sufficiently long, lasting for at least 4–5 years, and must not involve susceptible crops.

*T. basicola* has been recorded on numerous plants: the vegetable crops referred to above, major crops, and industrial production crops (cotton, peanut, soya, alfalfa, lupin), and various ornamental plants (chrysanthemum, pelargonium, and so on). Strains of *T. basicola* isolated from l.s.v. seem to all appearances polyphagous, but we still do not have any specific knowledge about their range of hosts. The species referred to only provide an indication. Cereals are a good choice as crops to precede lettuce, as is rice and sorghum.

**Soil disinfection** could be envisaged under shelters using a fumigant (methyl bromide**, dazomet, and so on).

Tools used to work the soil in contaminated plots must be carefully cleaned before being used in healthy plots. The same is true for tractor wheels. Careful rinsing in water and disinfecting of this equipment is often sufficient to clean them of soil and *T. basicola*.

**Draining of plots, fertilization,** and irrigation must be properly controlled; soil pH must be kept at around 6.0 and you must not add too much lime. You should be vigilant about certain organic materials appearing in the soil.

**Differences in sensitivity** between lettuce cultivars have been observed in Australia (Monaro, Kirrali, Centenary are not likely to be very susceptible).

* In order to give the proposed control methods a 'universal' nature, we have produced a fairly comprehensive list of the methods and fungicides indicated in the various grower countries. **It is clear that these proposals must be modified according to the country concerned and the pesticide legislation in force.**

** **Methyl bromide will be prohibited after 2005. Its use, which is no longer considered to be good plant health practice in Europe, involves risks with residues; abundant rinsing of the soil in water is therefore essential.**

# Other principal fungi acting as parasites on the roots and/or crown

| | ***Athelia rolfsii* (Curzi) Tu & Kimbrough** (Telomorph : Eukaryotes, Fungi, Basidiomycetes, Stereales, Atheliaceae)<br>Sclerotium rolfsii Sacc. **(Anamorph form Eucaryotes, Fungi, Mitosporic fungi)** (Sclerotium stem rot, Southern blight) | ***Pyrenochaeta lycopersici* Schneider & Gerlach** (Eukaryotes, Fungi, Mitosporic fungi)<br>**(Cryptogamic corky root)**<br>**('Fungal corky root')** |
|---|---|---|
| **Distribution and damage** | This Basidiomycete attacks mainly in hot tropical and subtropical regions. It affects an extremely large number of plants, both cultivated and uncultivated, over 100 mono- and dicotyledons. It is rarely very serious to lettuce.<br>In France it is present in the soil of some market gardens located in the Basque country. It is particularly serious in the overseas territories. | This fungus, responsible for tomato corky root, also colonizes lettuce roots. It has only been recorded on the latter on very rare occasions mainly in England and France. Its effect on lettuce is only serious now and again, causing reduced vigour of plants with consequent reduction in lettuce size at harvest-time. |
| **Symptoms** | Damp lesions develop on the leaves which are in contact with the soil. Rotting sets up and progressively takes over the crown, certain roots, the base of petioles; it spreads to the head. Affected tissues become brown, to varying degrees of intensity. We also see wilting and yellowing of leaves as well as collapse of plants followed by their progressive decomposition. Affected plants are distributed in sources of varying sizes. A white mycelium, sometimes abundant, covers the base of the plants and the soil. It is sometimes surmounted by numerous small smooth sclerotia, fairly spherical (1–3 mm), of a beige to reddish-brown colour (See **370–373**). | *Pyrenochaeta lycopersici* causes brown damage with a corky surface surrounding the roots in the form of bands. This damage can be found on the taproot and secondary root. A number of rootlets disappear. Although it can be confused with symptoms associated with corky root, damage due to this fungus is never extensive. However, lettuce displays limited development and looks rather stunted. |
| **Biological information** | It survives in the soil on plant debris in the form of a mycelium mass or sclerotia. It is also capable of surviving on a range of organic substrates. It infects a large number of hosts, particularly vegetables: bean, beetroot, cabbage, members of the Solanaceae, Cucurbitaceae, and so on. It penetrates the tissues directly, invading and decomposing them using its powerful enzymatic system. Hot wet periods encourage its spread. It develops successfully between 25–35°C. It is transmitted by contaminated soil, working the soil, water, and plants produced in infested nurseries. | It survives in the soil on plant debris. Several cultivated hosts used in rotation with lettuce are capable of harbouring and multiplying it; this is the case with tomato, melon, eggplant, bean, and so on. The same is true for several types of weed. It penetrates the root tissues then progressively causes them to become corky and rot. Its dissemination basically takes place via seedling and/or contaminated substrates, ploughing tools, and so on. There seem to be at least 2 types of strain: 'cold' strains with a thermal optimum of 15–20°C and hot strains which can still be pathogenic at 30°C. Growing susceptible crops in the same plot encourages it to spread to the soil. |
| **Protection*** | A number of control methods recommended for combating *Rhizoctonia solani* or *Sclerotinia* spp. can be used, both in the nursery and in the crop (destruction of crop residues and weeds, disinfecting the soil, 'solarization', crop rotation with maize and other cereals). Some fungicides have been experimented with; the cost of treatments makes them impractical in several situations. | This soil-based fungus requires the implementation of most control methods recommended for combating *Sclerotinia* spp. Implement crop rotation excluding the susceptible plant indicated above. In some cases the soil must be disinfected – the best option being solarization. Eliminate as many roots as possible once the crop has finished growing – to reduce the amount of contaminated plant debris returning to the soil. Differences in susceptibility between varieties have been observed. |

476

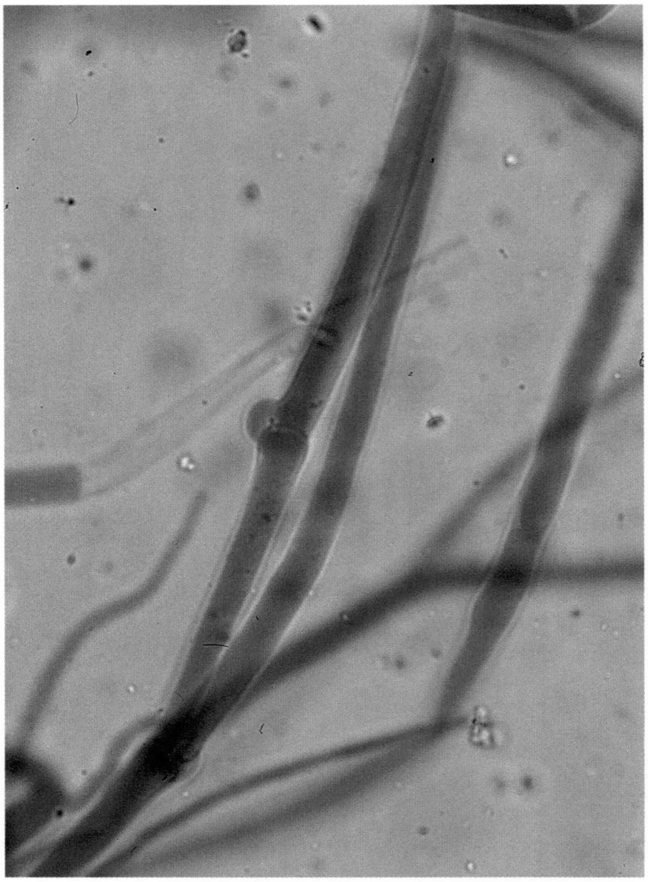

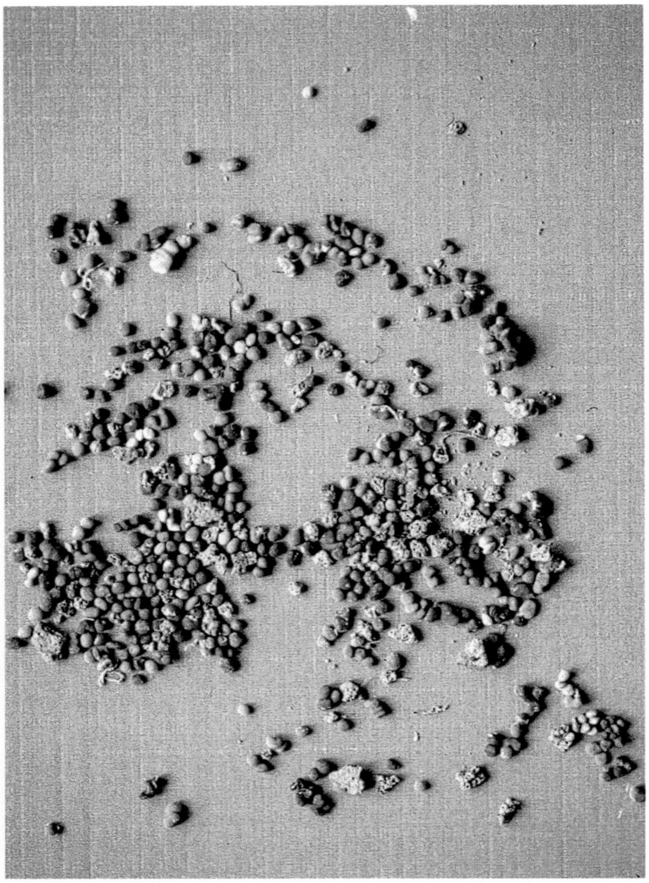

477

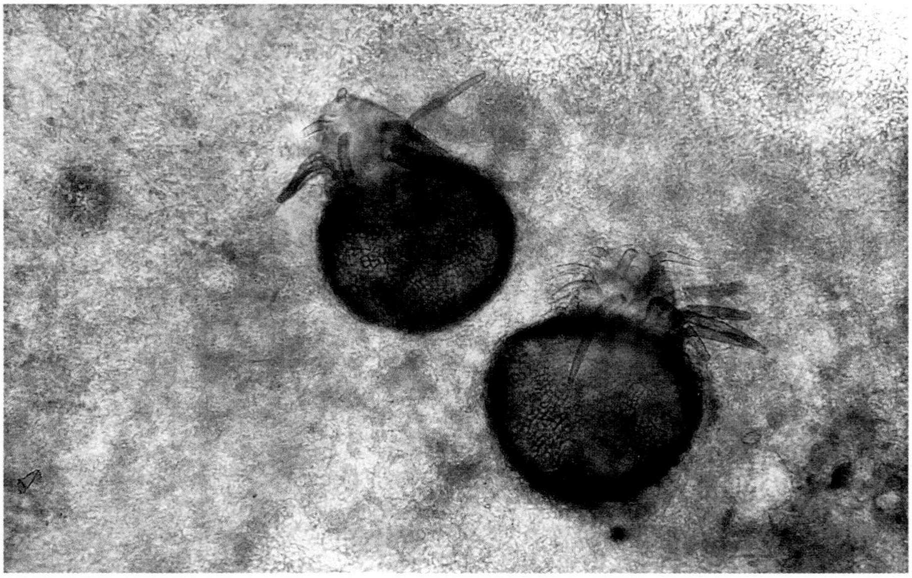

478

**476** Like the majority of Basidiomycetes, the mycelium of *Sclerotium rolfsii* displays the anastomosis loops characteristic of this group of fungi.

**477** It also produces sclerotia, white at first then becoming brown, comparable to radish seeds (1–3 mm in diameter).

**478** Pycnidia of *Pyrenochaeta lycopersici* have brown bristles (setae) irregular in size. They produce ellipsoid conidia (4–6 × 1–1.5 μm).

| | ***Phymatotrichopsis omnivora*** **(Duggar) Henneb. = *Phymatotrichum omnivorum* Duggar**<br><br>**('Texas root rot')** | ***Plasmopara lactucae-radicis*** **Stang. &Gilbn.**<br>(Prokaryotes, Pseudo-fungi, Chromista, Oomycota, Oomycetes, Peronosporales, Peronosporaceae)<br><br>**('Root downy mildew')** |
|---|---|---|
| **Distribution and damage** | This fungus is particularly predominant in semi-desert regions. It has been detected fairly recently in avocado and mango orchards in semi-tropical zones. It is very polyphagous, affecting almost 2,000 cultivated dicotyledons (cotton, alfalfa, vines, apple trees, and so on) as well as uncultivated varieties. It affects lettuce only very rarely; it has mainly been recorded in the southern USA where it is considered to be a minor pathogenic agent of lettuce. | This fungus, a systemic pathogenic agent of lettuce, is fairly rare; it has mainly been observed on the roots of lettuce grown in soil-less culture (NFT – Nutrient Film Technique System) in Virginia. It constitutes one of the rare types of mildew, with that of the sunflower, which can sporulate on the roots of plants. |
| **Symptoms** | Rotting of the root system can be seen associated with its colonization by the mycelium of this fungus. It disturbs sap movements; leaves quickly become yellow and become tinged with bronze. Plants suddenly wilt and die. Downy mycelium and mycelium strings are present on the roots, in the surrounding soil, and close to the soil surface. We also see small isolated or massed sclerotia, fairly spherical in shape, 1–2 mm in diameter, reddish-brown once they have matured. | This ground mildew causes necrosis on the roots where it forms numerous sporangiophores which can carry sporangia to the extremity of the sterigmata. The zoospores are liberated by an apical pore. It seems incapable of sporulating on lettuce leaves. Affected l.s.v. are reduced in size. |
| **Biological information** | *Phymatotrichopsis omnivora* can survive for several years in the soil thanks to its sclerotia which can be found down to 2 m in depth. It can also survive on numerous other cultivated plants (vines, cotton, alfalfa, apple trees, and so on) and uncultivated plants. Initially it sets up in the root cortex then spreads to the vessels. It appreciates chalky, alkaline soils cont-aining a low content of organic matter. Its thermal optimum is 28–30°C. | Its method of survival is not known. The oospores present in the roots are probably able to contribute to this. Once they have formed cysts, the zoospores germinate and penetrate the roots directly. The fungus possesses intercellular development. Subsequently, sporangiophores are produced on the surface of the roots. The zoospores liberated help to disseminate the disease. Aplerotic oospores are also formed in the tissues. This fungus seems extremely well adapted to NFT lettuce growing conditions. Its thermal optimum is 22–28°C. It does not seem to cause damage at temperatures below 18°C. |
| **Protection*** | This soil fungus does not normally require the implementation of special control methods as far as growing lettuce is concerned. The level of organic matter in the soil can always be increased and crop rotation can be carried out using monocotyledons. In addition you can base your treatment on the measures recommended for controlling the 2 *Sclerotinia* spp. | Control methods that can be used to combat it are identical to those recommended for the Pythiaceae (see fact file 8). You should be aware that certain cultivars are resistant to this new root mildew of lettuce. |

* In order to give the proposed control methods a 'universal' nature, we have produced a fairly comprehensive list of the methods and fungicides reported in the various grower countries.

**It is clear that these proposals must be modified according to the country concerned and the pesticide legislation in force.**

479

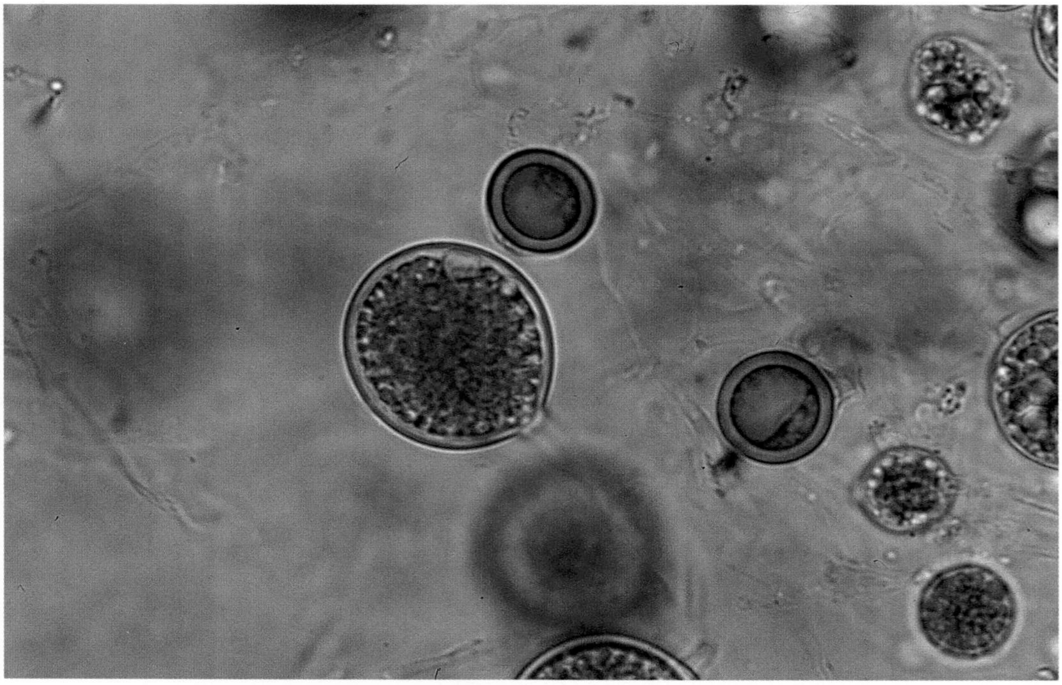

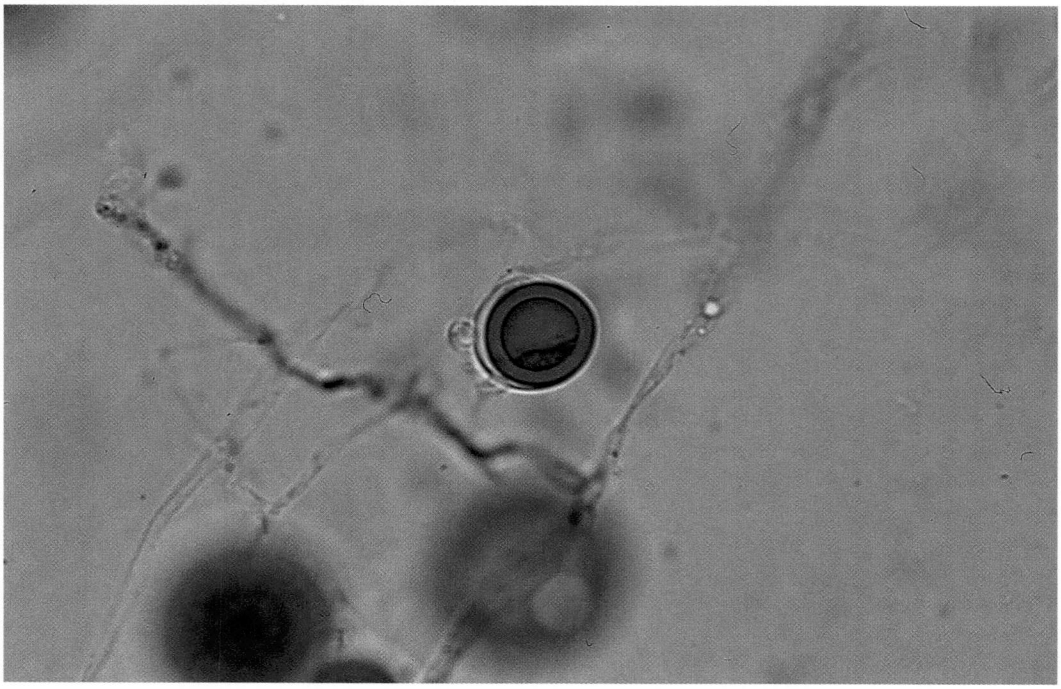

480

**479** *Pythium tracheiphilum* produces globular to subglobular sporangia, intercalary or terminal, whose diameter varies between 22–34 μm.

**480** - Its spherical, smooth-walled oogones are often fertilized by 1–2 antheridia, which are mostly monocline (14–17 μm). They produce thick-walled plerotic oospores.

# Vascular fungi

## *Pythium tracheiphilum* Matta

(Pseudo-fungi, Chromista*, Oomycota, Oomycetes, Pythiales, Pythiaceae)

## Vascular *Pythium* ('Pythium wilt and leaf blight', 'Lettuce stunt')

## Principal characteristics

### • Frequency and extent of damage

Nothing was known about the origins of the damage caused by vascular *Pythium* for many years, undoubtedly because of the difficulty in isolating this fungus, which is often masked by other micro-organisms. Initially it was described mainly in Europe (Italy, France, Netherlands, Germany, Sweden, and so on) and then in the USA. It may lie at the origins of major damage such as that reported in Italy and the USA in particular.

In France it has serious effects on certain plots which have grown repeated crops of lettuce. Many growers now dread it.

Strains attacking in the USA are probably different from those in Italy or the Netherlands.

### • Principal symptoms

The growth of affected lettuce is reduced; if attacks are very early, plant growth may be completely stopped and may remain dwarfed. Their reduced size contrasts with that of apparently healthy surrounding plants. The leaves of certain lettuces may become yellow and wilt. Initially, this wilting occurs at the hottest times of the day and plants recover at night. Subsequently this sometimes becomes irreversible, leading to the drying out and death of some plants.

The taproot can sometimes be deformed, rough on the surface, and greyish in colour; there are not many secondary roots.

A longitudinal section of the taproot of several diseased plants shows that the vessels are more or less brown. In some cases, the contiguous tissues are also affected. A longitudinal section of several secondary roots confirms that the vascular system as a whole is affected, given its marked brown colouration.

In addition, some authors have reported the presence of necrotic spots on leaves of lettuce attacked by *P. tracheiphilum*.

(See **12, 325, 427, 429, 434, 437–440**)

### • Biology, epidemiology

**Survival, sources of inoculum:** Although the biology of this *Pythium* is not very well understood, there is reason to believe that it is capable of living in a saprophytic state at the expense of organic matter. It seems capable of attacking or being harboured by a certain number of hosts apart from lettuce, that also guarantee its multiplication and survival: various *Lactuca*, the sunflower, artichoke, cardoon, salsify, fennel, pea and some weeds, including groundsel. Artificial inoculation, carried out on cucumber, cauliflower, tomato, and pea, have shown that this pseudo-fungus could attack these hosts. The strains encountered in the field probably do not have the same pathogenic power against these hosts. It can survive in the soil thanks to its resistant structures, such as its oospores and, to a lesser degree, its sporangia.

**Penetration and invasion:** It penetrates the epidermal tissues, undoubtedly through areas of damage, then spreads to the vascular system which it progressively invades, particularly the xylem, where we can see mycelial filaments.

* For many years the oomycetes, which contain almost 550 species, have been classified as Fungi. Their ultrastructure, biochemistry, and molecular sequences indicate that they belong to the Chromists including, in particular, algae (green and brown), diatoms, and so on.

**Sporulation, dissemination:** Subsequently it produces sporangia and oospores inside damaged tissues or on their surface (**479** and **480**). It is perfectly adapted to the aqueous phase of soils. Water is mainly responsible for its dissemination. Aerial dispersion has been recorded in the USA, following splashing of contaminated particles of soil during sprinkler irrigation or heavy rainfall. They produce broad areas of leaf necrosis which have not been observed in France.

**Conditions favourable to its development:** As with numerous Pythiaceae, its development is above all encouraged by the presence of water. High soil moisture (90% of retention capacity), and reduced gaseous exchange constitute an ecological advantage for this fungus. Temperature has less effect on its growth as contamination is possible between 5–43°C. Nevertheless it appreciates temperatures of around 20–24°C. Young plants and succulent tissues are more susceptible, which is confirmed by the fact that this *Pythium* often attacks early.

## Protection*

### • During cultivation

If the disease is observed in an established crop, unfortunately there is no curative method of control which can rectify this. Several **fungicides** are currently used to combat the Pythiaceae. Propamocarb HCl is the only product approved for lettuces in France; its effectiveness seems fairly limited. Some of the systemic fungicides used to combat mildew (*Bremia lactucae*) may be fairly effective against this *Pythium*.

As far as possible you should eliminate diseased plants as soon as the first symptoms appear so as to prevent the fungus from multiplying from the top.

### • Subsequent crop

In affected gardens, **nurseries** must be carefully monitored in order to check for the presence of *P. tracheiphilum*. If it does appear, soil disinfection must be considered. Subsequently, plants must be produced on a healthy substrate. Sowing blocks or trays must not be placed on the soil, especially if it has not been disinfected. It will be preferable to place them on a plastic film or shelf (also see page 357).

During **planting**, plants must not be placed in soil which is too wet or too cold; irrigation carried out at this stage of cultivation must not be excessive. You must be vigilant about checking the healthy quality of irrigation water coming from channels or pools which may have been contaminated.

Lettuce grown in the open field and in the greenhouse show **differences in susceptibility between varieties**.

A large number of **control methods recommended for combating Pythiaceae** can be implemented in the case of this *Pythium*. We recommend that you read about these in fact file 8.

* In order to give the proposed control methods a 'universal' nature, we have produced a fairly comprehensive list of the methods and fungicides reported in various grower countries.
**It is clear that these proposals must be modified according to the country concerned and the pesticide legislation in force.**

# *Verticillium dahliae* Kleb.

(Eucaryotes, Fungi, Mitosporic fungi)

# Verticilliosis ('Verticillium wilt')

## Principal characteristics

Groups of compatibility have been revealed with *V. dahliae*. Certain strains probably have a certain parasitic specificity.

### • Frequency and extent of damage

This very polyphagous fungus is present in numerous countries. It is not considered to be a serious pathogenic agent on lettuce. It has only been recorded in a very limited number of countries, especially in the USA where it is a serious problem in the Monterey and Santa Cruz counties.

This disease does not seem to attack in France.

### • Principal symptoms

The lower leaves of the crown of diseased lettuce go yellow, starting around the periphery of the lamina and becoming fairly sectorial. Eventually the leaves partially or completely wilt and become necrotic. At this stage, they remain strongly adherent to the head. Microsclerotia can sometimes be seen along a few veins.

A longitudinal section of the roots, the taproot and stem reveals that the vascular system has been affected. It is brown or green in colour, which may lead to confusion with damage caused by Fusarium (see fact file 13).

### • Biology, epidemiology

Very little is known about the epidemiology of this fungus on lettuce. It seems highly unlikely that it behaves in a different way on lettuce compared with other crops, particularly vegetable crops.

**Survival, sources of inoculum:** *V. dahliae* is a very poor competitor at ground level. In spite of this, its survival is guaranteed by its microsclerotia (**482**) which allow it to survive for around fifteen years in the soil. Its great polyphagous qualities allow it to attack numerous cultivated host plants (pepper, eggplant, tomato, Cucurbitaceae, cauliflower, tobacco, cotton, potato, and so on) or wild hosts (black nightshade, amaranth, and so on) which encourage its multiplication and survival.

**Penetration, invasion:** Contamination takes place either by direct penetration by the mycelium, or through various areas of damage in the roots. In addition, it may be encouraged by attacks of root knots and *Pratylenchus* spp. Once established, this fungus spreads to the vascular system, which it progressively colonizes. The plants react to this vascular invasion by forming a gum, or tylose, which prevents its progression through the veins. These mechanisms of defence contribute to lettuce wilting. *V. dahliae* produces microsclerotia and fragile whorls of condiphores, forming ovoid conidia in the tissues (**481**).

**Dissemination:** This is possible via compost or agricultural equipment dirtied with contaminated soil and plant debris. Dust from soil harbouring microsclerotia and/or conidia is easily disseminated by air currents, as well as by splashes of water and soil-based insects.

**Conditions favourable to its development:** There seem to be a certain number of strains with varying thermal requirements. Their thermal optima are probably located between 20–32°C. Short periods of light and poor lighting make plants susceptible to the disease which is probably more severe in neutral to alkaline soils. Monoculture of susceptible plants, excessively short rotation periods, or poor selection of rotational crops help to increase the effects of this disease in certain plots.

481

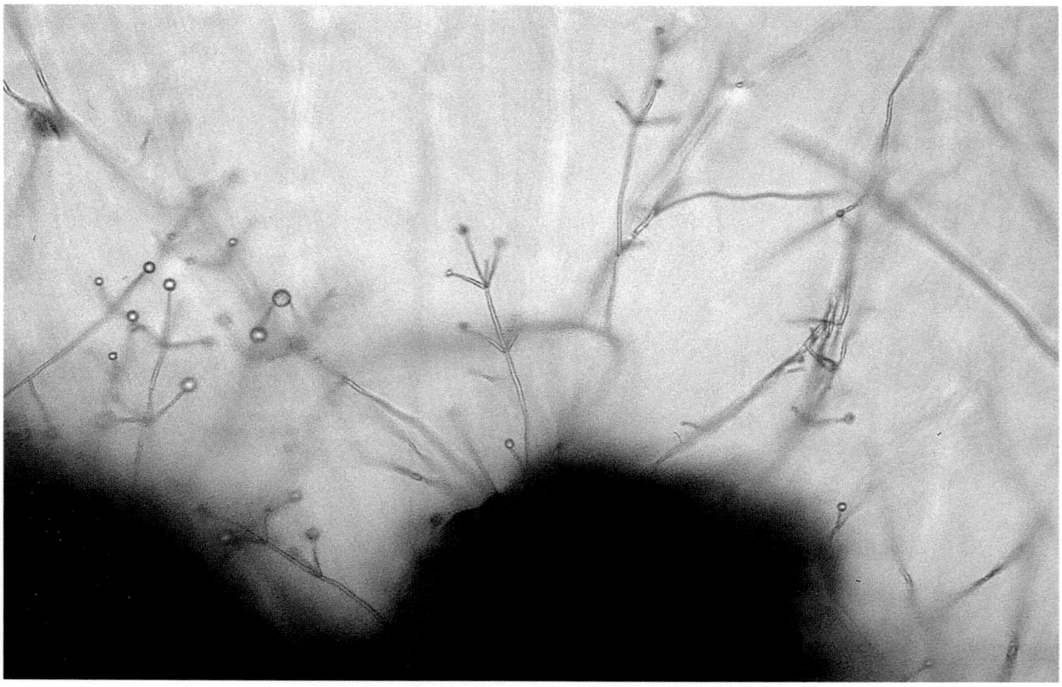

482

**481** The condiphores of *Verticillium dahliae* are branched in a whorl pattern; they produce ellipsoid, hyalin conidia, grouped together into droplets at their terminal or lateral extremities.

**482** This fungus also produces black microsclerotia (70–200 μm) which are globular or elongated, these differentiate them from *Verticillium alboatrum*.

## Protection

### • During cultivation

No method is capable of controlling this disease once the crop is growing.

At the end of the growing season it is fairly usual to bury **crop residues** in the soil. Plant tissues which have been buried are extensively colonized by *V. dahliae*, which produces numerous microsclerotia in this material. Disposing of plants restricts this phenomenon and helps to reduce the inoculum left in the plots.

### • Subsequent crop

The tools used in contaminated plots must be carefully cleaned before being used in other plots which are still healthy. The same applies to tractor wheels. Careful rinsing of this equipment in water will often be sufficient to remove earth and *V. dahliae*.

**Crop rotation** will prevent or delay the appearance of this disease. To be effective, this rotation must be sufficiently long and must not involve susceptible crops such as tomato, eggplant, potato, and so on. Control of certain weeds, such as black nightshade and amaranth, must be taken into account. The monocotyledons and cereals in particular do not seem to be affected by this vascular fungus.

Disinfecting the soil with a **fumigant** will not be useful for lettuce. It only has limited effectiveness and its cost may be high. **Solar disinfecting** of the soil, recommended for controlling other soil-based fungi such as *Rhizoctomia solani*, can reduce the effects of *V. dahliae* on cotton plants. The same may be true for lettuce. **Immersion** of contaminated plots probably helps to limit the number of microsclerotia present in the soil by reducing the quantity of oxygen available and increasing the quantity of $CO_2$. This measure does not seem to have reduced the incidence of the disease enough to be adopted. The remaining propagules are sufficient to allow *V. dahliae* to develop on its host.

The most effective solution is likely to involve using resistant varieties. **Partial resistance** exists in certain lettuce cultivars; it has not yet been clearly identified.

* In order to give the proposed control methods a 'universal' nature, we have produced a fairly comprehensive list of the methods and fungicides reported in various grower countries. **It is clear that these proposals must be modified according to the country concerned and the pesticide legislation in force.**

# *Fusarium oxysporum* Schl. f. sp. *lactucum* Hubbard & Gerik

# Fusariosis ('Fusarium wilt')

(Eukarytotes, Fungi, Mitosporic fungi)

## Principal characteristics

• **Frequency and extent of damage**

This soil-based fungus has mainly been reported in the USA, in California and in Arizona. Earlier it was identified in Japan in the districts of Nogano and Hokkaido as well as in Taiwan.

Its distribution in Europe is very limited. It may have caused damage in Germany and Italy. It has never been seen in France.

• **Principal symptoms**

*F. oxysporum* f. sp. *lactucum* can attack seedlings in the nursery which then wilt and die. We notice a reddish brown area of damage extending into the vessels of the taproot and certain roots, then spreading into the cortex.

Adult plants which have been affected display root and vascular symptoms comparable to those observed on seedlings. A longitudinal section reveals brown vessels. As a consequence, leaves located on one side of lettuce plants become yellow and become necrotic around the lamina periphery. The growth of certain plants may be reduced and the head may not form.

(See **446a** and **b**)

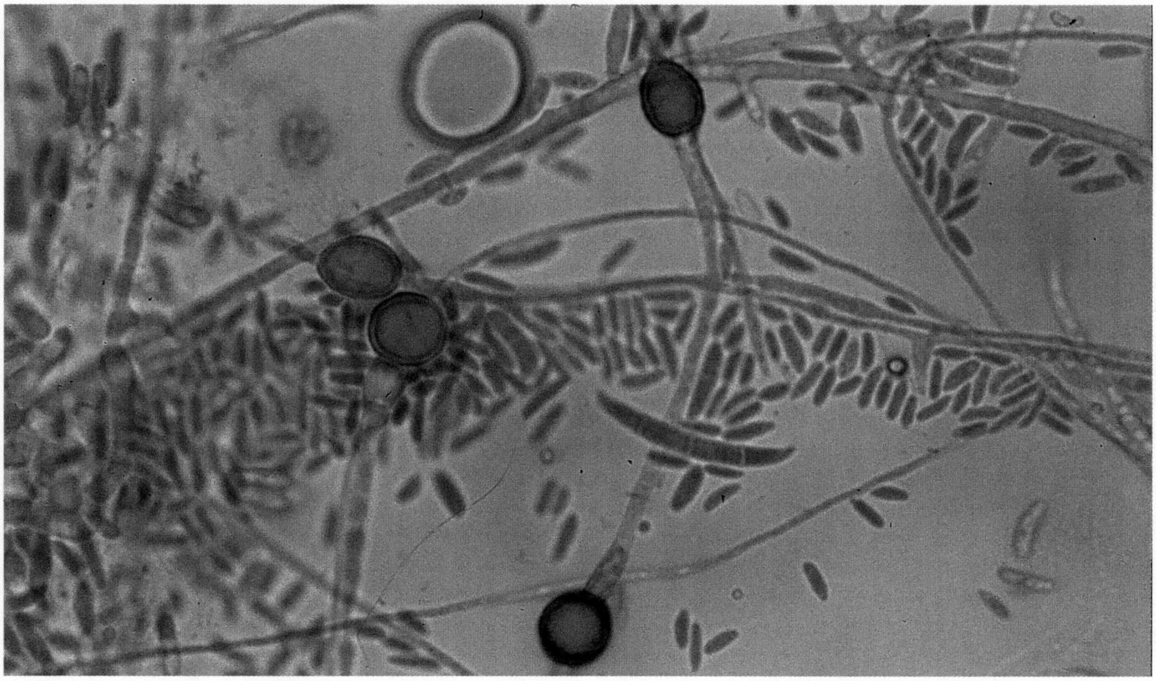

**483** *F. oxysporum* f. sp. *lactucum* produces crescent-shaped unicellular microconidia and macroconidia with a maximum of 3–4 partitions. It also forms thick-walled chlamydospores, intercalary or terminal, isolated or in chains.

• **Biology epidemiology**

The epidemiology of this fungus is not well understood. There is reason to believe that it has numerous similarities with other specialist forms of *F. oxysporum*, vascular parasites on various cultivated plants.

**Survival, sources of inoculum:** This fungus must be capable of surviving in the soil thanks to plant debris and its chlamydiospores with thick resistant walls. It is certainly well equipped for life as a saprophyte which allows it to colonize and survive on various organic compounds. No other plants capable of harbouring it, either cultivated or uncultivated, are known.

**Penetration, invasion:** After germination of its chlamydiospores, this *Fusarium* must be able to penetrate the lettuce either through natural areas of damage, such as those present at the emission point of secondary roots, or via various kinds of damage. Once inside the plant, it spreads to the vessels and invades them with its mycelium. As in the case of Verticilliosis, plants may react to this vascular invasion by a forming gum or tylose which will prevent its progression, but will also be a contributory factor in wilt.

**Sporulation, dissemination:** This fungus produces chlamydospores and microconidia in the vessels and on colonized tissues. Dissemination can certainly take place via agricultural equipment dirtied with contaminated earth and via plant debris. Dust from soil containing chlamydospores is easily disseminated by currents of air as well as by splashes of water.

**Conditions favourable to its development:** *F. oxysporum* f. sp. *lactucum* mainly prefers high temperatures. Its thermal optimum is probably located at around 28°C. In the USA, this disease is mainly observed in autumn.

## Protection*

• **During cultivation**

No method or product is capable of controlling this disease while the crop is growing.

After harvesting, it is fairly usual to bury the **crop residues** in the soil. Plant tissues which have been buried are abundantly colonized by *F. oxysporum* f. sp. *lactucum* which produces numerous chlamydospores on this material. Disposing of plants with their root system limits this phenomenon and helps to reduce the quantity of inoculum left in the plots.

• **Subsequent crop**

Given the limited extent and incidence of this disease on l.s.v., it will not normally require the implementation of any special measures.

The recommendation is to avoid growing lettuce in a plot which has already been affected. **Crop rotation** will help to prevent the appearance of this disease. Rotational periods must be sufficiently long in order to be effective.

**Soil disinfection** has short-lived effectiveness because, as with all types of *Fusarium*, it very rapidly re-colonizes soil which has been disinfected.

**Tools** used to till the soil in contaminated plots must be carefully **cleaned up** before being used in other healthy plots. The same is true for tractor wheels. Careful rinsing of this equipment in water will often be sufficient to remove infested earth.

The most effective method of controlling this vascular disease probably consists of using **resistant varieties**. Unfortunately, to our knowledge, no commercial varieties are yet available where the disease is important.

* In order to give the proposed control methods a 'universal' nature, we have produced a fairly comprehensive list of the methods and fungicides reported in the various grower countries.
**It is clear that these proposals must be modified according to the country concerned and the pesticide legislation in force.**

# Root colonizing fungus and virus vector

## *Olpidium brassicae* (Woronin) P.M.Dang.

(Eukaryotes, Fungi, Chytridiomycetes, Spizallomycetales, Olpidaceae)

## Virus fungal vector

### Principal characteristics

• **Frequency and extent of damage**
O. *brassicae* is certainly found worldwide. It is not considered to be pathogenic for the majority of plants on which it has been observed, with the exception of tobacco. It has been reported on the roots of l.s.v. grown both in, and out of the soil. It is particularly feared as it is the vector of two serious lettuce viruses: **Lettuce big vein disease (Mirafiori lettuce virus (MiLV)) and Lettuce ring necrosis agent (LRNA) (see fact files 34 and 35)**.

We are now aware of a strain of O. *brassicae* which predominates on the Crucifereae, and a second which favours lettuce and numerous other plants; this 'race' is a particularly strong virus vector.

• **Structures present in the root cells**
The presence of O. *brassicae* in the roots can easily be confirmed by observing them under the microscope. This fungus, an obligate aquatic parasite, forms resting spores or chlamydospores and the sporangia which produce numerous zoospores in the cells of the cortex (**484**).

(See **397**)

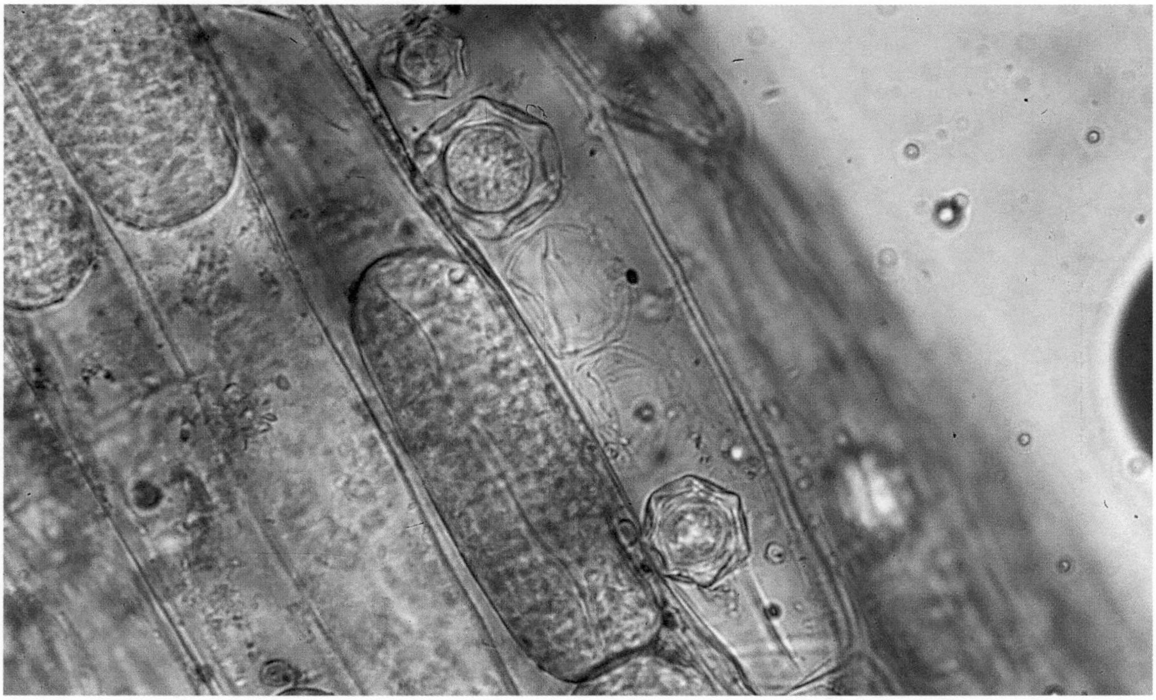

**484** O. *brassicae* is characterized by the presence of zoosporangia, filling the root cells of the cortex, and star-shaped resting spores.

• **Biology, epidemiology**

**Survival, sources of inoculum:** *O. brassicae* produces 'resting spores' or chlamydospores which guarantee its survival for several years in the soil and on plant debris. It can also survive on several intermediate hosts given its polyphagous qualities, particularly several market garden species: tomato, cucumber, and so on.

**Penetration, invasion:** Contamination takes place via uniflagellate zoospores which are produced from resting spores or sporangia. They are attracted by the roots, form cysts on their surface, then penetrate the cells directly. Here the fungus produces one of several massed sporangia which will generate the zoospores, guaranteeing secondary contamination.

**Dissemination:** The zoospores are liberated on the outside of the root cells thanks to release tubes. Once in the aqueous phase of the soil or in the nutrient solution, a flagellum allows them to move and spread to other plants. *O. brassicae* is undoubtedly spread by soil dust, contaminated seedlings, and circulating water.

**Conditions favourable to its development**: This fungus is perfectly adapted to aquatic life and rapidly spreads in hydroponic crops. It develops very successfully at temperatures between 10–16°C.

## Protection*

Combating this soil fungus can only be justified in order to limit the damage from the two redoubtable viruses which it transmits to lettuce.

Implementation of **crop rotation** is not effective because this fungus survives for a very long time in the soil and in addition it is very polyphagous. Damp heavy soils must be drained. They may be disinfected. Disinfecting with a **fumigant** may be envisaged. Only methyl bromide** seems to have any real efficacy against the vector *O. brassicae*. Other fumigants and/or methods of disinfecting produce very sporadic results. You should be particularly vigilant about ensuring that you do not re-infect disinfected soils. See fact files on MiLV and LRNA.

Certain **products** have been reported as being effective against *O. brassicae* in crops grown out of the soil: Agral (Alkyl phenol-ethylene oxide) and fungicides belonging to the Benzimadazole family. Studies have shown that the simultaneous use of methyl-thiophanate and zinc succeeded in reducing the damage from viruses transmitted by this vector. These results are controversial.

In order to reduce the risks of this parasitic fungus developing in your fields, it is vital to use a healthy **substrate** and **seedlings** of the same quality. Seedlings grown on mini sowing blocks must not be placed straight on to the soil, particularly if it has not been disinfected; they must be insulated with a plastic film and planted on raised beds in order to avoid water retention in the roots.

You must be particularly vigilant about the **water quality** as this may lie at the origins of contamination. UV rays probably offer a degree of efficacy in partially ridding the water and recycled nutrient solutions of the *Pythium* spp. and *O. brassicae*.

Given the numerous resting spores present in the roots, it will be essential **to eliminate as many root systems** as possible while the crop is growing and after it has been harvested.

* In order to give the proposed control methods a 'universal' nature, we have produced a fairly comprehensive list of the methods and fungicides reported in the various grower countries. **It is clear that these proposals must be modified according to the country concerned and the pesticide legislation in force.**

** **Methyl bromide will be prohibited after 2005. Its use, which is no longer considered to be good plant health practice, involves risks with residues; abundant rinsing of the soil in water is therefore essential.**

# Bacteria

## General information

**Bacteria** are relatively simple unicellular organisms, small in size (0.5–1 μm wide, 1–3 μm long), which can be distinguished under a light microscope with strong magnification. They do not have a differentiated nucleus, which classifies them among the **prokaryotes**, but a **circular chromosome without a nuclear membrane**. This chromosome, which contains the genotype of the bacterium, consists of double stranded DNA (deoxyribonucleic acid). **Plasmids**, much shorter strands of cytoplasmic DNA, also contain genetic information which can be exchanged with other bacteria, particularly after sexual reproduction which produces bacterial conjugation. Their reproduction more frequently involves simple cell division, resulting in the formation of two daughter bacteria (vegetative reproduction by **cleavage**). This very rapid method of multiplication (varying in time from about 20 minutes to a few hours) allows populations of bacteria to increase exponentially if conditions are favourable.

The bacterial walls vary, giving them different shapes: spherical, ellipsoid, filamentous, and so on; but the majority of those which attack l.s.v. are **rod shaped**. They are mobile thanks to the presence of flagella arranged around the body of the bacterium (*Erwinia* spp.), or at their extremity (*Pseudomonas cichorii*, and so on). The presence and location of the flagella allow us to identify the bacteria. Other criteria can also be used to distinguish them:

- the composition and structure of the bacterial wall (which allows us to split the bacteria into 2 groups: Gram$^+$ and Gram$^-$);
- the speed of growth and appearance of colonies on an artificial medium, any production of diffusable pigments;
- numerous biochemical characteristics;
- serological and molecular tests, and so on.

Approximately 250 species, biovars and pathovars, representing 17 bacterial genera, have been listed for bacteria. Four species have a serious effect on l.s.v.: *Pseudomonas*, *Xanthomonas*, *Erwinia*, and *Rhizomonas*. The ***Pseudomonas*** can be subdivided into two groups: the '**fluorescents**' (*P. cichorii*, *P. marginalis* pv. *marginalis*) and the '**non-fluorescents**' depending on whether or not they form a diffusable pigment which is fluorescent under UV light on the King medium. Among ***Erwinia***, we can essentially list *E. carotovora* subsp. *carotovora* which is damaging to l.s.v., both when they are growing and while they are being transported and stored.

Bacteria which predominate on l.s.v. are relatively independent and they can live as parasites or in a saprophytic state in the phyllosphere (**epiphyte** life) and the rhizo-sphere. They are found on other plants, in the soil, in organic matter, and on numerous types of plant debris on which they can easily survive. Their walls give them a certain degree of resistance.

All the organs of the plant can be affected:

- the leaves (*P. cichorii*, *X. campestris* pv. *vitians*, *E. carotovora* subsp. *carotovora*, and so on);
- the stem and the taproot (*E. carotovora* subsp. *carotovora*, *P. marginalis* pv. *marginalis*);
- the roots (*R. suberifaciens*);
- the vascular system (*E. carotovora* subsp. *carotovora* in certain cases).

They penetrate l.s.v. in several ways:

- through natural openings such as the stomata and the hydathodes (*P. cichorii*, *X. campestris* pv. *vitians*);
- through various areas of accidental **damage** (damage from hail or insects, and so on) or damage caused by growing operations such as, for example, those occurring during harvesting (*E. carotovora* subsp. *carotovora*, *P. marginalis* pv. *marginalis*).

Once in the tissues they develop first of all between the plant cells which they kill fairly quickly by means of various enzymes. This is particularly the case of *Erwinia* spp. but also of *P. marginalis* var. *marginalis* and *P. viridiflava*, which, thanks to their **pectinolytic enzymes**, are responsible for soft rot. They sometimes produce **toxins** which cause a yellow halo to form around the damage. This does not seem to be the case for l.s.v. Colonized tissues harbour large quantities of bacteria which can be spread in various ways:

- During **rainfall** generating splashing drops of water or by running water;

- via **contaminated seeds** (*X. campestris* pv. *vitians*);
- by man and his tools;
- by certain animals and insects.

The water present on the plants, or soaking the soil, is initially the factor conditioning bacterial epidemics. It can come from rainfall, irrigation, morning dew, and so on. It is essential for the effective development of contamination, infection, and dispersion of bacteria.

Temperature is less influential in the development of bacteria attacking l.s.v.

# Bacteria which affect the leaves and head

## *Pseudomonas cichorii* (Swingle) Stapp.

(Prokaryotes, Bacteria, Gracilicutes, Pseudomonadaceae)

## Bacterial leaf spot, Varnish spot

### Principal characteristics

**• Frequency and extent of damage**

This very polyphagous Gram⁻ bacterium is found extensively throughout the world in virtually all continents (Japan, USA, Italy, Portugal, Spain, and so on). It is found on both lettuce and endive. Sometimes it attacks in combination with one or several of the other bacteria described as affecting l.s.v.

In France, as in numerous other countries, it is found particularly in autumn and winter, mainly on mature plants.

**• Principal symptoms**

The symptoms caused by *P. cichorii* can be very varied. They mainly appear on the internal leaves, as harvest time approaches. Generally we see small chlorotic spots, which quickly become dark brown to black as they become necrotic. They are generally shiny, in circles or polygons, sometimes star-shaped, limited by the secondary veins which also become brown. The presence of a yellow halo is fairly rare.

In some cases, damage is restricted to the principal vein which also becomes brown to black in colour. Contiguous tissues display the same type of damage, with the taproot normally remaining healthy. Other secondary bacteria can colonize the tissues and produce soft rot. When the tissues dry, they are lighter in colour and have a paper-like consistency.

Other leaf damage sometimes occurs on the internal leaves of the head as harvest time approaches. These areas are often extensive and necrotic, brown and shiny, without any modification in consistency (sometimes known as '**Varnish spot**').

*P. cichorii* can also cause spots on the leaves of l.s.v. in the category of ready-prepared, pre-packaged, refrigerated fruit and vegetables while they are being stored.

(See **93, 94, 146, 160, 166, 247, 257,** and **276**)

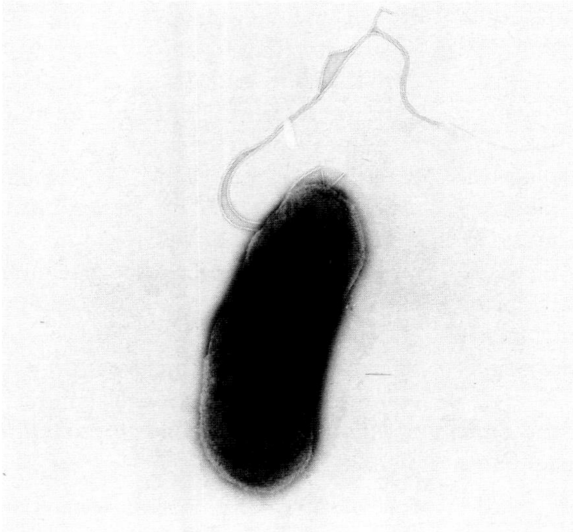

**485** *P. cichorii* is a rod-shaped, Gram⁻ bacterium with polar flagella.

**• Biology, epidemiology**

**Survival, sources of inoculum:** This bacterium establishes itself very easily in the soil and on plant debris. It has been isolated directly from the soil in Japan. It colonizes the rhizosphere of a large number of hosts, both cultivated and uncultivated. It is found in the pathogenic state on several vegetables, such as cabbage, chicory, celery, tomato, eggplant, several other leguminous plants, tobacco, flowers (gerbera, chrysanthemum, pelargonium, and so on) and numerous weeds: *Sonchus oleracea*, *Veronica* sp., *Solanum nigrum*, *Portulaca oleracea*, *Poa annua*, *Setaria* sp., *Senecio vulgaris*, *Capsella bursa-pasteuris*, and so on.

It can survive for several months on seeds and may survive in certain sources of contaminated dirty water. Certain tools and harvesting crates also constitute sources of contamination.

**Penetration:** Contamination may take place at the end of the rosette stage, assisted by rainfall or irrigation. *P. cichorii* does not seem to have any significant parasitic potential on lettuce. It mainly penetrates lettuce leaves through natural openings, such as the stomata or areas of damage (bursting during growth, damage due to wind, insects, or growth related damage, and so on). Numerous bacteria can be observed in the guard cells and in the intercellular spaces of the epidermis.

Subsequently they spread to the mesophyll. Direct contamination across the cuticle is rare and occurs when the leaves are covered and saturated with water. In fact, this bacterium is present on the surface of leaves throughout the growth cycle of l.s.v. Population density fluctuates during the growth cycle depending on climatic conditions and possibly on the salad variety. External leaves constitute a significant source of inoculum, allowing subsequent contamination of leaves in the head. As harvest time approaches, the bacterial population increases and inoculum pressure intensifies. The first spots appear rapidly after contamination (approximately 30 hours). Browning of the tissues progresses from the epidermis to the mesophyll. In heavily infested soils, contamination via the roots or the taproot may take place. The bacterium, behaving like a vascular micro-organism, is probably responsible for browning of the principal veins.

**Dissemination:** This certainly takes place as a result of splashes of water, running water, and normally via the seeds. This is particularly the case with certain host plants attacked by this bacterium. It can be acquired and transmitted by the leaf miner *Liriomyza trifolii* on the chrysanthemum. Plants which are contaminated in the nursery are also responsible for its dissemination.

**Conditions favourable to its development:** *P. cichorii* develops at temperatures between 5–35°C, with its optimum being located around 20–25°C. The bacterium is killed at temperatures in excess of 53°C. It particularly favours damp environments. This is why it mainly attacks during prolonged rainy spells, during which the water deposited on the leaves encourages contamination and dissemination.

## Protection*

### • During cultivation

Combating the bacterial diseases affecting l.s.v. is fairly tricky as they often strike late, which makes the use of any chemical means of control difficult. The use of bacteriostatic products such as copper is only relatively effective. In addition, you should be aware that strains of *P. cichorii* resistant to copper and streptomycin have been reported on celery in Florida. In addition, we should point out that l.s.v. are very susceptible to copper (risk of chemical injury). This is why the various forms which can be used must be applied at lower doses than normal.

It seems obvious, given the biology of this bacterium, that sprinkler irrigation must be avoided whenever possible, or this must be carried out in the morning rather than the evening, so that the plants dry out quickly during the day. This type of irrigation can be particularly dangerous as harvest time approaches, especially if it is not carefully managed in time.

It is advisable to avoid working on the plots while the plants are wet; risks of transmitting bacteria through contact are high.

Protective structures must be well-ventilated to allow the plants to dry out, particularly after sprinkler irrigation. As much plant debris as possible must be eliminated at harvest-time and this must not be buried in the soil as the bacterium can survive there relatively successfully.

### • Subsequent crop

This disease represents a significant risk for future crops. Given the relative efficacy of copper, growers will be well advised to set up prophylactic measures which may help to reduce the establishment and spread of this bacterium.

Crop rotation, implemented as a preventative measure in order to delay the appearance of soil-based pathogenic agents, will be able to limit its survival in the soil. In fact, rotations lasting for 3 years, recommended for other crops, seem to produce good results. Soil preparation must be particularly thorough. Good drainage must be encouraged. Preference must be given to plots located in well-ventilated positions as these conditions will help the lamina to dry. The proximity of susceptible crops and/or those which

* In order to give the proposed control methods a 'universal' nature, we have produced a fairly comprehensive list of the methods and fungicides reported in the various grower countries. **It is clear that these proposals must be modified according to the country concerned and the pesticide legislation in force.**

have already been contaminated must be avoided. Susceptible plants, particularly weeds, must be eliminated.

In growing areas where this is possible, crops should be planted during drier periods of the year.

Plant fertilization must be balanced, avoiding excessive nitrogen. Growing conditions conducive to the appearance of 'tip burn' must be avoided (see page 133).

L.s.v. intended for the range of ready-prepared, pre-packed, and refrigerated fruit and vegetables must be carefully harvested and rapidly refrigerated. Consignments of harvested crops which need to be kept for a relatively long period, or those which are intended for the range of ready-prepared fruit and vegetables, will be selected from plots which are clear of the disease.

In France, an anti-bacterial treatment based on copper is recommended for endive, at the 8–10 leaf stage.

Differences in sensitivity to bacterial disease between l.s.v. types and between cultivars have been observed in the field. These behavioural differences do not seem to have been evaluated during selection. In fact, at the moment, there are no varieties of l.s.v. resistant to *P. cichorii*.

# *Xanthomonas campestris* pv. *vitians* (Brown) Dye

(Prokaryotes, Bacteria, Gracilicutes, Xanthomonadaceae)

# Bacterial leaf spot and head rot

## Principal characteristics

Two new names have been attributed to this bacterium: ***Xanthomonas axonopodis* pv. *vitians*, *Xanthomonas hortorum* pv. *vitians*.**

### • Frequency and extent of damage

Numerous bacteria have been isolated from necrotic lesions observed on l.s.v. Among these, only this Gram$^-$ bacillus, with a polar flagellum, is considered to be pathogenic to these plants, together with *Pseudomonas cichorii*.

This bacterium, known to attack mainly in hot regions, is now very widespread. It has been recorded in numerous countries, on lettuce and endive, particularly in Australia, South Africa, USA, Canada, Venezuela, Japan, and several European countries such as Germany, Italy, France, and so on.

It seems to have been increasing over recent years and is causing extremely significant damage. Entire plots are sometimes invaded within a few days and almost all of the damaged lettuce is unfit for sale.

### • Principal symptoms

Following rainy periods or excessive irrigation, we observe minuscule oily spots forming mainly on well developed leaves and around the lamina periphery. Subsequently these spots rapidly spread; they are fairly circular in shape, damp in appearance and transparent, progressively becoming brown as they age until they become dark brown to black. These spots spread and converge; ultimately entire leaves are damaged. V-shaped wilting beginning at the lamina edge has also been reported. Lesions become progressively brown in the centre and become necrotic, particularly when the weather becomes more clement and the air is drier.

Other symptoms are associated with the presence of this bacterium:

- rotting of the head can sometimes be observed if climatic conditions are particularly favourable; it seems to be exacerbated by various secondary micro-organisms;
- damp damage on the stem, on young plants;
- symptom of hollow stem. In this case, plants may be stunted and/or wilted.

In addition, **X. campestris pv. vitians** may be responsible for the damping-off of seedlings.

(See **159**, **162**, **165**, and **167**)

### • Biology, epidemiology

**Survival:** The biology of this bacterium is rather poorly defined; it probably behaves in a way fairly similar to *Pseudomonas cichorii*. It survives and is transmitted by seeds from cultivated lettuce. It is found in the epiphyte state on wild lettuce (*Lactuca serriola*) and on some weeds, especially compositae, *Matricaria matricarioides*, and so on. In Germany, it has been reported as affecting tomato and pepper. This bacterium can survive from one year to the next in an infested plot.

**Penetration:** If seeds are contaminated the bacterium can be found very rapidly on the initial leaves of some seedlings in the nursery and, subsequently, on most of them if sprinkler irrigation is applied. In the open field, its epiphyte establishment on l.s.v. may occur at an early stage. Once on the surface of the leaves, *X. campestris* pv. *vitians* mainly penetrates in the presence of free water standing on the lamina via natural entries: through the stomata and the hydathodes.

**Dissemination:** Once the tissues have been invaded, numerous bacteria exude from the areas of damage. They are dispersed towards other leaves or other healthy plants via splashes of water during rainfall or sprinkler irrigation, and by the effects of wind and air currents in shelters. The seeds, which conserve the bacterium, certainly help to disseminate it to other establishments and new plots.

**Conditions favourable to its development**: As with the majority of bacteria, its development is always preceded and accompanied by wet periods. Morning dew, rainfall and copious sprinkler irrigation are particularly favourable. It develops at temperatures between 0–35°C, with the thermal optimum being located between 26–28°C. This bacterium is probably killed at temperatures in excess of 52°C. Areas of damage which occur following frost probably make the leaves in lettuce heads susceptible to infection.

## Protection

- **During cultivation**

Combating bacterial disease in l.s.v. is fairly tricky as they often strike late and there are no high-performance products. Treatments based on copper are not very effective and only have a preventive action. Copper hydroxide seems to be the most effective form, in combination with zineb or mancozeb. We ought to point out that some strains of *X. campestris* pv. *vitians* tolerant to copper have been reported in published works. In addition you must remember that l.s.v. are very susceptible to copper. This is why the various forms of copper that can be used must be applied at lower doses than on other crops.

As free water encourages the growth of this bacterium, sprinkler irrigation must be avoided as far as possible. It should be carried out in the morning rather than the evening, so that plants dry out quickly during the day. This type of irrigation is particularly dangerous as harvest time approaches, especially if applications are not properly managed in terms of time. The quality of water used must be checked, sprinkling with water in which plant debris has decomposed must be avoided.

Work must not be carried out in plots while plants are wet; risks of transmitting bacteria through contact are high.

Greenhouses must be well-ventilated to allow the plants to dry after water has condensed in the morning or after sprinkler irrigation.

As much plant debris as possible must be eliminated during harvesting and if possible not buried in the soil as the bacterium can survive there relatively well. If there are no alternatives, this debris must be deeply buried.

- **Subsequent crop**

As with other types of bacterial disease affecting l.s.v., growers are advised to set up prophylactic measures which can limit the establishment and spread of this bacterium.

As it cannot survive for very long in the soil, you would be well advised to implement crop rotation which will also be effective against soil-based pathogenic agents.

Extreme care must be taken with preparing the soil. Good drainage is desirable. In growing areas where this is possible, crops should be planted during drier periods of the year. Preference must be given to plots located in well-ventilated areas, as these conditions will help to dry the lamina. Proximity of susceptible crops and/or those which have already been contaminated must be avoided. Weeds, particularly those from the Asteraceae family, must be destroyed.

Bacteria-free seeds must be used. It is difficult to envisage disinfecting them using bleach given the fact that numerous seeds are now coated. You must avoid producing seeds in conta-minated regions, and sprinkler irrigation in seed-plant crops must be avoided. Plant fertilization must be balanced, without excessive nitrogen.

In France, an anti-bacterial treatment based on copper is recommended at the 8–10 leaf stage on endive.

Currently there are no varieties of lettuce resistant to *X. campestris* pv. *vitians*. Differences in sensitivity between varieties have nevertheless been observed in nature. In fact, the intensity of damage varies between types of lettuce and cultivars. In the USA we have seen more severe attacks on romaine and butterhead lettuce than on crisp batavia.

# *Pseudomonas marginalis* pv. *marginalis* (Brown) Stevens

# Marginal leaf spot

(Prokaryotes, Bacteria, Gracilicutes, Pseudomonadaceae)

**(Example of a bacterium which behaves as a secondary, opportunist, or saprophytic colonizer)**

## Principal characteristics

***P. marginalis* pv. *marginalis***, a Gram$^-$ bacillus, is present in numerous European countries, India, South America, USA, Japan, New Zealand, Australia, and so on. Usually it is only responsible for slight damage to lettuce. This bacterium has been reported on other vegetable species such as cabbage, cucumber, onion, potato, green bean, pea, and so on. It has been isolated from lettuce, endive, and chicory.

### • Principal symptoms

As its species name suggests, it causes damage which develops around the lamina periphery. Affected areas fairly quickly display a few spots which go brown, black and become necrotic. Subsequently tissues may dry out and become paler in colour with a paper-like appearance. If conditions are very damp, oily, black rot takes over the leaves. Some secondary veins sometimes appear darker.

During particularly wet periods, the bacterium is capable of spreading to the crown and stem; it is then associated with soft, olive coloured rotting of the pith, comparable to that caused by *Erwinia carotovora* subsp. *carotovora* (known as 'butt rot').

*P. marginalis* pv. *marginalis* probably causes damage when l.s.v. are being transported and stored under unsatisfactory conditions, particularly in terms of heat, sometimes in association with *E. carotovora* subsp. *carotovora*.

(See **161** and **168**)

### • Biology, epidemiology

This bacterium behaves in a fairly similar way to the two previously described bacteria.

**Survival**: It survives from one year to the next in the soil on plant debris from lettuce and also on several alternative cultivated hosts such as onion, cucumber, or cabbage.

**Penetration:** Contamination takes place during wet periods following splashing from rainfall or sprinkler irrigation. Once on the leaf surface, *P. marginalis* pv. *marginalis* penetrates the lamina through areas of damage and natural entries: the stomata, and especially the hydathodes which are frequently located around the leaf edges.

**Dissemination:** Once tissues have been invaded, numerous bacteria exude from the areas of damage and are dispersed to other leaves or other healthy plants as a result of splashing, which occurs during rainfall or sprinkler irrigation.

**Conditions favourable to its development:** The appearance of symptoms is always preceded and accompanied by wet periods. Morning dew, rainy periods, copious sprinkler irrigation, and more particularly the presence of free water on the lamina are especially conducive to *P. marginalis* pv. *marginalis*.

## Protection

The measures to be implemented in order to control this bacterium are identical to those recommended for controlling *Pseudomonas cichorii* and *Xanthomonas campestris* pv. *vitians*. We therefore advise you to consult fact files 15 and 16.

As *P. marginalis* pv. *marginalis* behaves as an opportunist, you must avoid any damage to plants. They must be carefully harvested at the right stage, and before they have become over-mature. In addition, temperatures for transport and storage must be kept close to 1°C.

# Bacteria which mainly affect the stem or roots



## *Erwinia carotovora* subsp. *carotovora* (Jones) *Bergey et al.* = *Pectobacterium carotovorum* (Jones) Waldee

(Prokaryotes, Bacteria, Gracilicutes, Enterobacteriaceae)

## Bacterial soft rot

### Principal characteristics

**• Frequency and extent of damage**

This Gram$^{-}$ bacterium is present worldwide, and mainly attacks various types of vegetable. It has been reported in numerous countries where l.s.v. are grown (Japan, USA, Hawaii, numerous European countries, and so on), on lettuce, endive, and chicory. It can cause serious damage both in the field and under protection (in and out of the soil), during transport and storage.

In France, the incidence of this bacterium is relatively low. Its attacks are mainly sporadic.

Various species of *Erwinia*, *E. aroideae* and *E. chrysanthemi*, have been recorded on lettuce in Japan and Swaziland.

**• Principal symptoms**

Salad crops affected by *E. carotovora* subsp. *carotovora* initially wilt fairly quickly. This symptom appears after rainy periods or in plots which have been excessively irrigated, close to harvest time. It is actually caused by this bacterium invading the vessels of the stem and taproot. A longitudinal or cross section of these organs reveals that the vessels are pink in colour, rapidly becoming brown. The pith is particularly badly affected. It becomes vitreous and gelatinous and takes on a greenish hue before going black and liquefying entirely under the effect of pectinolytic enzymes ('butt rot', 'jelly butt').

Dark spots are sometimes visible on the leaves. Subsequently, damp, viscous, dark brown to black rotting starts on a few leaves, then spreads to the whole head ('slime head'). Ultimately affected salad crops may liquefy completely. In contrast to the parasitic fungi responsible for wilting and rotting of leaves (*Botrytis cineria*, *Sclerotinia* spp.), no particular structure can be seen on the tissues.

*E. carotovora* subsp. *carotovora* also causes damage to salad crops kept in poor thermal conditions during transport and storage, sometimes in combination with *Pseudomonas marginalis* pv. *marginalis*.

(See **168, 297, 377, 436, 441–445**)

**• Biology, epidemiology**

**Survival, sources of inoculum:** *E. carotovora* subsp. *carotovora* is ubiquitous; it is present in numerous soils where it can survive without problems for several years, especially in plant debris in the aqueous phase. It is also found in the floral parts of l.s.v., which tends to indicate that it is relatively widespread within the plant environment. It is polyphagous and therefore likely to establish itself on a fairly large number of hosts, both cultivated and uncultivated, especially herbaceous dicotyledons.

**Penetration:** It penetrates salad crops essentially through areas of damage (mechanical damage, frost damage, 'tip burn', acid fog, and so on) or senescent tissues and, more rarely, through natural openings. It is a weak parasite which may attack secondarily to other pathogenic agents. Once in place, its cellulolytic and pectinolytic enzymes make an active contribution to its rapid spread throughout the tissues, which quickly rot and sometimes take on a nauseating odour.

**Dissemination:** As with numerous types of bacteria, it is easily disseminated by water during splashing and by running water. Insects, as well as workers with their tools, can help to spread it.

**Conditions favourable to its development:** It is essentially favoured by hot, damp, climatic conditions. Cloudy and rainy periods increase the risks of this *Erwinia* proliferating. It finds it much easier to contaminate salad crops which have reached or gone past their maturity and/or which present numerous areas of damage and a few senescent leaves. It seems capable of developing at temperatures of 5–37°C, with its optimum being located between 25–30°C. In dry soils, whose humidity is lower than 40%, the development of *E. carotovora* subsp. *carotovora* seems to decrease; in certain situations it disappears.

Poor control of storage temperature, the presence of numerous areas of damage, and the use of dirty water when washing or cooling salad crops encourage the damage caused by this bacterial disease during storage and transport.

## Protection

- **During cultivation**

It is practically impossible to control the development of this *Erwinia* once it is present in the taproot and stem of l.s.v. When attacks are noted (often at the end of the growing period) diseased lettuces must be rapidly eliminated.

It is also a good idea to lower the humidity of plant growth and to make every attempt to prevent the soil becoming too damp. This remark seems obvious if we are aware of the biology of this bacterium. To achieve this, sprinkler irrigation must be avoided where possible or be carried out in the morning rather than the evening so that plants can dry off quickly during the day. This type of irrigation can be particularly dangerous close to harvest time, especially if sprinkling is not carefully managed with regard to time. Greenhouses must be properly ventilated in order to dry the vegetation following nocturnal condensation or sprinkler irrigation.

No work should be carried out in the plots when the plants are wet as risks of transmitting bacteria through contact are high. Great care must be taken to handle the plants carefully during harvesting and to refrigerate them rapidly. Temperatures close to 0°C must be maintained during the transport and storage of salad crops, but make sure that they do not freeze.

As much plant debris as possible must be eliminated during harvesting and this must not be buried in the soil as the bacterium can survive there relatively well.

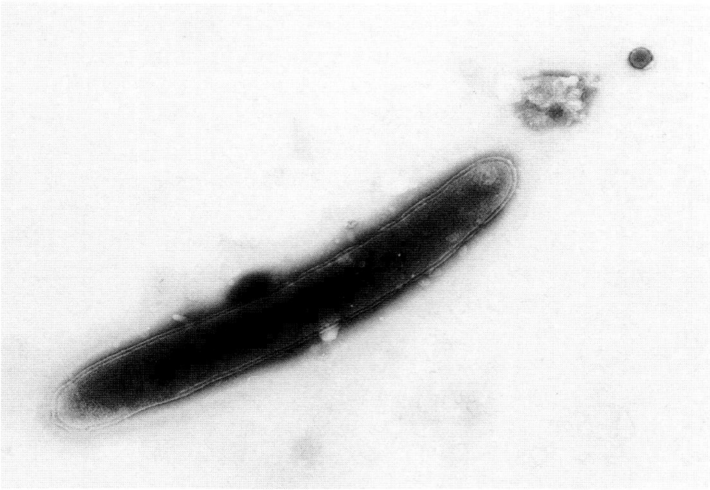

**486** *E. carotovora* subsp. *carotovora* is a rod-shaped Gram⁻ bacterium, without any peritrichous flagella.

- **Subsequent crop**

As this bacterial disease represents a fairly significant risk to future crops, growers will be well advised to set up prophylactic measures to limit the establishment and spread of this bacterium.

Crop rotations, implemented as a precaution in order to delay the appearance of soil-based pathogenic agents, will be able to limit its establishment in the soil. The choice of species to be grown is rather limited because this bacterium attacks numerous types of vegetable. By preference you should choose non-susceptible crops such as cereals, Gramineae, or maize. The proximity of susceptible crops and/or those which have already been contaminated must be avoided. Susceptible weeds must be eliminated.

If possible, in order to help the lamina to dry, choose plots located in well ventilated areas. In production areas, or where possible, crops must be planted during a fairly dry period of the year.

This soil must be prepared particularly carefully. Good drainage must be a priority. Plant fertilization must be balanced, avoiding excessive nitrogen. Growing conditions which encourage the appearance of 'tip burn' or various types of damage must be avoided.

L.s.v. intended for the class of ready-prepared, pre-packed, and refrigerated fruit and vegetables must be carefully harvested and rapidly refrigerated. Consignments of l.s.v. which need to be kept for a relatively long period or intended for the range of ready-prepared fruit and vegetables will be selected from plots which are clear of the disease.

Currently there are no varieties on the market which are resistant to *E. carotovora* subsp. *carotovora*. Lines of leaf lettuces which present a certain degree of resistance have been reported in Korea.

# *Rhizomonas suberifaciens* Van Bruggen, Jochimsen and Brown

(Prokaryotes, Bacteria, Gracilicutes, Sphingomonadaceae, *Sphingomonas*)

# Corky root

## Principal characteristics

Recently this bacterium has been renamed; the name ***Sphingomonas suberifaciens*** **(Van Bruggen, Jochimsen & Brown 1990) comb. nov.** has now been proposed. In the Netherlands, a *Rhizomonas* sp. which responds negatively to monoclonal serum specific to *R. suberifaciens* has been isolated. It also shows a weak DNA sequence homology with *R. suberifaciens*.

### • Frequency and extent of damage

*R. suberifaciens* (a Gram$^-$ bacillus which is difficult to isolate and therefore identify) has been reported in several countries, initially in the USA, especially in plots where lettuce crops are grown repeatedly. In this country, its attacks are serious, especially in California and Florida, causing **infectious corky root** of lettuce. It has also been reported in Australia, New Zealand, and Europe (in England and the Netherlands on iceberg lettuce; in Spain on *Lactuca serriola*; in Greece on *Sonchus oleraceus*). Symptoms of corky root have also been reported in Italy.

In France, the aetiology of lettuce 'corky root' has never been determined, apart from cases of attacks of *Pyrenochaeta lycopersici*, a fungus which is also responsible for corky damage on roots (see fact file 10). Corky roots and taproots have also been observed in crops without it being attributable to this fungus; in this case it seems likely that they could be attributed to *R. suberifaciens*. Their frequency and the seriousness of the damage recorded has never justified the aetiology of this disease.

We should point out that certain authors have associated the effects of **plant toxins,** emitted into the soil during decomposition of green plant debris under very wet conditions, with these symptoms. This situation has very seldom been reported. Excessive **nitrogenous fertilization** has also been implicated; it probably leads to the release of too much ammonia and nitrites into the soil. In this case we refer to **non-parasitic 'corky root'** or **'non-infectious corky root'**.

It is quite clear that in some plots the joint action of *R. suberifaciens* and excessive nitrogenous fertilization may be involved. We will not make any systematic differentiation between parasitic and non-parasitic corky root as their symptoms, as well as the methods of controlling them, are fairly comparable.

### • Principal symptoms

Generally speaking, diseased lettuce crops stand out in the plot because of their stunted size and partial head formation. The lower leaves may wilt at the hottest times of the day, sometimes becoming yellow and necrotic during severe attacks.

If several plants which have not grown well are pulled up, root damage, which can sometimes be considerable, can be noted. Roots presenting yellowish lesions rapidly become brown. In places, on the roots and taproot, we can see surface corkiness and longitudinal cracks. These also tend to swell. Their surface becomes rough and cracked as a result and we can see grooves, crests, and corky ridges on the root system.

The taproot is often badly affected; it is extremely corky and hypertrophied and becomes brittle. Ultimately, as numerous damaged roots have disappeared, only a small section of the taproot remains on the plants as well as a few adventitious roots which have newly formed around the crown.

Non-infectious corky root, associated with nitrogenous toxicity, is characterized by a pink to red colouring of the roots and central cylinder.

(See **4, 385, 405–408, 428, 430–433**)

• **Biology, epidemiology**

Very little is known about the biology of this bacterium.

**Survival:** *R. suberifaciens* can survive for several years in the soil, especially on plant debris. In fact, it can survive for at least 3 years in the soil without any presence of l.s.v. It has a fairly narrow range of hosts, mainly belonging to the Asteraceae family (endive, wild lettuce, *Sonchus oleraceus*, and so on). It can also be found on the roots of various plants which may allow it to be perpetuated in a plot (broad bean, tomato, melon, barley, and so on). This bacterium forms part of their rhizoflora, although it does not always cause apparent symptoms on their roots.

**Penetration:** The first stages of its infectious process are unknown.

**Dissemination:** This takes place via water, ploughing tools polluted by contaminated soil, and so on.

**Conditions favourable to its development:** The expression of this disease is favourably influenced by the excessive application of nitrogenous fertilizers. The dose applied and the nature of the fertilizer are fairly important. Nitrate forms are probably more conducive to corky root than ammoniacal forms. Heavy and asphyxiating soils, as well as those which have received too much irrigation, especially furrow irrigation, and a rise in temperature, predispose plants to this disease. Lettuces growing in light soils can also be affected if they are watered too much.

Temperature also influences the development of this bacterium. This is optimum at 31°C, but is relatively effective from 10°C. Its effects cease at 36°C.

## Protection

• **During cultivation**

Once symptoms are noted when lettuces are growing in a plot, unfortunately it is too late to intervene. We do not have any effective measures to prevent the progress of the disease: the damage is done.

We can, however, recommend reducing irrigation if it is too copious.

During harvesting, and when the crop has finished growing, make sure you do not bury too much plant debris in the soil.

• **Subsequent crop**

As the symptom of 'corky root' mainly occurs in plots where lettuce crops are repeatedly grown, if possible, fairly long periods of crop rotation must be implemented. For example, following a rotation with sugar cane we noticed a reduction in the damage caused by *R. suberfaciens*. Burying green manure can improve the structure of the soil and temporarily reduce infectious corky root. These results have been obtained in the USA by burying rye 3 weeks before planting. When the next crop of lettuce is planted, the soil must be dug over deeply in order to improve its texture; drainage must also be improved.

Disinfecting the soil with a fumigant is effective. Dazomet and methyl bromide have proved very effective in California.

L.s.v. must be planted on raised beds and their irrigation carefully monitored in order to ensure deep root development of the plants. It is preferable to use plants rather than sowing seeds directly. Excessively prolonged irrigation, especially in the first stages of seedling growth, must not be carried out as this saturates the soil with water. The disease establishes itself during these periods. Sprinkler or drip irrigation seems less conducive to the disease than furrow irrigation.

Soil analyses must be carried out in order to check that it does not contain high residual concentrations of nitrogen. It is preferable to use types of fertilizer which release nitrogen slowly.

A recessive monogenic resistance has been encountered in the genus *Lactuca* ('*cor*' gene). This resistance has been introduced in the USA in certain cultivars. Unfortunately, strains capable of circumventing this resistance have appeared. It is not present in French cultivars of lettuce.

# Phytoplasmas

## General information

The **Mollicutes** are the smallest and simplest known living organisms. They are prokaryote micro-organisms, like bacteria, but smaller in size. They also **lack a wall**, as they do not have the genetic information required for its formation. Nevertheless, mollicutes probably share a common ancestor with two Gram$^{+}$ bacteria from the *Clostridium* genus. Some of these can be grown *in vitro* (**spiroplasma**), whereas others cannot (**phytoplasma**, e.g. **mycoplasma** or MLO, *mycoplasma-like organism*). These are essentially **obligatory parasitic** micro-organisms of tissue cells conducting the elaborated sap (**phloem** vessels). They are susceptible to certain antibiotics, especially tetracycline which blocks their development but does not kill them.

Only the phytoplasma from the Aster yellows group currently pose problems for crops of l.s.v. Observation under an electron microscope reveals that they are fairly **polymorphous**. Their symptoms are relatively typical: yellowing and deformation of leaves, accumulation of latex, and so on.

# *Candidatus phytoplasma* sp. Aster yellows group

(Prokaryotes, Bacteria, Mollicutes)

# Aster yellows phytoplasma

## Principal characteristics

Aster yellows affect over 350 different plant species, both cultivated and uncultivated, over among 50 botanical genera. Several strains of phytoplasma have been recorded on l.s.v., and these may be responsible for somewhat different symptoms.

### • Frequency and extent of damage

This micro-organism has only been identified on l.s.v. (essentially on lettuce) in a limited number of countries, mainly the USA, Canada, and Italy. It has been causing damage for numerous years in several American states (New York, Wisconsin, Oklahoma, Texas, and so on). It may vary considerably in severity from one region and/or one year to another. This was the case in Texas during the 1980s, where differences of 5–46% of affected plants were observed from one experimental site to another. In other states, severe attacks affecting 100% of plants were reported. The same situation is found in the Canadian province of Ontario.

During the same period, and at the beginning of the 1990s, epidemics (which were sometimes severe) were reported in Italy, in the open field and sometimes even under protection, in several regions located in the centre and south of the country as well as on the Italian Riviera. In certain cases, this phytoplasma was found mixed with another phytoplasma responsible for chrysanthemum yellows or with viruses (BWYV, LMV, and so on). The phytoplasma is probably also present in Germany.

The incidence of this phytoplasma in other lettuce growing countries is not known. This also applies to France, where attacks of this micro-organism have been suspected in the presence of l.s.v., mainly displaying latex accumulation all along the leaf veins. The laboratory investigations carried out (electronic microscopy, serology, and so on) have never been able to reveal a particular phytoplasma, or any other micro-organism.

### • Principal symptoms

Infected plants present fairly varied symptoms. Young leaves from the heart may be affected first. They are chlorotic and only partially develop. In addition to being small in size, they are twisted and rolled.

If attacks take place on plants which are already well developed, we note chlorosis of the young leaves as well as 'tip burn' on the lamina. Thick axillary buds sometimes develop in the heart of the plant. Leaves from the crown become yellow and deformed (often smaller, rolled, and so on).

Latex deposits, pinkish to brown in colour, are visible under the veins of diseased leaves. Intense necrosis of the phloem can be seen.

If infection takes place at an early stage, plants remain dwarfed, and while their external leaves become yellow. If it takes place at a late stage, deposits of latex are visible only on a few leaves. In many cases, their head fails to form and consequently the vegetables cannot be marketed.

The symptoms described in Italy involve reddening and deformation, particularly on the lower leaves. The latter are also thicker. We also note the absence of head formation, intense necrosis of the phloem, and dwarfism, which means that the plants cannot be marketed.

### • Ecology, epidemiology

**Survival, sources of phytoplasma:** Phytoplasma survive in leafhopper vectors and on various hosts, cultivated (carrot, spinach, celery, and so on) and uncultivated, which constitute significant phytoplasma reservoirs. In the case of the phytoplasma responsible for Aster yellows, it survives the winter in the body of its vectors. This phytoplasma is polyphagous, and is mainly found on several weeds (*Salsola tragus*, several species of *Sonchus* and *Plantago*, *Taraxacum officinale*, *Lactuca serriola*, and so on). It has also been reported on Brussels sprouts, *Senecio cruentus*, *Argyranthenium frutescens*, Dahlia, *Hydrangea macrophylla*, *Lavandula officinalis*, olive tree, *Primula* sp., *Spartium junceum*, *Violeta odora*, tomato, and so on. These plants harbour it, multiply it, and also act as sources of inoculum.

The cycle of the leafhopper vector involves eggs which do not play any role in the survival of the phytoplasma over the winter, but guarantee the persisting presence of the insect from one season to the next.

**Transmission, dissemination:** As we suggested above, this phytoplasma is transmitted persistently, when several species of leafhopper: *Macrosteles laevis*, *M. quadripunctulatus*, *M. severini*, *M. quadrilinaetus* (syn. *M. fascifrons*) puncture the plant in search of nutrition. The last of these species seems most frequently involved in transmitting this disease. Leafhoppers, insects which pierce and suck sap, contaminate l.s.v. in the course of their migration. Once they come into contact with the leaf, they puncture the phloem vessels in order to obtain their nutrition, passively injecting or withdrawing phytoplasma (**487**). The phytoplasma involved multiply in these insects, affecting various organs, including the salivary glands, therefore making the leafhoppers infectious. *M. quadrilinaetus* may remain in this state for at least 100 days.

These insects occasionally visit infected plants. They are capable of flying long distances or remaining in one place. Five generations may succeed each other on different hosts in the course of one season. The date on which symptoms appear, generally 30–45 days after contamination, depends on the migration period of the vector or vectors. Migration is a complex phenomenon which is the result of insect populations moving from place to place in mass flights. Causes of migration are poorly understood. They seem to be associated with local conditions which are unfavourable to leafhoppers. Among the factors which influence migration and the nature of flights are: hunger, overpopulation, deterioration of the host, endocrine deficiency in insects or genetic effects, light availability, temperature, wind, and so on. Insects prefer young plants with succulent tissues. In drought conditions they move more readily from wild plants to irrigated crops. Impulsive movements are sometimes noted, but poorly understood. Cold winters help to reduce winter populations.

The Aster yellows phytoplasma cannot be transmitted by seed.

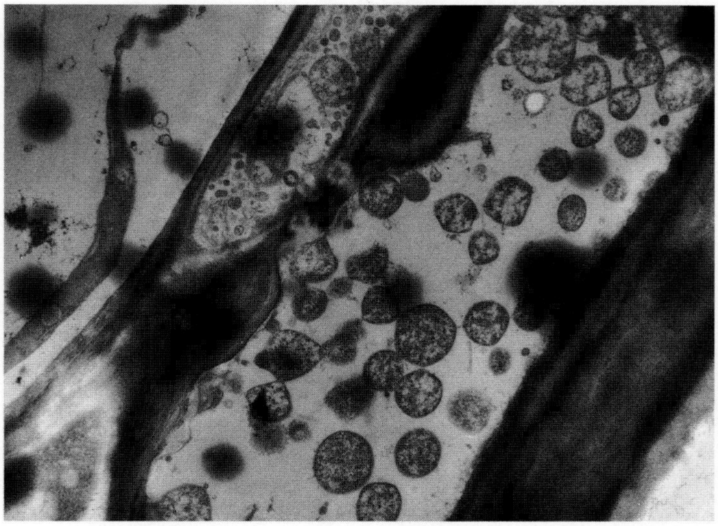

**487** Several rather spherical structures characterizing the presence of *Phytoplasma* sp. in phloem vessels.

## Protection*

- **Subsequent crop**

In countries where attacks occur in the nursery, plants must be protected. The best solution consists of using agrotextiles (unwoven mesh, woven fabric) constituting a mechanical barrier which is more effective than insecticide treatment. The effectiveness of insecticides is very controversial. Although a number are extremely effective against leafhoppers, they will not always prevent contamination. On the other hand, insecticide treatments employed against other insects will sometimes be sufficient to limit populations of leafhoppers. We have seen that aluminized mulching has

* The control of insect populations on a crop often involves using insecticides. In the case of l.s.v., their use may be questioned and legislation varies depending on the country. In order to give a 'universal' character to the proposed control methods, we recommend carrying out treatments which have been reported in publications.
**It is clear that these proposals must be modified according to the country concerned and the pesticide legislation in force.**

been successful in reducing the number of vectors and the incidence of the disease.

Careful weeding of nurseries, plots and their surroundings (edges of hedges and paths, and so on) must be carried out in order to eliminate the weeds referred to above. It is advisable to use healthy plants and not to plant crops of l.s.v. next to other crops which have already been affected.

Currently there are no varieties of lettuce resistant to the Aster yellows phytoplasma. A few differences in behaviour between cultivars have been reported in the USA, as well as tolerance in the Trianon Cos genotype.

# Viruses

## General information

Viruses are infectious entities which are infinitely small and invisible under an optical microscope. Their structure is particularly simple. It is limited to a protein shell, the **capsid**, containing a **nucleic acid** which is usually **RNA** (ribonucleic acid) in phytopathogenic viruses rather than DNA (deoxyribonucleic acid). This nucleic acid takes the form of one or several molecules in a chain formed of hundreds, or even thousands of units known as **nucleotides**, each of which is composed of **bases** (adenine, guanine, cytosine, uracil). It is therefore protected by the capsid which is itself made up of repeated sub-units (**capsomers**). This protein, which is very antigenic, is frequently used to produce specific serums which can detect plant viruses. Some of these, although fairly rare, also have a lipid envelope for example the tomato spotted wilt virus (TSWV).

The use of an electron microscope shows that the viral particles, or virions, affecting the l.s.v. take the form of symmetrical particles whose size can be measured in nanometres; they vary in shape: elongated rods (TRV), serpentine rods (LMV, TuMV, ENMV), bacilliform structures (AMV), spherical (TSWV), isometric (CMV, BBWV, DaYMV), and so on.

Numerous viruses can be artificially inoculated into l.s.v.; over 20 or so naturally attack these plants in the various production areas. Each of these viruses attacks a number of hosts, varying in size. L.s.v. are sometimes attacked by several viruses at one time; this is the case in France, for example, with CMV and LMV. Viruses only develop in living cells, and they are referred to as **obligate parasites**. These agents use the cellular 'biochemical machinery' to their own benefit in order to ensure their multiplication and the production of certain molecules (specific enzymes, lipids). They cause a range of symptoms which may lead to confusion and make their identification difficult. Consequently, various laboratory techniques should be used in order to detect them and identify them with certainty: indexing on various host plants, electron microscopy, serology and immuno-enzymatic methods, molecular techniques (polymerase chain reaction (PCR), molecular hybridization, and so on).

In general, if the plant cells of l.s.v. die, the virions share the same fate, with the exception of some very stable viruses which retain their infectious power in soil or in plant debris. If they are to perpetuate themselves it is essential for them to spread to the cells of healthy plants and penetrate them; to do this, the viruses have various means of disseminating themselves in nature.

A few rare viruses are easily **transmitted by contact**, particularly in the course of various growing operations. Indeed the grower, as he works with a crop, causes injuries to plants (breaking epidermal hairs, for example) which contaminate his hands, his tools and even his clothes. If healthy plants are damaged in this way the virus will be able to penetrate the plant cells. This method of transmission does not seem to affect l.s.v.

The vast majority of viruses affecting l.s.v. are transmitted by biting insects. **Aphids**, apart from their biological characteristics, are redoubtable virus vectors in these plants, particularly in a **non-persistent mode**. In this case, the viruses are acquired or transmitted by aphids in a few seconds in the course of brief trial punctures which allow these insects to 'taste' the plant on which they have landed. Aphids will remain viruliferous, that is capable of transmitting viruses, over a short period of time (from a few minutes to a few hours) (LMV, CMV, AMV, BBWV, DaYMV, TuMV, and so on). These viruses are categorized as **non-circulating**.

There are also other modes of transmission:

- the **semi-persistent** mode, viruses are transmitted to the plants or withdrawn from them in the course of longer punctures lasting around 30 minutes;
- the **persistent mode** (BWYV, LNYV, SYNB, and so on); the acquisition or transmission of viruses takes place in the course of extended feeding punctures (acquisition can last from a few hours to 1–2 days) located on the phloem vessels. Once absorbed, the **viruses complete** their cycle in the insect's body before they can be transmitted again; these are known as circulating viruses. They move through the digestive tract, the general cavity, and then concentrate in the salivary glands. Once they are viruliferous, aphids will remain in this state for several days and sometimes for the whole of their life.

Other insects also able to effectively transmit several viruses to l.s.v.: **whiteflies** (BPYV, LCV, LIYV) and **thrips** (TSWV, TSV).

A number of organisms, particularly fungi and nematodes, are also virus vectors. LRNA, MiLV and TNV are transmitted by a little evolved aquatic **fungus**: *Olpidium brassicae*. This can survive easily in the soil thanks to its resting spores. It has mobile, flagellated zoospores which, by contaminating the roots of healthy plants, partially guarantee the dissemination and transmission of these viruses. Several **nematode** genera (*Xiphinema*, *Trichodorus*, *Paratrichodorus*) ensure transmission of tobacco ring spot virus (TRSV), or tobacco rattle virus (TRV) in the course of their parasitic activities on the roots.

Among the viruses seriously affecting l.s.v., only LMV is transmitted by **seeds** in lettuce. This method of transmission by descent is sometimes found in certain other plants, particularly wild ones, which act as hosts for these viruses.

# Viruses transmitted by aphids

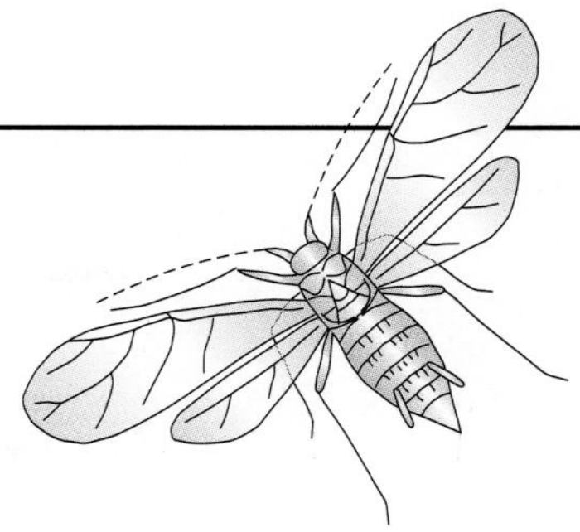

## Lettuce mosaic virus (LMV)

(*Potyviridae*, *Potyvirus*)

### Principal characteristics

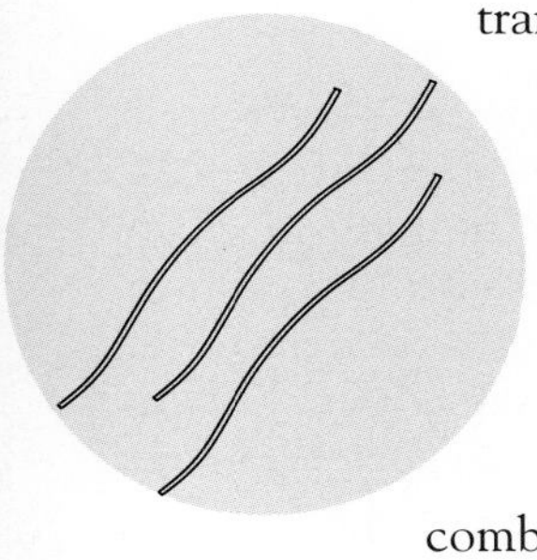

LMV is one of the most serious viruses attacking lettuce. It is a potyvirus which, in this species, is transmitted by seed. It infects escarole and endive (frisée) equally. There are numerous strains which are differentiated from each other by their biological and serological properties. For example, very aggressive strains have been isolated from wild composites, such as *Helminthia echiodes* (Picris), and also from endive and lettuce.

For several years, strains capable of combating the resistance gene or genes used most frequently in selection ($mo1^1$ gene, formerly '*g*' in Europe, $mo1^2$, formerly '*mo*', in the USA) have been isolated in numerous countries. A molecular characterization of the variable regions of the genome of this virus has allowed us to establish phylogenetic relationships in about 20 isolates which have been split into three groups.

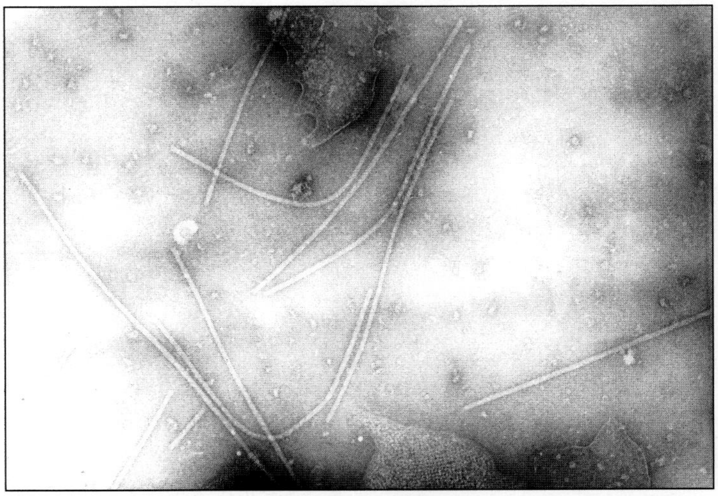

**488** Serpentine viral particles of 15 nm in diameter and approximately 750 nm in length (potyvirus type).

- **Frequency and extent of damage**

First recorded in the USA (in Florida), then in Europe, LMV has now been identified in every region where l.s.v. are grown, from the most northern zones to the warmest zones.

In the **open field**, it is still the major virus in countries where the production and use of controlled 'virus free' seeds is not general practice among lettuce growers. The same is true in countries where there are no resistant varieties. Occasionally, in spite of healthy seeds, we may observe serious local epidemics if susceptible varieties are grown. This situation may be due to changes in managing the growing medium, for example the introduction or inadequate control of virus reservoir plants.

In growing zones where resistant varieties are cultivated, the emergence of **virulent strains** which can combat the genes used can also result in serious local or regional damage. The frequency of LMV is generally low in crops grown under protection.

- **Principal symptoms**

If young plants are grown from infected seeds, on the leaves we very quickly observe lightening of veins, mosaic, or even crinkling of the lamina, and sometimes a few necrotic holes. Following these early infections, growth of plants may be greatly reduced. Ultimately, plants are stunted, fail to form a head, and therefore cannot be marketed.

Plants infected at a later stage reveal light green to yellow mottling, leaf deformation, and particularly rolling of external leaves. Plant development may be reduced to varying degrees. In certain cases of late infection, the expression of symptoms may be very limited; and only the dull appearance of plants indicate that they have been contaminated with LMV.

Symptoms are particularly clear in plants during the active growth stage; they are generally more characteristic and more pronounced in butterhead lettuce, romaine or leaf lettuce than in batavia or iceberg types.

You should be well aware that symptoms of LMV can be very variable, depending on the type and variety of lettuce grown, the strain of virus involved, the stage of plant development at the time of infection, and environmental conditions. For example, on certain varieties of lettuce and in the presence of particular strains of LMV, yellow areas or yellowing of veins may appear on the oldest leaves. Areas of necrosis may be dotted over the leaves or broader areas may appear, generally located around the edge of the lamina and are visible under contrasting temperature conditions.

On seed-plants, seed production is affected by the susceptibility of the variety and how early the infection strikes.

(See **14, 23, 24, 65–76, 309,** and **311**)

## • Ecology, epidemiology

**Survival, sources of virus:** This virus is transmitted by seeds in cultivated lettuce and also by *Lactuca serriola*. Infection of the embryo may originate in the ovule and, more rarely, in the pollen. The transmission rate varies depending on the genotype, strain and conditions of the medium; and may reach 15%. In varieties with a recessive resistance gene, the rate of transmission has been recorded as zero or lower than one per thousand. However, certain strains capable of combating resistance and transmissible at rates which may exceed 10% have appeared in several countries.

When using the seed lots which have not been checked or those which have become infected, it is the young seedlings from seeds housing the virus which are the initial source of inoculum. In this situation, epidemics in the field are very serious. In addition, LMV has a fairly broad spectrum of hosts likely to be infected and act as sources of inoculum. Numerous wild plants constitute reservoirs for potential viruses and act as source or relay plants. Among the Asteraceae, *Lactuca serriola*, *Lactuca virosa*, groundsel, sowthistle, dandelion, *Helminthia* sp. shepherd's purse, starwort, *Lamium amplexiacaule*, chenopodium, and so on, also harbour LMV. Amongst susceptible cultivated plants we may also list the pea, chickpea, and safflower; but their role in epidemics of LMV on l.s.v. seems minor. Over recent years, several ornamental species have been recognized as being hosts of LMV: *Osteospermum*, gazania, petunia, china aster, lisianthus. The favourable influence of the first two species in epidemics of LMV on lettuce has been clearly shown in California. Crops of endive, and also spinach, help to maintain LMV over winter.

Growing repeated crops of lettuce in the same place often results in a gradual increase in the infection rate in plots. This occurs above all when the varieties grown are susceptible to native strains and if numerous plants which are susceptible to the virus persist in the plot environment, acting as reservoirs for the virus.

**Transmission, dissemination:** LMV is transmitted by aphids in a non-persistent way. The vector aphid may acquire the virus from an infected plant or transmit it to a healthy plant in a few seconds, in the course of very brief punctures, known as 'test punctures'. These punctures allow it to verify that the plant is a host which will be favourable to its development. The aphid is capable of transmitting the virus immediately once it has been acquired, and remains capable of this for around 10 minutes. It loses this ability if it makes other test punctures or feeding punctures which last for a longer period. The proportion of plants it may contaminate is at its maximum in the vicinity of the source plant. Transported by the wind, it may be a formidable vector, effective over relatively long distances.

Extremely numerous species of aphids, especially *Myzus persicae*, *Macrosiphum euphorbiae*, *Aphis gossypii*, and *Aphis craccivora*, *Hyperomyzus lactucae*, *Nasanovia ribisnigri*, *Pemphigus bursarius*, and so on, are likely to transmit LMV with varying degrees of success.

The extremely high effectiveness of the transmission method of this virus means that very rapid dissemination of the disease may be observed in the plot without significant swarms of aphids having been observed.

## Protection*

### • During cultivation

Unlike the situation with cryptogamic diseases, there is no curative method which can effectively control viruses, particularly LMV, while the crop is growing. An infected plant will remain in this state throughout its life, even if symptoms sometimes tend to subside.

If attacks take place in the nursery and are detected at an early stage, the few plants which present symptoms of LMV must be rapidly eliminated, and under no circumstances planted at a later date.

Aphid treatments are essential for controlling aphid populations affecting l.s.v. Unfortunately, they are frequently ineffective in controlling viral epidemics. In fact, aphid vectors often come from outside the plot and transmit the virus in the course of brief punctures, even before the aphicide has had time to act. In addition, the difficulties currently being encountered in relation to controlling aphids on l.s.v., sometimes associated with phenomena of insecticide resistance, do not improve this situation.

### • Subsequent crop

Combating lettuce mosaic is mainly based on using seeds tested as virus-free, and on using resistant varieties.

Depending on the country, and the conditions for producing seeds and l.s.v., the virus thresholds tolerated in seeds may vary. For example, in numerous European countries, the seeds used by specialists within the field are produced under conditions which minimize the risks of infection from seed-plants. Seed lots are nevertheless checked for absence of LMV. Consequently, the threshold of tolerance, established a few years ago on the basis of epidemiological studies, is 1 per 1,000. ELISA tests, carried out by selection companies, can guarantee this threshold.

In the USA, where almost all varieties cultivated are susceptible to LMV but where growing methods are different (large plots where the same crops are not grown repeatedly), the imposed threshold is lower, in the range of 0 infected seeds per 30,000. This policy has allowed a significant global reduction in the damage caused by this virus. Serious epidemics have nevertheless been observed locally in certain years. They are due, not to the poor quality of seeds, but to the presence of unsuspected virus hosts (*Osteospermum*, gazania) grown around the edge of plots of l.s.v. in some fields.

With regard to the use of **genotypes resistant to** LMV, some resistant varieties of iceberg and romaine (possessing the $mo1^{2}$ gene) exist in the USA, but their use is not yet widespread. In France, and progressively in other European countries, the introduction of the $mo1^{1}$ resistance gene in most varieties of butterhead lettuce, batavia, romaine and cutting lettuce, can result in a fairly full range of resistant varieties.

Unfortunately, strains capable of combating the resistance conferred by the $mo1^{2}$ and $mo1^{1}$ genes have been identified in France, in other European countries and, more rarely, in the USA. The strains cannot be transmitted by seeds. For a dozen or so years other strains, virulent on resistant varieties and likely to be transmitted by the seeds at high levels on all varieties, have been isolated in France and South America (Chile, Brazil). This situation means that it is essential to check seeds from all varieties, whether susceptible or resistant. Our knowledge of LMV **resistance genes** used in lettuce has developed. The $mo1^{2}$ gene, identified in the USA in some *Lactuca serriola* from Egypt, and the $mo1^{1}$ gene identified in France in the Gallega de Invierno variety, have for many years been considered identical. Recently, it has been shown that these genes were alike, or very closely linked. The $mo1^{2}$ gene seems to be effective against certain strains which circumvent the $mo1^{1}$ gene. Another dominant resistance gene $mo1^{2}$ has been revealed in certain varieties of lettuce. It is very effective against certain isolates identified in Greece and the Middle East, but this gene is circumvented by most strains typically found in Europe, the USA, or Australia. Its use in selection programmes therefore seems of limited interest.

* The control of insect populations on a crop often involves using insecticides. In the case of l.s.v., their use may be questioned and legislation varies depending on the country. In order to give a 'universal' character to the proposed control methods, we recommend carrying out treatments which have been reported in publications.
**It is clear that these proposals must be modified according to the country concerned and the pesticide legislation in force.**

In lettuce no new source of resistance to the isolates combating the *mo1*[1] and *mo1*[2] genes has been found.

On the other hand, a dominant gene, *mo1*[3], effective against all the pathotypes identified, has been discovered in *Lactuca virosa*. The introduction of *Mo*[3] into lettuce is under way, starting from inter-specific crosses. In fact, ultimately several genes controlling different mechanisms of resistance will need to be brought together in order to obtain long-lasting resistance.

You should also be aware of all the measures which aim to prevent or, at the very least, limit as much as possible the introduction of LMV and its spread in l.s.v. plots. Consequently, in countries where contamination occurs at a very early stage, nurseries and young seedlings must be protected; unwoven mesh may be used to achieve this. The mechanical barrier created in this way will delay contamination. Careful weeding of nurseries, plots, and their surroundings (edges of hedges and paths, and so on) must be carried out in order to eliminate sources of virus and/or vectors. You must avoid setting up a nursery or salad crop next to crops which have already been affected or are susceptible to LMV, particularly pea, chickpea, safflower during summer periods, endive (frisée), and spinach in the winter.

We should point out that spraying mineral oils on l.s.v. has reduced the percentage of plants infected and therefore the damage associated with LMV.Il convient de signaler que des pulvérisations d'huiles minérales sur salade ont permis de réduire le pourcentage de plantes infectées et donc celui des dégâts liés au LMV.

# Cucumber mosaic virus (CMV)

*(Bromoviridae, Cucumovirus)*

## Principal characteristics

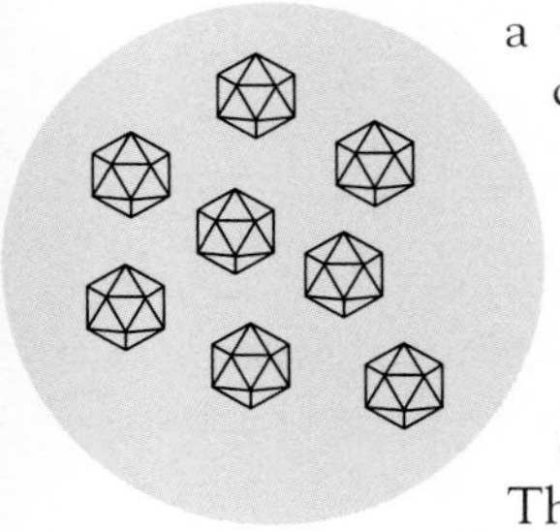

CMV is a very polyphagous ubiquitous virus. It has the largest recorded range of hosts. It has a very serious effect on market garden crops grown in the open field, particularly the Cucurbitaceae, and also tomato, pepper, spinach, and celery. A large number of strains have been described on the basis of their biological properties (symptomatology, thermosensitivity). There are two main groups of serotypes. Certain isolates possess a satellite, which can modify the expression of their symptoms on certain hosts.

### • Frequency and extent of damage

CMV has a serious effect on numerous cultivated species on every continent; its incidence on lettuce crops grown in the open field seems very variable. In Northern European regions (Great Britain, Germany, Belgium), it is considered to be a serious virus affecting output by 10–40% depending on variety. This is also the case in New York state, in the USA, where it can cause significant losses in autumn crops. In other European countries, such as France, Spain, or Italy, where the virus is extremely frequent in market garden areas, the damage caused to lettuce is considered to be fairly low. It is often isolated from all types of lettuce during the summer, but does not cause very marked symptoms during this season except for slight mottling and reduced growth. Consequently, its incidence may be underestimated. When the contrast between day and night temperatures is more extreme, as is the case with autumn crops, symptoms are more marked. This may result in a reduction in the number of plants suitable for marketing.

Combined CMV-LMV infections are common; in this case, we note more serious damage and stunted growth than is the case with a mono-infection by either of these viruses.

### • Principal symptoms

Symptoms vary considerably depending on climatic conditions, the stage of infection, and the lettuce variety. When plants are grown during the summer, in Mediterranean areas, symptoms are often limited to slight mottling or unobtrusive mosaic. Symptoms often tend to diminish as the plant grows. In periods of extreme contrasts in temperature, growth of young plants may be greatly reduced and yellow mottling and necrosis may be observed on leaves.

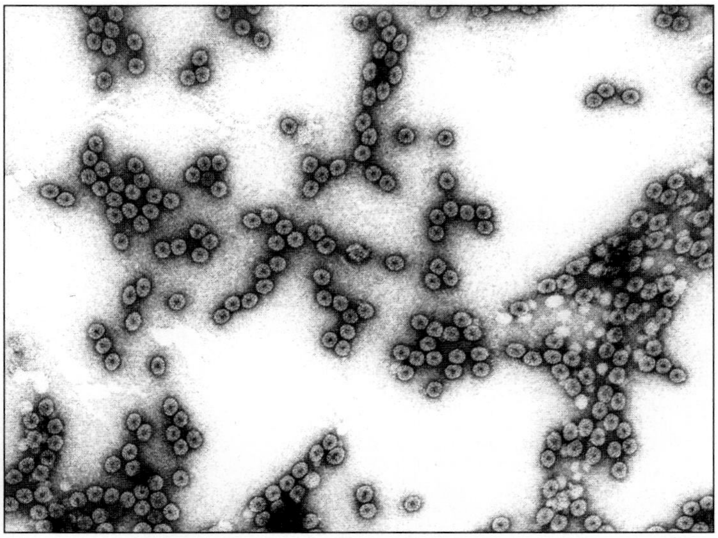

**489** Isometric viral particles; they measure approximately 29 nm in diameter.
**CMV**

The effects of CMV on lettuces are so slight that evaluations of sensitivity in cultivated varieties can only be established on the basis of the percentage of weight loss on harvesting. Isolates belonging to both serotypes have been isolated from lettuce, but no difference in aggressiveness between these has been revealed.

(See **77–79,** and **311**)

- **Ecology, epidemiology**

**Survival, sources of virus:** CMV can infect over 800 different species representing 85 botanical families, from both the monocotyledons and the dicotyledons. Market garden and ornamental crops (hardy annuals and perennials) are particularly badly affected.

Among spontaneous plants, species as common as purslane, black nightshade, groundsel, sowthistle, wild lettuce, dead nettle, speedwell, madder, and mercury are infected.

During the summer period, crops such as the Cucurbitaceae, tomato and pepper, and in winter spinach and celery, as well as numerous wild plants, may play this role. Weeds above all guarantee the survival of the virus from one season to another on and in the vicinity of plots. In addition, some species, such as starwort, are capable of transmitting the virus through the seed to their descendants.

**Transmission, dissemination:** CMV is transmitted by aphids in a non-persistent way. In the course of very brief punctures, known as 'test punctures', the aphid vector can acquire the virus on an infected plant and transmit it to a healthy plant in a few seconds.

These punctures allow it to verify that the plant is a host which will be favourable to its development. The aphid is capable of transmitting the virus immediately once it has been acquired, and remains capable of this for around 10 minutes. It loses this ability if it makes other test punctures or feeding punctures which last for a longer period. The proportion of plants it may contaminate is at its maximum in the vicinity of the source plant. Transported by the wind, it may be a formidable vector, effective over relatively long distances. Around 60 species of aphids, including *Myzus persicae*, or A *phis gossypii*, are capable of transmitting CMV.

As with almost all cultivated species, the virus is not transmitted by seeds in l.s.v.

The extremely effective method of transmission means that dissemination of this virus can be very rapid without large swarms of aphids being observed. Several abiotic factors play an important role with regard to the biology and effectiveness of flights of aphids:

- wind influences their distribution;
- temperature affects lettuce growth, multiplication of the virus, and development of aphid colonies;
- the crop environment. The proximity of other susceptible crops and wild plants which have been contaminated particularly encourages contamination.

We must add regional parameters to these factors, such as arrangement of plots, hedges used to protect crops against prevailing winds, local climatic conditions and their influence on certain reservoir plants, and so on.

## Protection*

- **During cultivation**

No curative method can combat CMV. An infected plant will remain in this condition throughout its life.

Generally speaking, eliminating plants affected by the virus right at the start of an epidemic can help to slow this down; the difficulty in observing the characteristic symptoms of CMV in lettuce sometimes makes this approach impractical. In addition, as symptoms only appear around a fortnight after the plants have been contaminated, l.s.v. in the incubation stage may constitute a virus source and may play a part in spreading the epidemic.

Aphicide treatments are essential in order to limit swarms of certain aphids on lettuce. Generally, they are ineffective in controlling the spread of epidemics of different types of viral diseases. In fact, vector aphids often come from outside the plot and may transmit the virus even before the aphicide has had time to act.

* The control of insect populations on a crop often involves using insecticides. In the case of l.s.v., their use may be questioned and legislation varies depending on the country. In order to give a 'universal' character to the proposed control methods, we recommend carrying out treatments which have been reported in publicatons.
**It is clear that these proposals must be modified according to the country concerned and the pesticide legislation in force.**

- **Subsequent crop**

It is advisable to implement a set of measures aimed at preventing or, at the very least, limiting the introduction of the virus as much as possible, as well as its spread in plots of l.s.v.

In countries where contamination takes place at a very early stage, nurseries and young seedlings must be protected. To do this you may have recourse to agrotextiles (unwoven mesh, woven fabrics, and so on). The mechanical barrier created in this way will delay contamination.

Careful weeding of nurseries, plots and their surroundings (edges of hedges and paths, and so on) must be carried out in order to eliminate sources of virus and/or vectors. Crops of l.s.v. must not be planted next to crops which are more susceptible to CMV, such as tomato, spinach, the Cucurbitaceae, and so on. Crop rotation may be envisaged in production zones where the pressure from infected weeds in surrounding areas is significant.

In Great Britain and Germany, as well as in the USA, where the virus may result in significant losses, studies have revealed varieties which are less susceptible to CMV, but no really effective resistance has been discovered in *Lactuca sativa*.

On the other hand, a dominant monogenic resistance has been identified in an accession of *Lactuca saligna* and introgressed into lettuce in the USA. Unfortunately, this resistance is specific to a few strains and has already been circumvented. Other accessions of *Lactuca saligna* and *Lactuca serriola* have proved to be tolerant. The accumulation of the various genes involved in these tolerances is very likely to constitute a real solution for effectively controlling CMV. In addition, *Lactuca virosa* may be resistant to this virus.

# Beet western yellow virus (BWYV)

(*Luteviridae*, *Polerovirus*)

## Principal characteristics

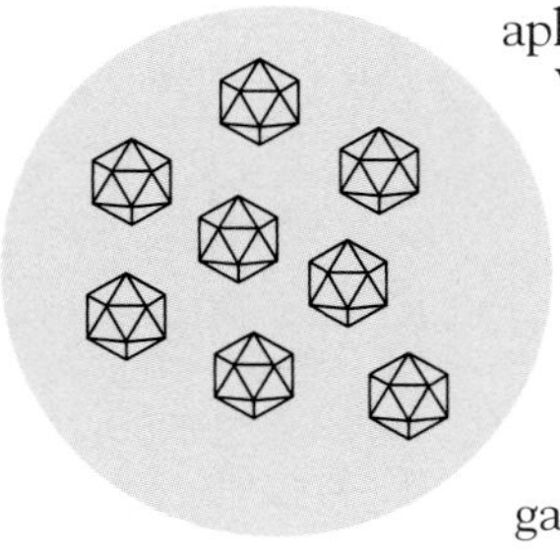

BWYV, like all the *Luteoviruses*, is transmitted by aphids and is restricted to the phloem. Worldwide, it affects major cultivated species such as beet, cabbage, rape, soya, and so on. It is one of the most serious viruses affecting lettuce grown in the open field in Europe.

Controlling this virus is currently very difficult, particularly in market garden crops.

A recent study has been carried out on beet in order to discover more information about the biological and molecular variability of BWYV, the virus responsible for the moderate yellowing in this species. It has actually revealed three distinct viral species, which have recently been submitted to ICTV:

- Beet mild yellowing virus (BMYV);
- Beet chlorosis virus (BChV);
- Brassica yellowing virus (BrYV).

Only BMYV is likely to infect lettuce; in fact it has already been described in Germany as affecting this plant.

### • Frequency and extent of damage

Identified on lettuce in the 1950s in the USA, BWYV was then isolated in Great Britain (1970), in France (1977), and throughout Europe, Israel, Australia, and Japan. Its difficult identification (virus not mechanically transmissible, tricky serological detection) delayed this proof. Consequently, for many years symptoms of the disease were attributed to other reasons, particularly to mineral deficiencies. Nowadays the virus attacks in the open field in all production zones, but its damage is particularly serious during very light periods of the year and/or in sunny regions.

The name '**summer yellows**' has often been given to the disease. All types of lettuce are susceptible, the 'butterhead' types, selected for their resistance to going to seed, are more particularly affected. In terms of plants which are suitable for marketing, in the batavia type, the consequences of the disease are less serious overall. The same is true for the red varieties if they have not become infected at too early a stage. Escarole and frisée endive, as well as chicory, are susceptible, but the damage caused to autumn crops usually appears to be minor.

### • Principal symptoms

After infection by aphids, symptoms generally appear between 15–25 days depending on environmental conditions (light intensity), the stage of infection, and varietal suscepti-bility.

Initially, chlorotic patches appear on the lamina of the oldest leaves, then on the intermediate leaves. Gradually these spots extend and become clearly yellow. Ultimately, we see virtually generalized interveinal yellowing; an area of tissue which is still green persists along the veins. Affected leaves, particularly lower ones, may have a necrotic zone around the edge of the lamina. In addition, they are thicker and break when folded, this allows us to distinguish this virus from premature physiological senescence. In addition, this disease has sometimes been confused with magnesium deficiency.

Early infections cause significant reduction in the weight of the lettuce head (up to 50%). The phenological stage of the plant at the time of infection and the relative susceptibility of the variety influence the extent of the damage. This damage, of varying degrees of severity, must be trimmed off lettuce varieties in order to make them suitable for marketing.

In infected seed-plants, inter-veinal yellowing is very characteristic on the bracts; in addition, fewer seeds are produced.

No change in the degree of strain aggression has been revealed on lettuce, except in the USA where a particular isolate, BWYV-ST9 with an ARN satellite, causes more serious symptoms. Mixed infections of BWYV+LMV or BWYV+ CMV cause much more significant damage. The synergistic effects of BWYV and LMV in lettuce may cause internal necrosis of veins in certain varieties of iceberg type batavia lettuce.

(See **8, 120–129**)

- **Ecology, epidemiology**

**Survival, sources of virus:** BWYV has an extremely broad range of hosts since it may infect approximately 150 species from 23 botanical families. Potential sources of virus are therefore very numerous, both in cultivated species and in spontaneous species. Beetroot, chard, spinach, all types of cabbage, rape, turnip, radish, soya, pea, chickpea, broad bean, a few of the Cucurbitaceae, potato, and so on, present symptoms varying in intensity.

Certain isolates differ according to their range of hosts. For example, some beetroot isolates do not infect lettuce; isolates found on lettuce or cabbage do not infect beetroot, or are not very serious. This behaviour can now be explained by the 3 viruses responsible for yellowing which have been described as affecting beetroot (BMYV, BChB, and BrYV).

A low proportion of weeds susceptible to this virus show symptoms of yellowing or inter-veinal reddening, but most act as healthy carriers (infected plants do not have any apparent symptoms). The most common are wild lettuce, groundsel and sowthistle (Asteraceae), shepherd's purse and garden cress (Crucifereae), and also purslane, plantains, mallow, amaranth, saltwort, and so on.

Growing successive crops of lettuce, leguminous crops (cabbage and cauliflower, rape, spinach) or other susceptible crops in the field during the winter, as well as the existence of numerous susceptible wild species, help the survival of an inoculum throughout the year.

**Transmission, dissemination:** This virus can be transmitted by several species of aphid in a persistent way, involving an extremely specific mechanism. In this way the virus integrates into the insect via the alimentary canal, then moves into the haemocoele in which a latency period of 12–24 hours is required before BWYV accesses the salivary canal and can be transmitted. In fact, it takes almost 48 hours from the phase of acquiring the virus from the plant's phloem vessels followed by its complete cycle in the aphid to its extremely effective transmission. The potential aphid vector thus remains infectious for over 50 days. The virus is retained by the aphids during their successive metamorphoses but is not transmitted to their descendants.

Among the dozen or so species capable of transmitting BWYV, *Myzus persicae* and *Macrosiphum euphorbiae* seem to be the most important vectors. *Aphis craccivora*, *Aphis gossypii*, and *Acyrthosiphon solani* also transmit it.

The virus is not transmitted by seeds.

The particular nature of the method of transmitting this virus via aphids, the broad range of plant species susceptible and capable of acting as a source of inoculum, mean that dissemination of BWYV is very extensive, and also means that this can take place over very long distances. In addition, the seriousness of epidemics will often be proportional to the level of aphid populations which are potential vectors.

## Protection*

Combating BWYV is very difficult, given the large number of plant species likely to be infected, the persistence of the virus in the vector, and the relative abundance of species of vector aphids.

- **During cultivation**

No curative method can combat BWYV. An infected plant will remain in this state throughout its life.

Aphicide treatments are essential to limit swarms of aphids on l.s.v. They are generally relatively effective in controlling the

* The control of insect populations on a crop often involves using insecticides. In the case of l.s.v., their use may be questioned and legislation varies depending on the country. In order to give a 'universal' character to the proposed control methods, we recommend carrying out treatments which have been reported in publicatons.
**It is clear that these proposals must be modified according to the country concerned and the pesticide legislation in force.**

development of epidemics of this virus if they are applied as a precaution, thus helping to reduce the aphid populations in crops and surrounding environment. Once summer yellows is firmly established in the crop, it is often too late to intervene effectively.

- **Subsequent crop**

It is advisable to implement a set of measures aimed at preventing, or at the very least limiting the introduction of this virus as much as possible, as well as its spread into crops of lettuce.

In countries where contamination takes place very early, nurseries and young seedlings must be protected. Agrotextiles may be used for this purpose (unwoven mesh, woven fabrics, and so on). The mechanical barrier created in this way will delay contamination.

Even if eliminating source plants may seem unrealistic, **careful weeding** in and around plots of lettuce will limit potential sources of viruses and vectors.

In production areas where l.s.v. plantings are staggered, in adjacent plots, the frequency of infection will increase progressively. Given the fairly long period of incubation (3–4 weeks depending on the stage of infection) that separates the time of infection from the appearance of symptoms, any delay in plant contamination will reduce the extent of the damage. In fact, plants infected at a late stage in the course of their growth cycle will only display symptoms on a small number of leaves. Gentle trimming will make these plants suitable for marketing.

Limiting aphid populations, thanks to the use of insecticides, outside and inside the plot or plots will help to reduce epidemics. On lettuce, a plant on which problems with residues occur particularly easily, the range of insecticides which are fairly persistent should be used carefully, following the manufacturer's instructions and those of your plant expert.

An important varietal screening study, covering the hundreds of lettuce genotypes, has been carried out by several research teams. It has not revealed any high-level resistance. Nevertheless this study has shown that batavias and icebergs are less susceptible than romaine or butterhead lettuce. It also seems that differences in sensitivity exist within each of these types.

In addition, we should point out that crosses between varieties resistant to going to seed (America type, very susceptible to the virus) and more light coloured spring varieties have led to summer varieties with lower susceptibility.

A recessive *bwy* gene, conferring good tolerance to BWYV, has been identified in one variety of batavia and one of butterhead lettuce. In addition, a dominant gene has been identified, giving quasi-immunity, in an accession of *Lactuca virosa*. Interspecific crosses have been carried out and a number of families of descendants are being evaluated.

# Alfalfa mosaic virus (AMV)

*(Bromoviridae, Alfamovirus)*

## Principal characteristics

### • Frequency and extent of damage

AMV is very polyphagous and ubiquitous; it has been recorded on l.s.v., particularly lettuce, growing in the open field in numerous countries. Even so, serious epidemics have only been reported in plots located close to fields of alfalfa, which is a preferred host of this virus.

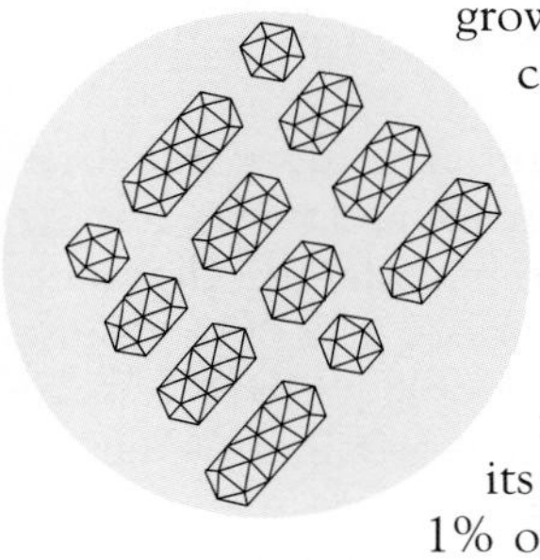

In France, AMV is regularly isolated from lettuce and endive, but its damage is insignificant, with less than 1% of plants generally showing symptoms. This situation is quite surprising as AMV can cause serious epidemics in other market garden species such as pepper, which registers considerable losses locally in some years.

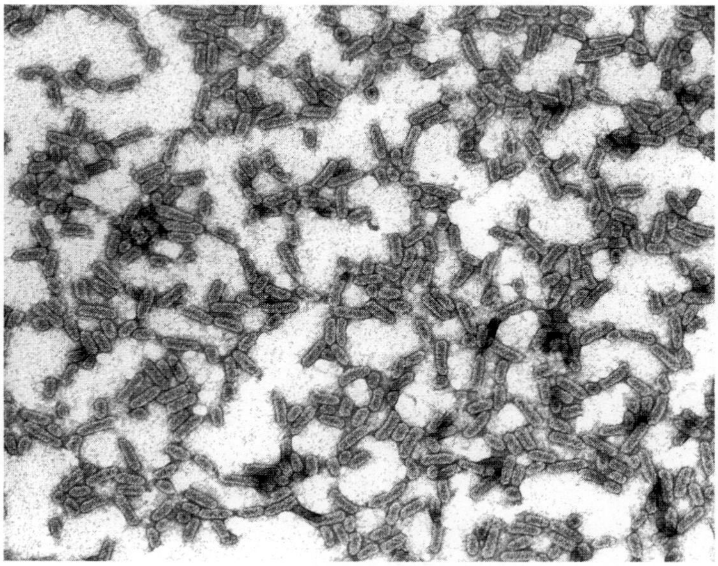

**490** Polyhedric, bacilliform viral particles, 18 nm in diameter.
**AMV**

### • Principal symptoms

This virus causes bright yellow or white spots, varying in size, generally angular, and located all along the veins. Sometimes we can see small whitish rings. Often this damage only affects one or two leaves from the skirt. More rarely, some leaves may display real mosaic with light yellow areas.

In the USA, where this virus has been given the name 'calico', affected plants are sometimes stunted and display leaf necrosis.

(See **106–109**)

### • Ecology, epidemiology

**Survival, sources of virus:** AMV is very polyphagous. It is capable of artificially infecting a large number of hosts, over 400 plant species from 50 botanical families. Approximately 150 herbaceous or woody plants can harbour it in nature. It can therefore survive easily from one season to the next on a few types of weed, and also on several cultivated plants. Excluding alfalfa, which seems to constitute the largest virus reservoir, a number of these may act as a source of inoculum during winter or summer periods (tomato, pepper, potato, celery, bean, pea, clover, and so on).

**Transmission, dissemination:** AMV is transmitted from infected plants to other plants via aphids, in a non-persistent way. These very quickly acquire viral particles absorbed in their stylets and the pediments of their mouthparts in the course of brief 'test' punctures.

They are capable of transmitting this immediately, over a short period of less than a few minutes to a few hours. Over a dozen species of aphids are capable of transmitting AMV to plants (in particular *Myzus persicae* and *Aphis craccivora* on lettuce).

Transmission by seed takes place in a few plant species, particularly in alfalfa and pepper, but never lettuce.

Dissemination of AMV is essentially carried out by aphids; it therefore depends on the nature of the movements of these insects. Several abiotic factors play an essential role in the biology and effectiveness of flights of aphids:

- wind influences their distribution;
- temperature affects lettuce growth, multiplication of the virus, and development of aphid colonies on plants;
- the crop environment, the proximity of other susceptible crops (particularly alfalfa) and numerous weeds which have been con-taminated particularly encourages contamina-tion;
- finally, we must add regional cultivation situations and practices to these factors, such as arrangement of plots, how they are positioned in relation to the wind, any hedges used to protect crops, survival of plants which are virus reservoirs, and so on.

## Protection*

### • During cultivation

There is no curative method of effectively controlling AMV while the crop is growing. An infected plant will remain in this state throughout its life.

If attacks take place in the nursery and are detected early, the few plants showing symptoms of AMV must be rapidly eliminated; under no circumstances must they be planted at a later date.

Aphicide treatments are essential to control aphid populations on lettuce. Unfortunately, very often they are ineffective in controlling virus epidemics. In fact, vector aphids frequently come from outside the plot and transmit the virus in the course of brief punctures, even before the aphicide has had time to act.

### • Subsequent crop

Because of the low incidence of this virus and its mode of transmission, traditional measures must be implemented in order to prevent or, at the very least, limit the introduction of this virus as much as possible, as well as its spread in plots of l.s.v.

In countries where contamination takes place at a very early stage, nurseries and young seedlings must be protected. To do this you may have recourse to agrotextiles (unwoven mesh, woven fabrics, and so on). The mechanical barrier created in this way will delay contamination.

Careful weeding of nurseries, plots and their surroundings (edges of hedges and paths, and so on) must be carried out in order to eliminate sources of virus and/or vectors. Crops of l.s.v. must not be planted next to crops which are more susceptible to AMV, such as alfalfa, pepper, tomato, spinach, and so on.

No resistance to AMV has been reported in lettuce or other *Lactuca*.

* The control of insect populations on a crop often involves using insecticides. In the case of l.s.v., their use may be questioned and legislation varies depending on the country. In order to give a 'universal' character to the proposed control methods, we recommend carrying out treatments which have been reported in publications.
**It is clear that these proposals must be modified according to the country concerned and the pesticide legislation in force.**

# Broad bean wilt virus (BBWV)

(*Comoviridae*, *Fabavirus*)

## Principal characteristics

• **Frequency and extent of damage**

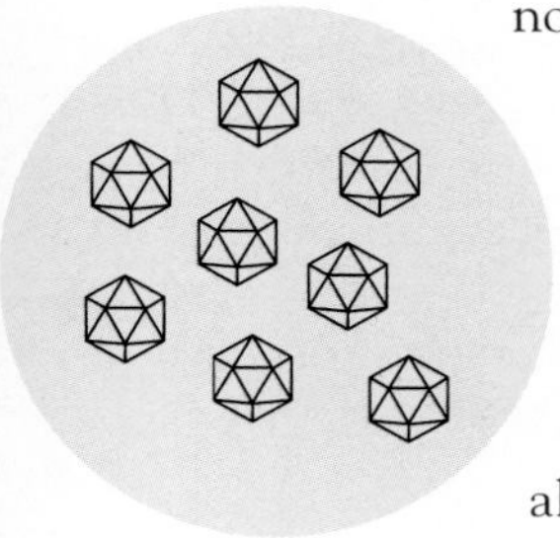

BBWV has been detected on lettuce growing in the open field in several countries: in the USA, northern Europe, Turkey, Japan, Australia, and so on.

In France, this virus has been found on fava bean, pepper, tomato, and artichoke. It was originally isolated on lettuce in 1976. However, no serious epidemic has been recorded on lettuce and the damage observed is always minor. In most cases, the virus is mainly seen on crops at the end of summer or in autumn.

• **Principal symptoms**

The symptoms caused by BBWV are relatively unobtrusive and can even lessen in time. On butterhead lettuce growing in the field we may notice discolouration of external leaves and mottling on the youngest of these. Under conditions of artificial transmission on young plants (batavia or butterhead), growth is sometimes greatly slowed down and chlorotic spots may appear, accompanied by generalized mottling.

In Germany, significant malformation has been attributed to BBWV. In the USA, this viral disease, serious mainly in the state of New York, causes malformation and leaf mottling as well as minuscule necrotic damage. It has also been shown that a very marked difference in temperature between day and night was directly associated with the gravity of symptoms. In addition, in the case of mixed infection with LMV, symptoms associated with the latter virus are exacerbated.

The various isolates collected on lettuce, belonging to two serological groups, do not seem to cause different symptoms on other l.s.v.

(See **80** and **81**)

• **Ecology, epidemiology**

**Survival, sources of virus**: BBWV is relatively polyphagous. Approximately 100 plant species, including around 20 cultivated crops (several market garden Solanaceae, fava (broad) bean, bean and pea, spinach, and artichoke) can be the natural hosts of this virus.

A few perennial ornamental species, such as the iris and begonia, are also likely to become infected. Among wild species, the plantains, *Plantago lanceolata* and *Plantago major*, may play an important role as virus reservoirs during the winter. This is also the case with the rhizome of *Linaria vulgaris*. Amaranth and sowthistle are common wild plants which may also harbour the virus.

**Transmission, dissemination:** BBWV is transmitted to other plants via aphids, in the non-persistent way. These insects very rapidly acquire viral particles, absorbed in their stylets and the teguments of their mouthparts, in the course of brief 'test' punctures. They are capable of transmitting these particles immediately, from a short period of no more than a few minutes to a few hours. Around 20 species of aphids are capable of transmitting BBWV, including *Aphis craccivora*, *Aphis fabae*, *Aphis nasturtii*, *Acyrthosiphon pisum*, *Hyperomyzus lactucae*, *Macrosiphum euphorbiae*, *Macrosiphum solanifolii*, and *Myzus persicae*. This last species seems to spread extremely effectively.

It is not transmitted by seeds in lettuce, or in any other plant species.

## Protection*

Combating this virus presents the difficulties inherent in all viruses transmitted by aphids in the non-persistent mode.

### • During cultivation

There is no curative method capable of effectively controlling BBWV while the crop is growing. An infected plant will remain in this condition all its life.

If attacks take place in the nursery and are detected at an early stage, the few plants showing symptoms of BBWV must be rapidly eliminated and under no circumstances planted at a later date.

Aphicide treatments are essential in order to limit swarms of certain aphids on l.s.v. Generally they are ineffective in controlling the spread of epidemics of disease. In fact, vector aphids often come from outside the plot and may transmit the virus even before the aphicide has had time to act.

### • Subsequent crop

Because of the low incidence of this virus and its mode of transmission, traditional measures must be implemented in order to prevent or, at the very least, limit the introduction of this virus as much as possible, as well as its spread in plots of l.s.v.

In countries where contamination takes place at a very early stage, nurseries and young seedlings must be protected. To do this you may have recourse to agrotextiles (unwoven mesh, woven fabrics, and so on). The mechanical barrier created in this way will delay contamination. The very first diseased plants must also be eliminated.

Careful weeding of nurseries, plots and their surroundings (edges of hedges and paths, and so on) must be carried out in order to eliminate sources of virus and/or vectors. Crops of l.s.v. must not be planted next to the crops previously listed as being susceptible to BBWV.

Reflective mulching (aluminized or black) may partially reduce contamination on lettuce.

Many varieties of lettuce grown in Europe show good tolerance to BBWV. This situation is quite surprising as this virus has only a low economic impact and has not therefore given rise to any specific selection programme.

It is also interesting to note that good levels of resistance to BBWV have been found in batavia type iceberg lettuce varieties, for example Vanguard and Vanguard 75, formerly grown in California (recessive monogenic resistance). Numerous varieties selected from Vanguard 75, the genotype used as the parent resistant to LMV, are often found to be resistant to both viruses. The gene involved in resistance to BBWV is not, however, associated with the gene of resistance to LMV.

In addition, several accessions of *Lactuca virosa* have proved to be resistant to BBWV.

* The control of insect populations on a crop often involves using insecticides. In the case of l.s.v., their use may be questioned and legislation varies depending on the country. In order to give a 'universal' character to the proposed control methods, we recommend carrying out treatments which have been reported in publications.
**It is clear that these proposals must be modified according to the country concerned and the pesticide legislation in force.**

# Dandelion yellow mosaic virus (DaYMV)

(*Sequiviridae*, *Sequivirus*)

## Principal characteristics

DaYMV was identified in Great Britain on *Taraxacum officinale* in 1944. It was isolated, from the 1970s onwards, on lettuce in most northern European countries and also in China.

Numerous isolates, collected in France from crops grown in the open field and more rarely under protection, have proved to be very close to DaYMV. Their characterization is under way in order to determine their exact identity.

- **Frequency and extent of damage**

Although DaYMV has been reported on lettuce in England, Scandinavia, Eastern Europe, the Netherlands, and Germany, we do not have any accurate data about its real incidence on crops of l.s.v. It seems that lettuce mottle virus (LMoV), identified in Brazil, is identical or extremely close to DaYMV. This virus has essentially been identified on lettuce resistant to LMV but nevertheless presenting growth anomalies and leaf discolouration.

Given the difficulties encountered in diagnosing this virus, it is highly likely that its economic importance in relation to lettuce, as well as to endive (escarole and frisée), has been underestimated.

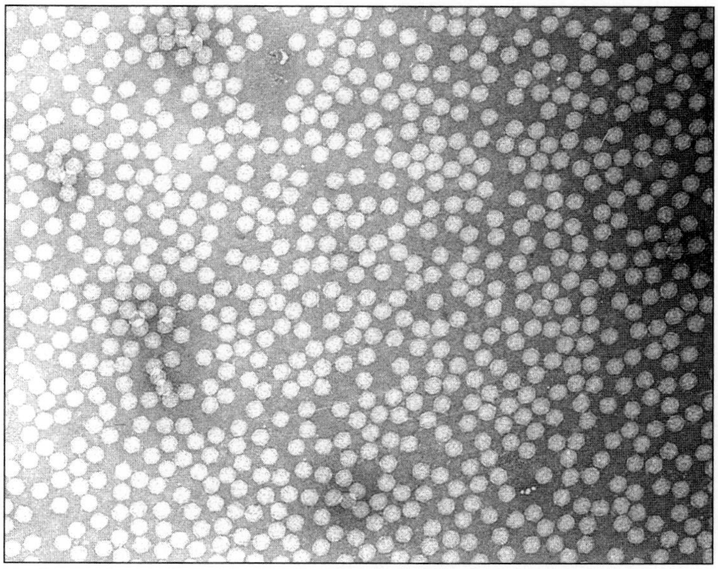

**491** Very high concentration of isometric particles of DaYMV.

- **Principal symptoms**

In the field, where attacks are the most destructive, lettuce appears stunted and sometimes extremely necrotic. Leaves are narrower, slightly rolled, or twisted. Sometimes young leaves display necrotic damage, more or less limited and located both close to and distant from the veins. Rings may also be observed. Chlorotic spots are visible on certain leaf stages. Fine necrosis sometimes appears, giving the lamina a slightly 'bronzed' appearance. The earlier the stage of infection the greater the severity of symptoms. We may also see vein lightening, more or less marked mosaic, necrotic spots, and so on.

Under protection, fairly similar symptoms are found. Initially young leaves seem bronzed, their veins lighten and, ultimately, fairly marked mottling is established on the lamina. A multitude of minuscule necrotic areas of damage sometimes cover the lamina.

(See **82–87**, and **119**)

• **Ecology, epidemiology**

**Survival, sources of virus:** The sources of contamination from this virus are not fully understood. To date, only lettuce and endive, in terms of cultivated plants, and dandelion (*Taraxacum officinale*), in terms of weeds, can be considered to be reservoir plants ensuring the survival of this virus. On the other hand, DaYMV can be artificially inoculated in numerous plant species belonging to five botanical families. Not all the isolates studied present the same range of hosts. Spinach, pea, and zinnia have proved to be susceptible. Most wild *Lactuca*, sowthistle (for certain isolates), and several chenopodia also become infected in the laboratory.

**Transmission, dissemination:** According to certain published works, DaYMV is probably transmitted by a few species of aphid, but the information we have available is largely contradictory. Artificial transmission carried out has proved to be negative on occasion or has resulted in low rates of transmission to l.s.v. Species of aphid such as *Myzus ornatus*, *Myzus ascalonicus*, and *Acyrthosiphon solani* probably transmit it. According to authors the ability of *Myzus perisicae* to transmit DaYMV is probably low or zero. Current experimental results suggest that its mode of transmission is probably semi-persistent, namely that punctures lasting for 10 or more minutes are required, both to acquire the virus from an infected plant and to transmit it to a healthy plant. In addition, it is quite possible that, as is the case with another sequivirus, the presence of a second virus referred to as an 'assistant' is necessary for DaYMV to be transmitted.

Current data do not allow us to understand why this virus spreads extensively, but does conclude that it is not transmitted by seeds in lettuce.

## Protection*

There is no curative method for effectively controlling DaYMV while the crop is growing. An infected plant will remain in this condition all its life.

The lack of knowledge about the aetiology and epidemiology of DaYMV does not allow us to suggest any specific preventative methods for controlling this virus. Consequently we suggest that you consult the section entitled 'Protection' in the other fact files dealing with viruses transmitted by aphids and that you adopt the measures proposed: eliminating weeds, protecting nurseries against vectors, eliminating the first diseased plants, and so on.

In addition, mechanical transmission carried out on around 20 varieties of butterhead lettuce, both resistant and non-resistant to LMV, have not revealed any varietal resistance.

* The control of insect populations on a crop often involves using insecticides. In the case of l.s.v., their use may be questioned and legislation varies depending on the country. In order to give a 'universal' character to the proposed control methods, we recommend carrying out treatments which have been reported in publications.
**It is clear that these proposals must be modified according to the country concerned and the pesticide legislation in force.**

# Turnip mosaic virus (TuMV)

*(Potyviridae, Potyvirus)*

## Principal characteristics

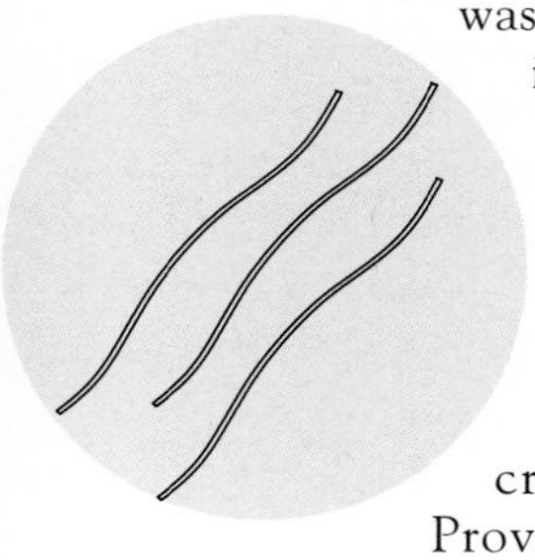

- **Frequency and extent of damage**

TuMV is found worldwide. It mainly attacks in temperate regions on crops of Cruciferae. It was isolated for the first time on lettuce in 1966 in the USA where damage has been described on a few specific varieties of iceberg or romaine.

In Europe, the virus is detected relatively frequently on escarole and frisée endive, and sometimes on iceberg lettuce in summer or autumn crops. It was recently isolated in Provence from a variety of batavia, but never from butterhead or romaine lettuce. Losses in lettuce crops can currently be considered to be negligible. On the other hand, it is likely that, in crops of iceberg lettuces, and in those of escarole and frisée endive, symptoms are sometimes attributed to AMV, resulting in an underestimation of the damage associated with TuMV.

In addition, we should point out that when breeding l.s.v. it is particularly important to make sure that resistance to this virus is not lost. In fact, numerous varieties of butterhead and romaine lettuce are resistant as you will note subsequently when reading about methods of protection.

- **Principal symptoms**

On iceberg type lettuce, mottling in broad, yellow, circular or irregular areas appears on old and intermediate leaves. Ultimately the oldest leaves often become bright yellow all over, with a few leaf sectors sometimes remaining green. The lamina frequently becomes necrotic. Unlike BWYV, veins do not remain green and leaves do not become brittle. Intermediate leaves display a yellow and green distorting mosaic, accompanied by small necrotic spots. Plant growth is affected in all cases. If plants are infected at a juvenile stage, they will remain stunted and deformed and may die.

At the seed-plant stage, the bracts may show a more or less necrotic yellow mosaic. Many flowers fall prematurely or are sterile. Seed production is significantly reduced.

(See **26, 50, 99–105, 110,** and **308**)

- **Ecology, epidemiology**

**Survival, sources of virus:** TuMV has a wide range of hosts including various cultivated Cruciferae (cabbage, cauliflower, Brussels sprouts, Chinese cabbage, radish, turnip, swede (rutabaga), mustard, and so on), but also species from 19 other types of dicotyledons. Numerous wild plants are capable of harbouring it, such as *Capsella bursa-pastoris* or *Stellaria media*. Some floral species, such as the anemone, petunia, or zinnia, are also susceptible.

**Transmission, dissemination:** TuMV is transmitted by over 40 species of aphids, in the non-persistent way, particularly by *Myzus persicae*, *Aphis craccivora*, and *Macrosiphum euphorbiae*.

The virus is not transmitted by seeds in lettuce, escarole, and frisée endive.

## Protection*

### • During cultivation

No curative method can control TuMV. An infected plant will remain in this condition throughout its life.

Generally speaking, eliminating plants affected by the virus right at the start of an epidemic can help to slow this down although the difficulty in observing the characteristic symptoms of CMV in lettuce sometimes makes this approach impractical. In addition, as symptoms only appear around a fortnight after the plants have been contaminated, plants in the incubation stage may constitute a virus source and may play a part in spreading the epidemic.

Aphicide treatments are essential in order to limit swarms of certain aphids on lettuce. Generally they are ineffective in controlling the spread of epidemics of disease. In fact, aphid vectors often come from outside the plot and may transmit the virus even before the aphicide has had time to act.

### • Subsequent crop

It is advisable to implement a set of measures aimed at preventing or, at the very least, limiting the introduction of the virus as much as possible, as well as its spread in plots of l.s.v.

In countries where contamination takes place at a very early stage, nurseries and young seedlings must be protected. To do this you may have recourse to agrotextiles (unwoven mesh, woven fabrics, and so on). The mechanical barrier created in this way will delay contamination.

Careful weeding of nurseries, plots, and their surroundings (edges of hedges and paths, and so on) must be carried out in order to eliminate sources of virus and/or vectors. New growth of mustard, rape, rutabaga must be destroyed in nearby plots. Crops of l.s.v. must not be planted close to crops which are more susceptible to TuMV, particularly crops of various types of cabbage. Susceptible varieties must not be grown in production areas which are particularly badly infested by weeds.

Given the mode of transmission of this virus and the large number of plant species which can potentially become infected, the use of resistant varieties is the essential method of control. Several types of resistance (immunity, tolerance, and lower susceptibility to infection) have been described in l.s.v. Resistance to the virus is conferred by a dominant gene known as '*Tu*'. It seems that the majority of varieties of butterhead and romaine lettuce have extremely strong resistance to TuMV.

We should point out that susceptibility to this virus may be closely linked to a resistance gene to mildew (gene *Dm 5/8*), isolated from a particular accession of *Lactuca serriola*. In the past, in breeding varieties of iceberg lettuce which were already old, breeders lost resistance to TuMV and at the same time introduced the resistance gene *Dm 5/8* to *Bremia lactucae*. We find the same situation in current American, European, and French varieties resistant to mildew and still widely grown today.

* The control of insect populations on a crop often involves using insecticides. In the case of l.s.v., their use may be questioned and legislation varies depending on the country. In order to give a 'universal' character to the proposed control methods, we recommend carrying out treatments which have been reported in publications.
**It is clear that these proposals must be modified according to the country concerned and the pesticide legislation in force.**

# Endive necrotic mosaic virus (ENMV)

(*Potyviridae*, *Potyvirus*)

## Principal characteristics

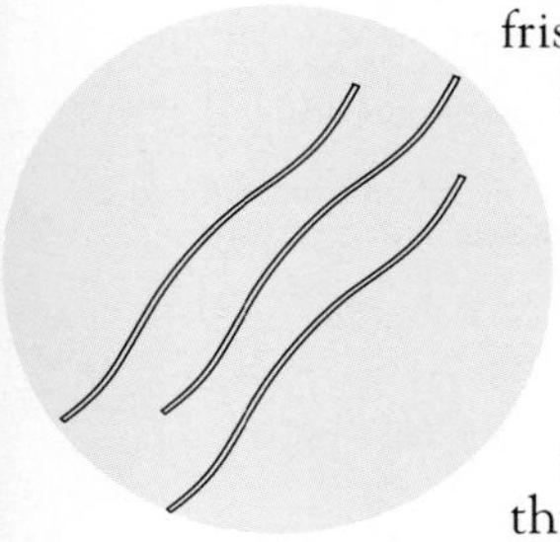

ENMV is a virus which has been recently and partially characterized. It was recorded for the first time in Germany in 1995 on escarole and frisée endive, but it seems that the disease had already been in existence for around 10 years. Although its name is not yet officially recognized, elements of characterization provide proof that it is definitely a new potyvirus. It does not have any serological relationship with the other potyviruses identified throughout the world on lettuce, such as LMV, TuMV, or BiMoV, as well as around 50 other potyviruses.

### • Frequency and extent of damage

Initially, the virus was isolated on frisée and escarole type endive in the Frankfurt region. Subsequently very similar isolates were detected on iceberg lettuce in several regions of Germany where crops were very badly affected.

In 1999 the virus was identified in France, in Provence, on several rows of butterhead lettuce at the time of selection. In the year 2000 we detected it on lettuce grown in the open field in the Ain.

At the moment it is difficult to evaluate the incidence of ENMV, but this is very likely to be underestimated, particularly in crops of endive and lettuce. In fact, symptoms of this virus may be confused with those caused by LMV, TuMV, or even, in certain cases, TSWV.

Fortunately, most current varieties of lettuce seem to be resistant to ENMV.

### • Principal symptoms

In the open field, lettuce displays serious necrotic mosaic. Plants which are affected when young are deformed and stunted and their lamina presents broad areas of necrosis.

On seed-plants the bracts are swollen to varying extents and have broad necrotic sectors. As is the case with TuMV, many flowers fall prematurely or are sterile. Consequently, seed production is significantly reduced.

Following artificial mechanical inoculation the symptoms observed on lettuce may, depending on the genotype, be varied: chlorotic to yellow spots appear progressively on all the leaves, as well as mosaic, and so on. In all cases, plant growth is affected.

(See **95–98**)

### • Ecology, epidemiology

**Survival, sources of virus:** Currently there are no data relating to the number of spontaneous or cultivated species harbouring ENMV. In fact, the extent of risks of epidemics will depend on the extent of the range of potential hosts of this virus. From now on crops of endive must be considered to be reservoirs for potential viruses, especially in autumn and winter.

**Transmission, dissemination:** As with all the types of potyvirus, ENMV is transmitted by aphids in the non-persistent way, potentially by a large number of species.

A few tests carried out on iceberg and butterhead lettuce have not concluded that it is transmitted by seeds in lettuce.

## Protection*

There is no curative method of effectively controlling ENMV while the crop is growing. Generally an infected plant will remain in this condition throughout its life.

The lack of data about the epidemiology of ENMV does not really allow us to propose specific preventative methods of control. As you will see from the fact files about other lettuce viruses transmitted by aphids, numerous measures are commonly implemented in order to limit their epidemics. Consequently, we suggest that you consult some of these, particularly those relating to potyviruses (LMV, and so on), in order to select and implement fairly general anti-virus methods (elimination of weeds, protection of nurseries against vectors, elimination of initial diseased plants, and so on).

In the future, the most successful method is likely to be the use of resistant varieties. An initial evaluation of the sensitivity of varieties belonging to several types of lettuce showed that most cultivars of butterhead lettuce were immune, as were the few romaine varieties tested. In iceberg lettuce, several varieties have proved to be very susceptible, whereas others are immune. Preliminary results have revealed a similarity between the susceptibility of icebergs to ENMV and TuMV. Research into the links between certain genes of susceptibility or resistance to this virus is under way.

* The control of insect populations on a crop often involves using insecticides. In the case of l.s.v., their use may be questioned and legislation varies depending on the country. In order to give a 'universal' character to the proposed control methods, we recommend carrying out treatments which have been reported in publications.
**It is clear that these proposals must be modified according to the country concerned and the pesticide legislation in force.**

# Other viruses transmitted by aphids

| Virus | Symptoms | Vectors and particle shape | Principal characteristics |
|---|---|---|---|
| **Bidens mottle virus (BiMoV)**<br><br>*Potyviridae*<br>***Potyvirus*** | Lightening of leaf veins, mottling sometimes turning into vein necrosis. Plants attacked at an early stage remain dwarfed. | Several aphid species, in the non-persistent way. *Myzus persicae* is particularly efficient.<br>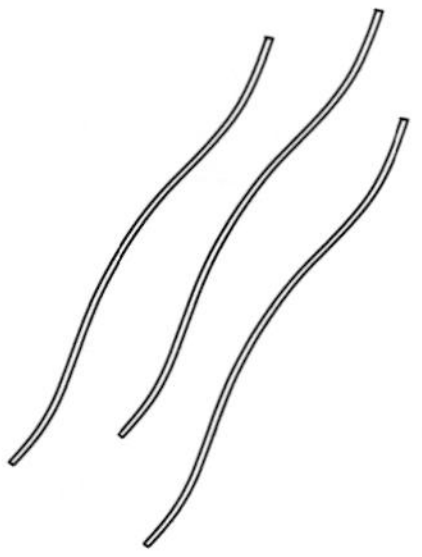 | The distribution of BiMoV seems to be limited to the USA (especially Florida). iThis virus, whose damage rarely seems serious on l.s.v. (*Cichorium endivia, Lactuca sativa*), is harboured by a wide range of hosts including cultivated plants, particularly ornamental ones (*Zinnia elegans, Ageratum conyzoides, Rudbeckia hirta hybrida*) and weeds (*Lepidium virginicum, Bidens pilosa, Erigeron* sp., *Senecio* sp.) on which the virus was recorded for the first time in 1968. It is also found on *Lupinus angustifolius*. It is not transmitted by seeds. Several genotypes of butterhead and romaine lettuce are probably fairly resistant to BiMoV and sometimes to LMV. In certain cases, resistance is probably conferred by a recessive '*bl*' gene. It is sometimes found in combination with LMV on the same plants. |
| **Beet yellow stunt virus (BYSV)**<br><br>***Closteroviridae***<br>***Closterovirus*** | Progressive yellowing of old leaves and abrupt collapse of plants which subsequently die. A longitudinal section of the taproot and stem reveals the presence of internal necrotic lesions. | Effectively transmitted by *Hyperomyzus lactucae*, in the semi-persistent way. *Myzus persicae, Macrosiphium euphorbiae* can be vectors in a more random way.<br>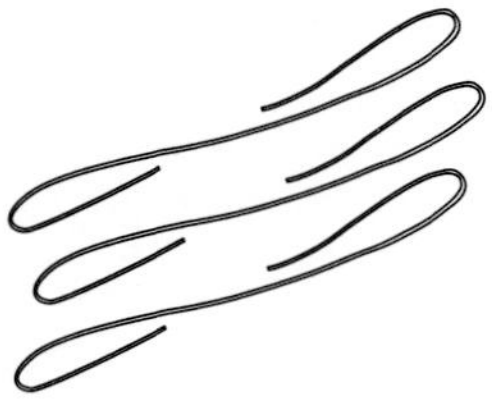 | BYSV attacks in the USA and California in particular, where it was recorded in 1963. In this country, it has been isolated on *Sonchus* and sporadically on lettuce. It has also been recorded in Great Britain. It has a fairly broad range of hosts, including species from 5 botanical families (Asteraceae, Chenopodiaceae, Geraniaceae, Portulaceae, and Solanaceae) and wild lettuce. Its cultivated hosts are beet and lettuce. The virus can be transmitted for 1, 2 and sometimes 4 days. It does not seem to be transmitted by seeds. |
| **Sowthistle yellow vein virus (SYVV)**<br><br>***Rhabdoviridae***<br>***Nucleorhabdovirus*** | Lightening of veins especially around the periphery of the lamina, which may lead to confusion with the symptoms of 'big vein'. Strips of dark green tissue all along the veins (vein banding). Plants infected at an early stage grow poorly. | Transmitted by the sowthistle aphid *Hyperomyzus*, in the persistent way.<br>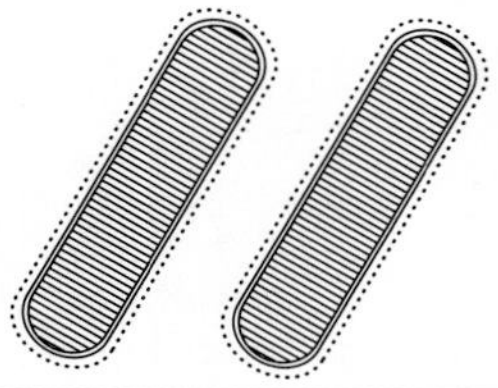 | This virus has been known since the beginning of the 1960s in the USA. It has essentially been recorded on l.s.v. in California and Arizona. However, it has been isolated on other plants in Italy, the Netherlands and France. The sowthistle (*Sonchus oleraceus*) seems to be the principal source of the virus and *Hyperomyzus lactucae* the only vector. It also infects *Sonchus asper*. |

To select the methods of protection to be implemented in order to control these viruses, taking into account their mode of transmission, we suggest that you consult the virus fact files relating to CMV (non-persistent) and BWYV (persistent).

| Virus | Symptoms | Vectors and particle shape | Principal characteristics |
|---|---|---|---|
| **Sonchus yellow net virus (SYNV)**<br><br>***Rhabdoviridae Nucleorhabdovirus*** | Lightening of veins and yellowing of leaves. Broad yellow, intervein spots on lower leaves. Lettuce infected at an early stage may remain dwarfed. | Transmitted by the aphid *Aphid coreopsidis,* in the persistent way.<br>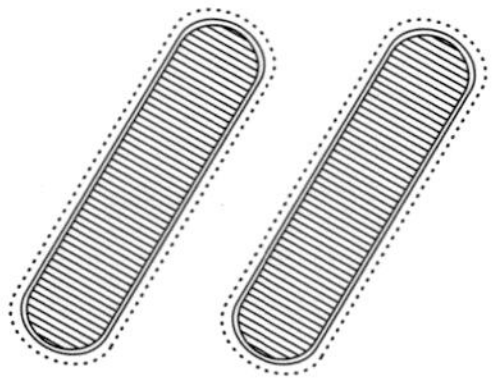 | SYNV is a virus which is similar to SYVV. Its action on lettuce appears to be limited to Florida where it affects romaine and crisphead lettuce. This virus is harboured by several weeds: *Bidens pilosa, Senecio glabellus, Sonchus* spp. |
| **Lettuce necrotic yellows virus (LNYV)**<br><br>***Rhabdoviridae Cytorhabdovirus*** | Pale green to chlorotic lettuces, with a rather flattened appearance. Mottling sometimes present on lower leaves. Plants which have been attacked before the head has formed may have internal necrotic damage and die. Sometimes they are also stunted. | Essentially transmitted by the sowthistle aphid *Hyperomyzus lactucae,* in the persistent way. *Hyperomyzus carduellinus* and *Nasonovia ribisnigri* are also capable of transmitting it naturally and artificially respectively.<br>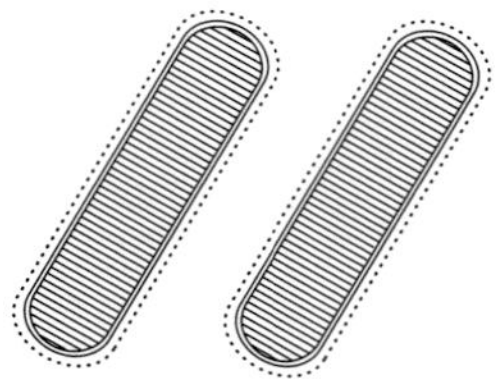 | This virus has a very low incidence on lettuce, on which it was recorded for the first time in Australia time in 1963. LNYV, or a similar virus, has probably been isolated from time to time in New Zealand, but also in Italy, Spain, and Great Britain. Its attacks are encountered above all in production areas where reservoir weeds are present in large numbers: *Sonchus oleraceus, Sonchus hidrophilus,* and so on, at a period of the year conducive to vectors. It is not transmitted by seeds. |
| **Lettuce speckles mottle virus (LSMV)**<br><br>***Umbravirus*** | This virus causes symptoms fairly similar to those of BWYV. Chlorotic, angular spots appear on old leaves. | Transmitted by several aphids: including *Acyrthosiphon (Aulacorthum) solani, Brevicorne brassicae, Myzus persicae,* in the persistent way.<br> | LSMV was recorded for the first time on beet, on lettuce and spinach in 1978 in the USA. It can be artificially inoculated into numerous hosts from several botanical families: (Asteraceae, Chenopodiaceae, Solanaceae). |
| **Lettuce mottle virus (LMoV) (Le MoV)**<br><br>**Non classé** | Mottling on lettuce leaves, preceded by vein lightening. The size of young leaves is reduced and they are deformed. Leaf necrosis is also noted and formation of the head may be affected. | Mainly transmitted by the sowthistle aphid *Hyperomyzus lactucae.*<br> | This virus has only been recorded in Brazil after 1986. Currently it is poorly characterized and therefore not yet classified. Recent data has shown that it is probably DaYMV or an extremely similar virus. Its range of hosts is not known; only lettuce seems to be affected. It does not seem to be transmissible by seed. Several cultivars have proved to be resistant in the course of varietal screening. |

To select the methods of protection to be implemented in order to control these viruses, bearing in mind their mode of transmission, we recommend that you consult the fact files for viruses relating to CMV (non-persistent) and BWYV (persistent).

## An incompletely characterized rhabdovirus

For several years, the symptoms observed on lettuce seed-plants, grown in France in the Rhone valley, have been associated with the presence of rhabdovirus. It has been possible to observe particles directly under an electron microscope in lettuce extracts. On affected plants we may see fairly broad spots of yellow mottling. On some varieties, small yellow smudges, more or less angular in shape, were distributed irregularly around the edge of leaves.

The virus implicated had been transmitted as an experiment, via mechanical inoculation, to a small range of hosts including several species of tobacco, petunia, *Datura stramonium*, and *Physalis floridana*. Cytopathology studies located the virus in the cytoplasm, in endoplasmic reticulum vesicles: it was therefore a *Cytorhabdovirus*.

Two *Cytorhabdoviruses* have already been described in publications as being capable of infecting lettuce: *lettuce necrotic yellows virus* (LNYV), exclusively present in Australia and New Zealand, and the *Sonchus virus* (SV), isolated in Argentina. On the basis of the symptoms observed on lettuce and the range of hosts, the virus isolated in France is probably similar to SV. The principal characteristics of LNYV are shown in the table opposite. The vectors of SV are not known.

The characterization of this *rhabdovirus* isolated in France will require additional research.

# Viruses transmitted by whiteflies

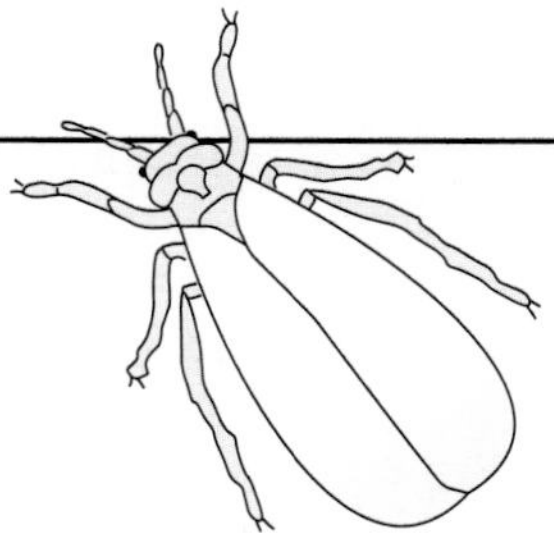

## Beet pseudo-yellows virus (BPYV)

(*Closteroviridae*, *Closterovirus*)

### Principal characteristics

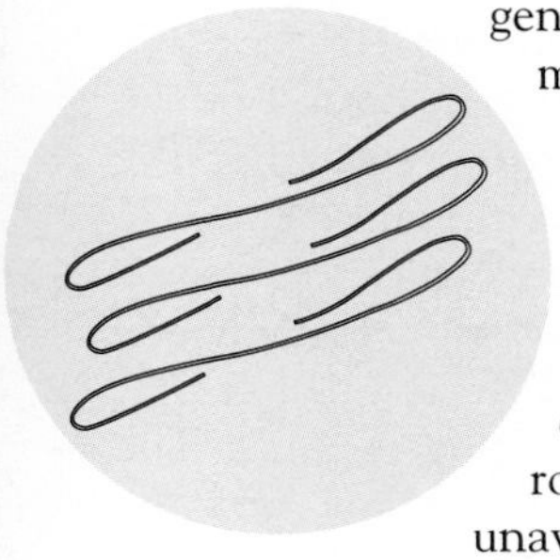

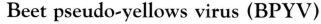

BPYV is currently located in the '*Closterovirus*' genus or 'clostero-like' viruses, but additional molecular studies may relocate it to the *Crinivirus* genus. Although it was isolated for the first time in 1965, it has still not been fully characterized. Located in the phloem cells, BPYV cannot be mechanically transmitted. It is very difficult to purify and at the moment routine diagnostic tools are still unavailable. It is now widely accepted that the viruses previously described under the names of *Cucumber yellows virus*, *Muskmelon yellows virus*, *Melon yellows virus*, *Cucumber infectious chlorosis virus*, and *Cucumber chlorotic spot virus* are, in fact, isolates of BPYV strains.

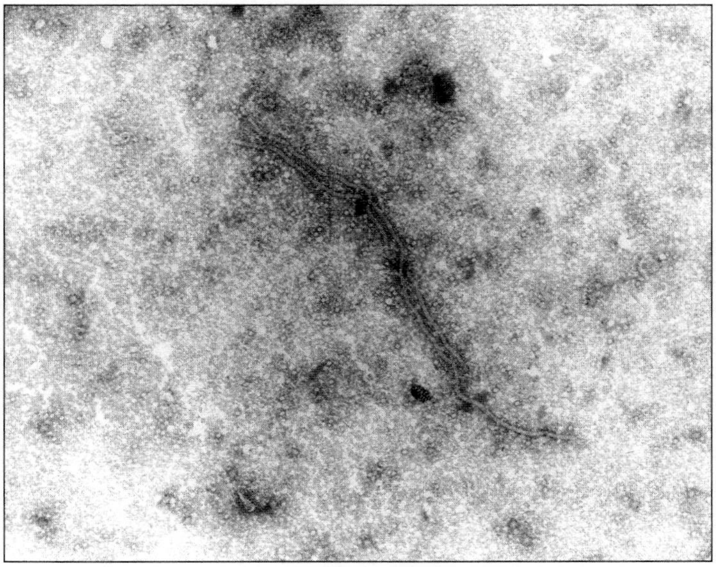

**492** A few serpentine and filamentous particles of BPYV..

**• Frequency and extent of damage**

BPYV mainly causes damage to lettuce being grown under cover. It is probably harboured by other plant species. Isolated in the USA in an experimental greenhouse, for some considerable time the virus has been considered to be trivial in that country; in fact, crops of lettuce are rarely grown in greenhouses in the USA. Subsequently it was observed in France, the Netherlands, the Czech Republic, Italy, and Turkey. This virus probably attacks throughout the Mediterranean basin. It is also present in Asia and Australia.

Although its symptoms on lettuce are relatively characteristic, it is undeniable that the yellowing it causes on this plant, as well as on other susceptible cultivated species, has been incorrectly attributed to nutritional disorders. In this situation and/or in the absence of diagnostic tools, we have certainly underestimated the effects of this virus in certain production areas.

In France and the Netherlands, BPYV has caused extensive damage in crops of lettuce grown under cover as it has not been possible to implement prophylactic measures suited to its biology and epidemiology. Currently, the impact of this disease on lettuce has lessened considerably. The most serious epidemics take place when there is crop rotation between lettuce and cucumber. The latter host is very susceptible to both greenhouse whitefly and the virus; significant losses are sometimes noted after this kind of crop rotation.

In Spain, its real impact is difficult to evaluate in comparison with other 'clostero-like' viruses and *Criniviruses* responsible for yellowing on lettuce and on the Cucurbitaceae. It is now considered to be a serious virus affecting Cucurbitaceae in the USA and in Asia.

Its wide range of hosts and the difficulty in controlling its vector, which moves from one crop to another, make this a highly threatening virus, especially on l.s.v.

- **Principal symptoms**

On l.s.v. grown in the autumn and winter we observe inter-veinal yellowing which is virtually identical to that caused by BWYV. This symptom first appears on a few lower leaves. It should be noted that only the part of the lamina exposed to light shows this discolouration and that veins remain green. Subsequently, yellowing spreads to all the lower leaves and to a few intermediate leaves. Old leaves become slightly thicker and brittle when pinched. Younger leaves do not show any symptoms.

As with most other types of viral diseases, the seriousness of the damage depends on how early the infection occurs. If plants contract the virus in the nursery, yellowing will affect numerous leaves by harvest-time and, in addition, the lettuce will be small in size. It is quite obvious that, under these conditions, their market value will be reduced as they will have to be trimmed; in some cases they will not even be marketable.

The symptoms caused by BPYV on other affected species are comparable to those observed on lettuce.

(See **114, 120–129**)

- **Ecology, epidemiology**

**Survival, virus reservoirs:** BPYV is capable of infecting and surviving on around 40 plant species, from 14 botanical families. Amongst these we can list cucumber, melon, marrow, beet, chard, escarole and frisée endive, spinach, carrot, and so on. Ornamental plants such as the marigold (*Tagetes* spp.), zinnia, *Callistepus* spp., *Aquilegia* spp., *Godetia* spp. are also potential hosts. Wild plants which may be located close to greenhouses, such as shepherd's purse, groundsel, sowthistle, dandelion, *Lactuca serriola*, chenopodium, purslane, mallow, and nightshade are also likely to be infected.

We must bear in mind that BPYV can also be easily introduced into the greenhouse via seedlings, especially those produced close to crops which have already been affected.

**Transmission, dissemination:** The greenhouse whitefly, *Trialeurodes vaporariorum*, is the only known vector of BPYV. It is transmitted in the semi-persistent mode. Generally speaking, the longer the acquisition period, the more effective the transmission of the virus, although some insects can already transmit BPYV after only one hour's contact with the plant. Transmission is at its optimum after 'a meal' lasting 6 hours. The latency period only lasts a few hours, and the whitefly is then capable of transmitting the virus. Subsequently, the whitefly can remain infectious for up to 6 days, even if it takes nourishment in the meantime from plants which have not been infected. The period of time between contamination and symptom expression is at least 15 days.

BPYV is not transmitted by seeds.

## Protection*

- **During cultivation**

There is no curative method of effectively controlling BPYV while the crop is growing. Generally speaking, an infected plant will remain in this condition all its life.

Given the duration of the incubation period of the disease (2–4 weeks), diseased seedlings are rarely observed in the nursery. If this were the case, we would advise you to eliminate them rapidly.

In certain cases insecticide treatments are essential to control whitefly populations on lettuce. Even if they are not very effective, they do restrict the development of this virosis.

Diseased lettuce must be rapidly eliminated after harvesting and not left in place as is often the case in numerous greenhouses. This measure will prevent the whitefly vectors from multiplying in this location and representing a danger for nurseries or future crops.

In the case of shelters, the building may need to be kept clear of plants for a few weeks.

* The control of insect populations on a crop often involves using insecticides. In the case of l.s.v., their use may be questioned and legislation varies depending on the country. In order to give a 'universal' character to the proposed control methods, we recommend carrying out treatments if they have been reported in publications.
**It is clear that these proposals must be modified according to the country concerned and the pesticide legislation in force.**

• **Subsequent crop**

In countries where contamination takes place at a very early stage, nurseries and young seedlings must be protected. To do this you may have recourse to agrotextiles (unwoven mesh, woven fabrics, and so on). The mechanical barrier created in this way will delay contamination.

Careful weeding of nurseries, plots, and their surroundings (edges of hedges and paths, and so on) must be carried out in order to eliminate sources of virus and/or vectors. A nursery or crop of l.s.v. must not be planted close to the previously listed crops which are more susceptible to BPYV, especially cucumber.

If there are no other options, insecticide treatment must be applied before pulling up cucumbers prior to setting up a nursery or crop of lettuce. In addition, greenhouse windows must be rendered 'insect proof' and preventative insecticide treatment carried out.

Currently there are no sources of resistance to BPYV in the lettuce.

# Other viruses transmitted by whiteflies

| Virus | Symptoms | Vectors and particle shape | Principal characteristics |
|---|---|---|---|
| **Lettuce chlorosis virus (LCV)**<br><br>***Closteroviridae***<br>***Crinivirus*** | Lightening of veins; yellowing, reddening, and rolling of leaves. Plants may remain dwarfed. Symptoms similar to BWYV. | Transmitted by *Bemisia tabaci* and *Bemisia argentifolii* with the same efficacy in semi-persistent mode.<br>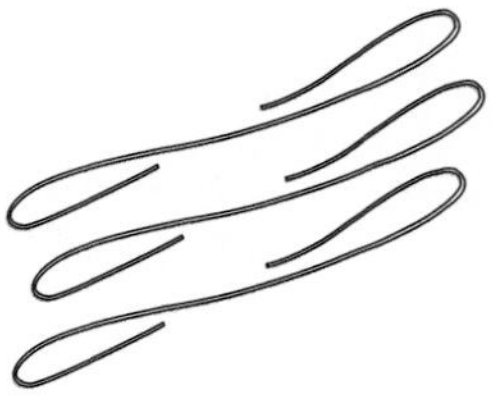 | This virus has mainly been reported in the USA (California, Arizona) where its incidence is low. It is capable of affecting over 27 plant species from in 12 botanical families. This remark is not insignificant since epidemics of LCV essentially seem to depend on the high presence of weeds affected by viruses in the plot environment (*Physalis wrightii, Conyza canadensis, Lactuca serriola, Solanum elaeagnifolium, Helianthus nutalli*). It has also been repor-ted on beet. Joint attacks of LCV and LIYV on the same plant have sometimes been noted in the field. Unlike LIYV, LCV does not affect the Cucurbitaceae. |
| **Lettuce infectious yellows virus (LIYV)**<br><br>***Closteroviridae***<br>***Crinivirus*** | Lightening of veins; yellowing, reddening, and rolling of leaves. Plants may remain dwarfed. Symptoms similar to BWYV. | Transmitted by *Bemisia tabaci* in semi-persistent mode.<br>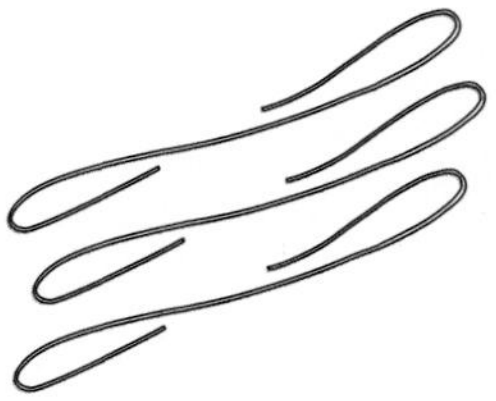 | This other *Crinivirus* is also mainly noted the on l.s.v. in the USA (California, Arizona) where it was described in 1980. It poses serious problems in the open field in California where levels of infestation of 100% and losses in the order of 50–70% have been recorded. It has even been reported in hydroponic crops in Arizona. This virus is currently in regression because of the disappearance of *Bemisia tabaci,* in affected regions, with the increase of *Bemisia argentifolii* (ineffective vector of this virus). LIYV affects a large range of cultivated hosts (cotton, Cucurbitaceae, beet, carrot, and so on) and weeds (*Malva* spp., *Chenopodium* spp. *Physalis* spp., *Ipomoea* sp., *Lactuca canadensis,* and so on). During the season, its vector passes easily from cotton to melon, ending up on lettuce, each time infecting the plants. |

The majority of control methods recommended for controlling epidemics of BPYV can also be used to control these two *Criniviruses*.

# Viruses transmitted by thrips

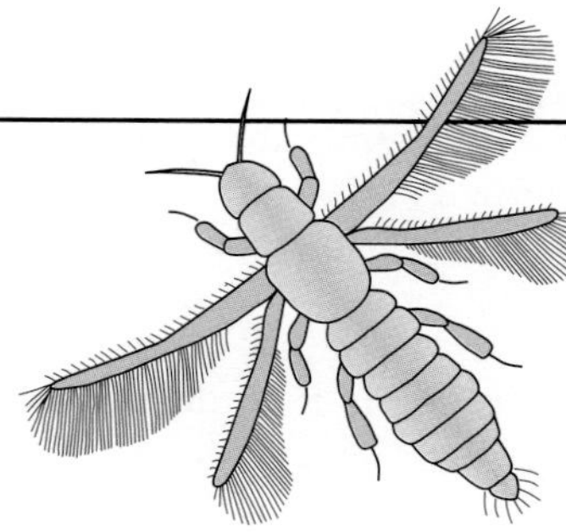

## Tomato spotted wilt virus (TSWV)

(*Bunyaviridae*, *Tospovirus*)

### Principal characteristics

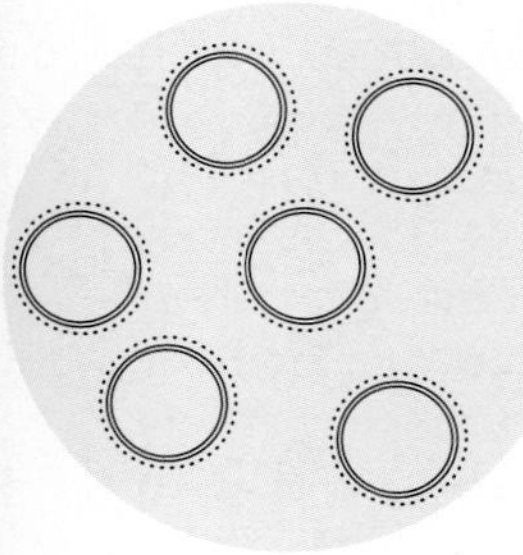

For many years two viruses have been un-wittingly grouped together under the name TSWV. In fact, studies carried out fairly recently on a large number of isolates have shown that these could be separated, initially, into two serologically distinct groups:

- the L-TSWV group (for lettuce) which appears to be the most widespread;
- the I-TSWV group (for *Impatiens*); the isolates of this group are now considered to belong to another virus (described in 1990), which has been given the name ***Impatiens necrotic spot virus*** (INSV). This virus can also be transmitted by *Frankliniella occidentalis*; for a long time it was considered specific to *Impatiens*. Currently, it is known to infect several ornamental crops (anemone, gerbera, begonia, and so on); it has recently been identified on lettuce in Italy.

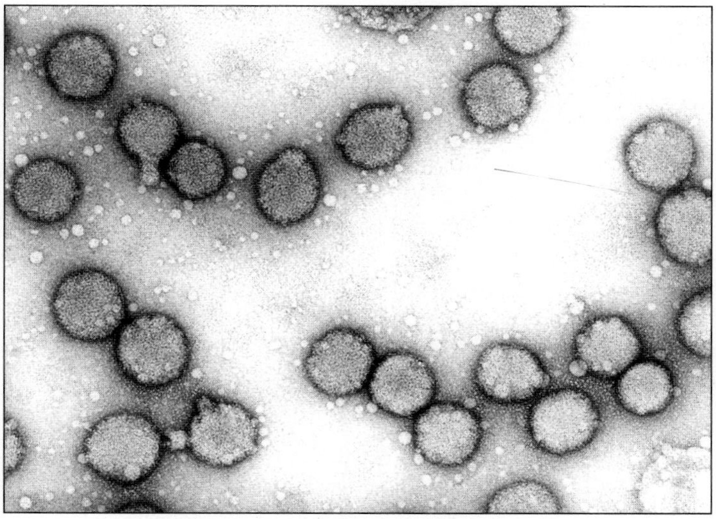

**493** Spherical viral particles, 85 nm in diameter. The presence of a lipid envelope gives them a specific appearance.
**TSWV**

- **Frequency and extent of damage**

TSWV is excessively polyphagous. It is present virtually all over the world and attacks with varying degrees of severity in some lettuce growing countries. Its damage can be very significant in several states in the USA, in various eastern European countries (Bulgaria, Hungary, Poland, and so on) and in Greece. It seems to be worsening in certain Asian and Oceanian countries.

In France, this disease, which affected several crops in the country before the last war, had almost completely disappeared. During the 1980s, the introduction of a new vector, the thrips *F. occidentalis*, greatly jeopardized this situation. It is now considered to be a concern in numerous horticultural establishments. In some production regions, the coexistence of other contaminated vegetable or floral crops with l.s.v. results in serious epidemics which cause significant damage. This is particularly the case in the south of France (Provence-Côte d'Azur and Languedoc-Roussillon regions) where the initial damage on butterhead lettuce, batavia, escarole and endive were described in 1989. Crops grown in the open field or under shelter may be affected, usually during the summer period.

- **Principal symptoms**

We can see a multitude of small light brown to black necrotic lesions on the young leaves and their petiole.

Old leaves display chlorotic spots which become necrotic and brown in colour; the lamina may also wilt around the periphery and go yellow. This damage then spreads; subsequently, broad sections of the lamina become necrotic. In certain cases, leaves are deformed to varying degrees.

Plants which become infected at an early stage grow poorly and often die. If they survive, their head fails to form. Their leaves are also tinged with 'bronze'. Secondary micro-organisms, especially bacteria, colonize the damaged tissues, resulting in their decomposition.

(See **191, 298, 312–317**)

### • Biology, epidemiology

**Survival, virus reservoirs**: As we stressed previously, this virus infects numerous hosts, both cultivated and uncultivated. Over 1,000 have been listed. It can therefore survive easily in the surrounding environment of l.s.v crops, especially on various weeds such as *Amaranthus* spp., *Anagallis arvensis*, *Capsella bursa-pastoris*, *Chenopodium amaranticolor*, *Convolvulus arvensis*, *Fumaria officinalis*, *Oxalis corniculata*, *Picris echioides*, *Poa annua*, *Solanum nigrum*, *Sonchus* spp., *Stellaria media*, *Taraxacum officinale*, *Veronica* spp.

Numerous cultivated plants harbour it; 107 have been listed in France:

- vegetable species, grown for aromatic or industrial purposes: aubergine (eggplant), tomato, pepper, bean, pea, melon, cucumber, marrow (zuccini), spinach, potato, cabbage, artichoke, beet, parsley, lavender, coriander, tarragon, basil, tobacco, and so on;
- ornamental species: marguerite, anemone, aromatic plants, begonia, marigold, chrysanthemum, dahlia, zinnia, cyclamen, gladioli, gerbera, lily, petunia, ranunculus, jasmine, and so on.

It is also found on certain food crops in hot zones: vine, peanut, chayote, pineapple, and so on.

All these plants are virus reservoirs which means that it is extremely difficult to control TSWV.

**Transmission, dissemination:** TSWV can be transmitted by several species of thrips in the persistent way (circulating - propagating). In France, only *F. occidentalis* seems to be involved. This is a very effective vector, much more so than *Thrips tabaci*, in which biotypes incapable of transmitting this virus to other plants have been identified. *Frankliniella shultzei* and *Frankliniella fusca* can also transmit TSWV in other countries where it is rife. Recently this virus was transmitted as an experiment to the peanut by two other thrips: *Thrips palmi* and *Scirtothrips dorsalis*. Only the larvae are capable of acquiring the virus within a minimum of 15 minutes. Inoculation takes place essentially when adults are feeding. They puncture the epidermal cells, injecting the saliva which causes the lysis of the cell content, then suck this up. They need 5–15 minutes to do this. The latency period lasts for at least 4 days. Thrips, which are small in size, can move over short distances. They are sometimes carried by ascending air currents that transport them passively over several hundred metres, or even further in certain situations. The adults, whose lifetime varies from 30–45 days, are very viruliferous until their death. Symptoms can appear 7–14 days after inoculation.

TSWV is also transmitted by seeds and plant reproductive organs in several plants, but not in l.s.v. Care must be taken with seedlings which may have become contaminated in the nursery.

## Protection*

In France, TSWV is classified as one of the quarantinable parasites which must be controlled.

### • During cultivation

There is no curative method of effectively controlling TSWV while the crop is growing. An infected plant will remain in this condition throughout its life.

If attacks take place in the nursery and are detected at an early stage, the few plants showing symptoms of TSWV must be rapidly destroyed and under no circumstances planted at a later date.

Insecticide treatments are essential for controlling thrip populations on l.s.v. A number of products can be used both in the soil and on aerial parts of the plant. In France, the number of products which can be used is very limited, or in fact none may be available. Certain aphicides have secondary effects on these insects. Generally, treatments are not very effective in controlling epidemics of this virus, especially in the open field. In fact, thrip vectors frequently come from outside the plot and transmit the virus while they are feeding, even before the

* The control of insect populations on a crop often involves using insecticides. In the case of l.s.v., their use may be questioned and legislation varies depending on the country. In order to give a 'universal' character to the proposed control methods, we recommend carrying out treatments which have been reported in publications.
**It is clear that these proposals must be modified according to the country concerned and the pesticide legislation in force.**

insecticide has had time to act. Strains of *F. occidentalis* resistant to several insecticides (dimethoate, acephate, oxamyl, fenpropathrine) have been recorded; these further reduce the efficacy of chemical control of these insects. This situation must make us aware of the benefits of alternating insecticides with different modes of action. To be effective, treatments must be carried out at an early stage, to allow good penetration of the drench, so that it can reach the places where the thrips are hiding.

Once the crop has finished growing, all diseased plants must be eliminated. It might also be a good idea to leave the plot empty for 3–4 weeks so that the larval stages still present on the ground and plant debris can develop, and the thrips then disperse elsewhere. Under protection, a comparable period of keeping the building empty may also be implemented. In some cases, fumigation of the soil is recommended. Shelters may also be disinfected.

### • Subsequent crop

In countries where contamination takes place at a very early stage, attempts must be made to prevent or limit contamination as much as possible. To achieve this, it will be advisable to:

- destroy weeds and nymphs found on the soil. Soil disinfection may also be used;
- protect young plants by having recourse to agrotextiles (unwoven mesh, woven fabrics, and so on). The mechanical barrier created in this way will delay contamination. All seedlings planted must be healthy;
- weed the nursery surroundings in order to eliminate sources of virus and/or vectors. This measure is particularly important in the case of this virus. Careful weeding must also be carried out in the open field.

It is also a good idea to incorporate, in rotation, those crops that are barely or not at all susceptible to the vectors. You must also avoid establishing a nursery or planting a crop of l.s.v. close to a crop which is susceptible to this disease and its vectors (especially ornamental species: anemone, chrysanthemum, ranunculus). As with aphids, it has been shown that covering the soil with an aluminized plastic film can reduce the number of thrips present in l.s.v. plots, as well as the incidence of this disease.

Insecticide treatments will be essential to limit thrip populations.

A model for predicting the incidence of this viruses has been developed in Hawaii. It showed that the initial frequency of diseased plants in the crop gave a better indication of the risks of damage at harvest time, rather than the numbers of thrips. For this reason it will be advisable to eliminate the very first diseased lettuces.

We should point out that a number of descendants of an interspecific cross between *Lactuca sativa* and *Lactuca saligna* showed some resistance to a Hawaian strain of TSWV. In addition, two partially resistant cultivars, Tinto and Ancora, had similar genes of a lesser sensitivity; this resistance seems to be partially dominant.

In addition, genotypes of the wild species *L. perennis*, incompatible with lettuce, have been identified as being resistant.

# Other viruses transmitted by thrips

| Virus | Symptoms | Vectors and particle shape | Principal characteristics |
|---|---|---|---|
| **Impatiens necrotic spot virus (INSV)**<br><br>***Bromoviridae***<br>***Tospovirus*** | Vein necrosis, concentric rings are visible on the lamina. Leaves may be deformed and lettuce growth is blocked. | Transmitted by *Frankliniella occidentalis* and *Frankliniella fusca*.<br>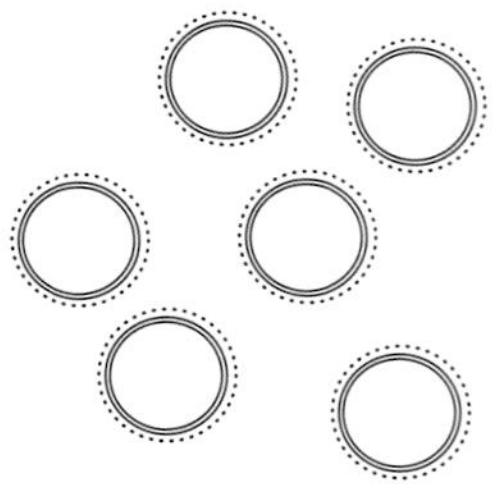 | INSV has been reported in western and eastern Europe and in the USA. It infects *Impatiens,* most of whose leaves do not show any symptoms. We find the same situation on several other ornamental crops. It was reported in 1997 on lettuce in Italy (Emilie Romagne). In France we now see its symptoms on lettuce and endive in the Alpes-Maritimes. |
| **Tobacco streak virus (TSV)**<br><br>***Bromoviridae***<br>***Ilarvirus*** | Numerous small necrotic spots on leaves; we also note chlorotic, concentric rings which become necrotic. Deformed, folded leaves give the plants a peculiar appearance. Growth may be slowed, and they sometimes become completely necrotic and die. | Several species of thrips (*Thrips tabaci, Frankliniella* spp.). It is also transmitted by seeds in beans and several weeds such as *Datura stramonium, Chenopodium quinoa, Helilotus alba,* and so on.<br>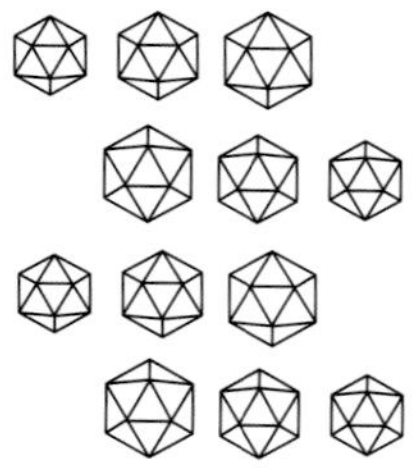 | Present worldwide, it infects numerous hosts naturally or in experiments, such as cotton, tomato, asparagus, bean, soya, vine, strawberry, alfalfa, tobacco, ornamental plants (*Dahlia* spp., *Rosa setigera,* gladioli), and several fruit species. It is found on several weeds (*Chenopodium quinoa Datura stramonium,*) growing around the edges of l.s.v. plots, which help to encourage its epidemics. Its attacks are fairly insignificant on l.s.v. Lettuce seems much less susceptible than endive. Some methods of control recommended for TSWV are applicable. |

# Viruses transmitted by nematodes

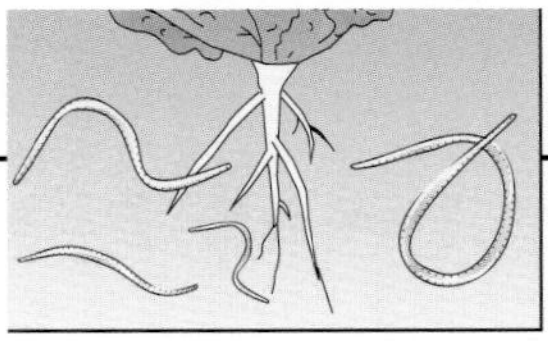

| Virus | Symptoms | Vectors and particle shape | Principal characteristics |
|---|---|---|---|
| **Lettuce necrotic spot virus (LNSV)**<br>***Comoviridae Nepovirus*** | Mosaic or mottling, concentric rings, arabesques, necrotic spots can be seen on leaves. | Unidentified vector. The disease seems to be associated with the presence of nematodes from the *Xiphinema* genus. Transmission by seeds has not been investigated.<br>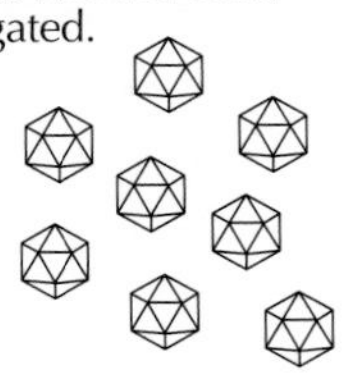 | This virus was identified in crops of lettuce grown under protection in Portugal, in 1998. |
| **Tobacco rattle virus (TRV)**<br>***Tobravirus*** | Mottling, spots, rings and irregular patterns, in varying shades of yellow. Plant growth is slowed and leaves are flatter than ordinarily. | Several species of nematodes from the genuses *Trichodorus* (*T. minor, T. primitivus, T. viruliferus)* and *Paratrichodorus*, (*P. christei, P. primitivus*, and so on). These nematodes remain infectious for several months, or even several years. (Weak transmission by seeds in a few weeds.)<br>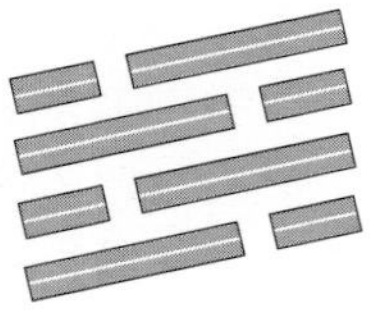 | This virus is extremely widespread in several continents. It causes damage to potato, tobacco, pepper, celery, spinach, and on ornamental crops (*Tulipa* sp., *Narcissus pseudonarcissus, Hyacinthus* sp., *gladioli)*, and it also affects other cultivated and uncultivated plants (*Capsella bursa pastoris,* and so on). It has already been recorded on lettuce in the USA (on romaine in Sacramento), in Denmark, and in Italy (Campania). Combating this virus consists of controlling its vectors. We recommend that you apply certain measures recommended in the fact files relating to nematodes. |
| **Tobacco ring spot virus (TRSV)**<br>***Comoviridae Nepovirus*** | Widespread yellow mottling of leaves from the skirt. Rings and irregular patterns in various shades of yellow, varying in intensity, comparable to those observed on plants infected by TRV. Plants which have been attacked are frequently stunted and very squat. | Transmitted by ectoparasitic nematodes from the *Xiphinema* genus, particularly *Xiphinema americana*. In certain hosts, it can be transmitted by grasshoppers and thrips while feeding. (Transmission by seeds is possible in lettuce.)<br> | In spite of being widespread throughout the world, TRSV has been reported very rarely on l.s.v., for example in Slovenia on lettuce. It has been recorded more frequently on soya, *Vigna unguiculata, Capsicum* spp., aubergine (eggplant), tobacco, tomato, cucumber, *Armoracia rusticana*. |
| **Tomato black ring virus (TBRV)**<br>***Comoviridae Nepovirus*** | Ring shaped spots on lettuce, identified at the time as a strain of TBRV. | Transmitted by several nematodes, such as *Longidorus elongatus*. Some hosts can transmit it via seeds.<br> | TBRV is extremely widespread throughout the world; it can be inoculated into a large range of hosts. It has been reported on leek, beet, bean, tomato, potato, and on numerous weeds. To our knowledge it has only been reported once on lettuce, in France, at the end of the 1970s. |

# Viruses transmitted by fungi

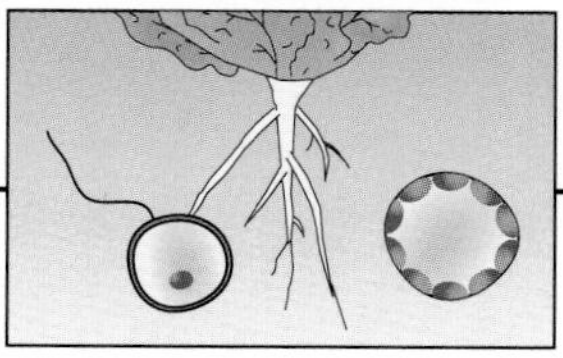

## Lettuce big vein virus (LBVV)
## Mirafiori lettuce virus (MiLV)
*(Ophiovirus)*

### Principal characteristics

Big vein is one of the most serious diseases affecting lettuce. It can also affect endive (frisée and escarole). The disease is particularly serious since there is no satisfactory method of combating it.

For many years its aetiology has been mysterious. Although it was described in 1934, in California, the causal agent (Big vein agent (BVA)) remained unidentified for many years. After various hypotheses, a viral type agent was suspected and its transmission by the fungus *Olpidium brassicae* was demonstrated (**see fact file 14**).

Several types of particles from different genera have been associated with the disease. Between 1983 and 1987, rod-shaped particles measuring 320–380 nm × 18 nm were implicated. The virus was called ***Lettuce big vein virus* (LBVV)**. Officially accepted as being responsible for the disease, it is considered to be the type member of the *Varicosavirus* family. In the absence of verification of Koch's postulate, doubts existed with regard to the involvement of this virus. Consequently other routes were pursued, mainly involving a *Tobamovirus*. Studies conducted in Italy showed that a new virus called **Mirafiori lettuce virus (MiLV)** was associated with big vein, in the absence of LBVV. Recently Koch's postulate was verified, demonstrating that MiLV alone, isolated from lettuces suffering from big vein, could be transmitted by zoospores of *Olpidium brassicae*, could cause the disease, and once again be isolated from inoculated plants. It therefore seems that LBVV is not the agent responsible for big vein, even if it is often isolated in a complex with MiLV.

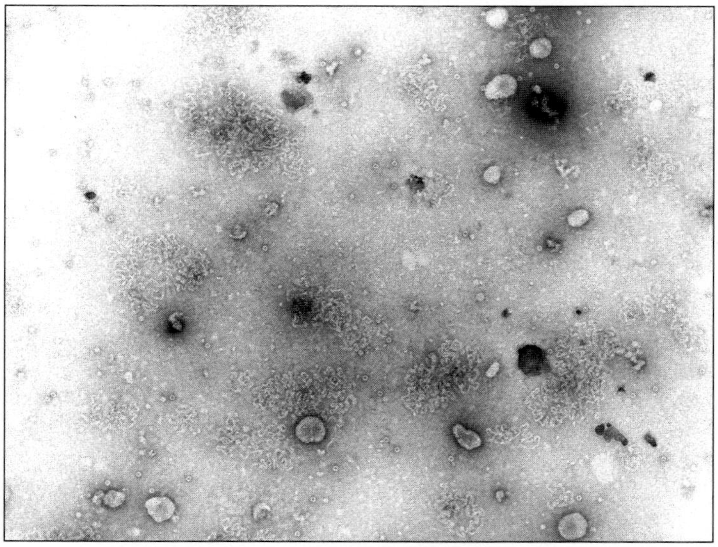

**494** Particles MiLV are difficult to observe under an electron microscope.

Particles of MiLV have a rare morphology, like that of other viruses from the *Ophiovirus* genus, with filaments measuring 3 nm in diameter, folded back to varying degrees to form masses of two different sizes with an indeterminate outline (**494**). Under an electron microscope, they are very difficult to observe in raw plant extracts.

#### • Frequency and extent of damage

Big vein was recorded for the first time in California on lettuce. Since then, its presence has been acknowledged in around 20 countries. Identified in most European countries, it is probably responsible for losses currently estimated at €40 million per annum. The disease appears in temperate and Mediterranean regions, and even in hot zones where there may be a significant contrast in temperature. It can also attack in Japan, Australia and New Zealand. Big vein can cause damage to crops grown under shelter and in NFT production; it is also frequent in the open field. The most serious attacks take place on soil where lettuce and endive are frequently grown. If l.s.v. are grown as a monoculture, the disease tends to worsen from year to year; thus attacks affecting 50–70% of plants have been recorded in certain plots. Damage generally occurs at the end of autumn and in winter.

• **Principal symptoms**

The first plants affected are quickly identified in crops as they appear to be standing up abnormally erect, depending on the type of lettuce; in addition, lettuce plants seem blistered, or even ruffled. Ultimately, plants are usually unsuitable for marketing. If they are infected at an early stage, they may remain stunted. All types of lettuce are affected, but slight differences in symptoms may be distinguished.

On butterhead, batavia, iceberg, and romaine lettuce we note, above all, lightening of chlorophyll tissues located along the veins, giving the impression that these are broadening (big vein symptoms).

On the other hand, on oak leaf and lollo lettuce, it is the stunted appearance of the plants and the deformation of the leaves which is the most remarkable sign.

Low temperatures encourage the appearance of symptoms. They are accentuated if the air temperature is around 14–16°C. They are reduced or disappear if the air temperature rises to 22–24°C or above, whatever the temperature of the soil at root level. The succession of cold and more clement climatic periods may explain the presence, on l.s.v., of normal leaf stages alternating with stages displaying deformed leaves.

(See **10, 17, 18, 25, 52, 88–94**)

• **Ecology, epidemiology**

**Survival, sources of virus:** The vector of MiLV is *Olpidium brassicae*, a chytridiomycete fungus which is an obligate parasite. This fungus is responsible for the survival and dissemination of the virus by very resistant resting spores and zoospores respectively. The resting spores may survive for many years in the soil and thus perpetuate the presence of the virus in plots.

*Olpidium brassicae* has a relatively large range of hosts, both cultivated and uncultivated. Their role in ensuring that the vector or fungus remains viruliferous in the soil is not well understood. Hosts showing symptoms, such as several species of *Lactuca*, *Cichorium*, and *Sonchus*, may increase the number of viruliferous resting spores and zoospores in the soil. The importance of this inoculum in an epidemic of this disease in lettuce is no doubt low if we consider the multitude of viruliferous spores associated with survival and dissemination formed when a crop of lettuce or endive is growing. The possibility of a reliable serological diagnosis of MiLV should allow the hosts of the virus to be studied.

**Transmission, dissemination:** In the presence of susceptible hosts, the case in question being lettuce, primary infections occur via viruliferous zoospores which are released by resting spores. These mobile spores infect the epidermal cells of young roots. Subsequently, the fungus remains in the roots. There it forms numerous zoosporangia which produce a significant number of zoospores, some of which are viruliferous. Subsequently numerous secondary infections occur if soil humidity is suitable.

*Olpidium brassicae* is encouraged by cool soils, usually heavy and badly drained, which remained saturated with water for several days. The expression of symptoms requires temperatures of below 18°C.

These favourable conditions largely explain the distribution of the diseases in terms of plots (humid zones) and in terms of time (end of autumn, winter). We should point out that exceptional attacks of big vein have sometimes occurred in April, when temperature conditions were favourable. Finally, damage has sometimes been observed in sandy soils in which permanent humidity was abnormally maintained.

## Protection

• **During cultivation**

Unlike the majority of cryptogamic diseases, there is no **curative method of effec-tively controlling** viruses, particularly MiLV, while the crop is growing. An infected plant will remain in this condition throughout its life.

During and at the end of the growing season, **plant debris**, particularly the root systems, must be eliminated from plots and destroyed, to prevent it from being buried at a later date in the soil alongside the resting spores of *Olpidium brassicae*.

- **Subsequent crop**

It is advisable to plant **healthy seedlings** in order to avoid introducing MiLV into soil which is still virus-free.

The soil in nurseries must be **disinfected** with methyl bromide (while this is still authorized by legislation) as this treatment is effective. It would also be advantageous to place the seedlings produced on sheets or on plastic film in order to prevent the roots, which have been formed, coming into contact with the soil.

Eradicating the disease from a given plot is virtually impossible since the resting spores of the fungus vector retain their capacity to transmit the virus for 15–20 years. Consequently, contaminated soil will remain in this condition for a long period. The **soil** of future plots of l.s.v. must be **carefully prepared** and drained in order to avoid the formation of puddles of water conducive to the development and dissemination of *Olpidium brassicae*. L.s.v. will preferably be planted on beds. Under protection, **fumigation** of the soil with methyl bromide can be done, particularly to limit the number of viruliferous resting spores. The efficacy of this treatment may last for 2 years, if carried out correctly. Once methyl bromide has been banned in 2005, we will probably have no means of controlling this virus; in fact, none of the soil treatments experimented with has presented satisfactory efficacy.

No resistance to big vein, at a high level, has been found in lettuce. In the USA, a few varieties of iceberg showing lower susceptibility have been marketed; the 'Pacific' variety is considered to be the most tolerant. Trials conducted in the open field, in Europe, have not confirmed this. Currently, the selection of resistant plants is made by simple observation of symptoms present on genotypes of lettuce cultivated and graded in contaminated soil. This method is very random, because of the strong influence of temperature and environmental conditions on the appearance of symptoms. No resistant material is available for other types of lettuce.

A good level of resistance has been discovered in certain accessions of *L. virosa*, but we do not know if this is resistance to the virus or the vector.

The availability of serological tests for detecting MiLV should allow us to re-evaluate resistance to this virus in *Lactuca sativa* and other species. In addition, studies relating to the level of receptiveness of lettuce roots to *Olpidium brassicae* are currently being carried out in Europe.

# Lettuce ring necrosis agent (LRNA)

*(Ophivirus)*

# Orange spot

## Principal characteristics

In France, '**orange spot disease**' owes its name to the primary symptoms it causes on batavia lettuce, on which it was observed for the first time. It was called '**necrotic ring disease**' in other countries.

Current research indicates that the causal agent is of a viral type. To date, the virus responsible has not been characterized; it is referred to in published works under the name of *Lettuce ring necrosis agent*. The mode of transmission of this virosis by *Olpidium brassicae* (**see fact file 14**) and its frequent association in the field with big vein have suggested a common causal agent, in spite of extremely different symptomatology. Indeed MiLV (ophiovirus) and LBVV (varicosavirus) are frequently isolated together in lettuces affected by orange spot disease. Recently, an isolate of *varicosavirus* showed that it was capable, after mechanical transmission, of partly reproducing the disease syndrome, but Koch's postulate could not be established.

### • Principal symptoms

On batavia, we initially observe rings, spots of an orange to brown colour, and a rather oily appearance. These symptoms are visible first of all on the lower face of the lamina. On batavias with narrow leaves, this damage develops fairly rapidly into paper-like necrosis.

On butterhead lettuce, yellow, punctiform or angular areas initially appear, sometimes giving the upper face of the leaf a marbled appearance. Rings and orange spots may also develop.

In fact these orange rings are visible on all types of lettuce, firstly in the axil of the leaves and all along the veins. They may spread to the whole lamina. It is always the lower or intermediate leaves which manifest symptoms. Although rigorous observation can identify the first symptoms at the 15–20 leaf stage, it is generally at the pre-head stage that growers begin to evaluate the extent of the damage.

(See **115–118, 152, 197, 225–230, 251, 252, 310,** and **376**)

### • Ecology, epidemiology

**Survival, sources of virus:** Knowledge of the biological cycle of the orange spot agent is fairly rudimentary. Initially the fact that this disease was transmitted in the soil was rapidly established. Subsequently, the involvement of *Olpidium brassicae* as vector was demonstrated in the course of experimental inoculations using zoospores. The frequency of LRNA transmission during these experiments was very low as if this virus had been 'lost' by the zoospores of the fungus during successive cycles and transfers from plant to plant. The vector of LRNA is therefore *Olpidium brassicae*, an obligatory parasitic chytridiomycete fungus. This fungus guarantees the survival and dissemination of the virus by very resistant resting spores and zoospores, respectively. As in the case of MiLV, the virus is 'carried' internally in the resting spores of the fungus. It is therefore likely that LRNA may remain infectious in these spores for several years.

We should remember that *Olpidium brassicae* has a very wide range of hosts, both cultivated and uncultivated. At the moment we do not know what influence they may have on the survival of LRNA in a given plot.

**Transmission, dissemination:** As in the case of big vein, we believe that primary infections occur on lettuce via viruliferous zoospores which are released by the resting spores present in the soil. These zoospores are mobile and infect the epidermal cells of young roots. Subsequently, the fungus remains in the roots. There it forms numerous zoosporangia which produce a significant number of zoospores, some of which are viruliferous. Consequently, secondary infections may occur if the humidity of the soil is conducive.

*Olpidium brassicae* is encouraged by cool soils which are usually heavy, and poorly drained, remaining saturated with water for several days. Temperatures below 18°C are required for symptoms to appear. These favourable conditions can be compared to those required for symptoms of big vein to appear; consequently it is not surprising to find the two diseases in the same plots at the same periods of the year.

In the Netherlands, it has been shown that drainage water, or water which has been trapped at very low depth, may be responsible for contamination and significant dissemination of the disease.

## Protection

### • During cultivation

There is no **curative method** of effectively controlling the viruses and particularly LRNA while the crop is growing. An infected plant will remain in this state all its life.

During and at the end of the growing period, **plant debris**, particularly the root systems, must be eliminated from plots and destroyed, to prevent them from being buried at a later date in the soil alongside the resting spores of *Olpidium brassicae*.

### • Subsequent crop

**Healthy seedlings** must be planted in order to avoid introducing LRNA into soil which is still virus-free.

The soil in nurseries must be **disinfected** using methyl bromide (while this is authorized by legislation) as this treatment is effective. It would also be advantageous to place the seedlings produced on sheets or on plastic film in order to prevent the roots, which have been formed, coming into contact with the soil.

Eradicating the disease from an infected plot seems as impossible as in the case of big vein; resting spores of the vector fungus probably retain their capacity to transmit the virus for several years. Consequently, contaminated soil will remain in this state for many years. The **soil** in future l.s.v. plots must be **well prepared** and drained in order to prevent the formation of puddles of water propitious to the development and dissemination of *Olpidium brassicae*. It will be preferable to plant varieties on beds. Under protection, **fumigation** of the soil with methyl bromide may be envisaged, particularly to limit the number of viruliferous resting spores. This treatment may be effective for 2 years, if carried out correctly. In Provence, and Roussillon, it appears that the spread of the disease was directly associated with the progressive abandonment, soon to be obligatory, of this method of disinfecting. Experiments with **solarization** (see page 248) conducted in the Lyon region showed that this method of disinfecting had a degree of effectiveness against this disease. However, the mechanism of action still remains unexplained. We should remember that solarization is only possible in production areas where the climate and production schedule allow it.

The search for **resistance** is made difficult by the fact that we do not know which agent is responsible and do not have any means of detecting this. Only observation carried out in the field, in heavily contaminated soil, has shown differences in susceptibility within lettuce types, with iceberg lettuce being the least susceptible to the disease. Lower susceptibility has been noticed within European-type batavia genotypes.

The combination of solarization and the use of less susceptible lettuce can delay the expression of symptoms; consequently, a larger number of plants remain marketable.

| Virus | Symptoms | Vectors and particle shape | Principal characteristics |
|---|---|---|---|
| **Tobacco necrosis virus (TNV)**<br><br>***Tombusviridae Necrovirus*** | Brown necrotic spots appearing close to the veins may coalesce.<br>Young seedlings sometimes die. On adult plants, symptoms are essentially located on lower leaves. | *Olpidium brassicae* is its natural vector.<br> | This virus is present worldwide. It mainly attacks in Europe, but does not often affect l.s.v. It has been reported on lettuce grown in NFT in Belgium; this growing system provides conditions which are totally favourable to vector dissemination as well as that of TNV. It infects numerous hosts in addition to lettuce, particularly market garden and ornamental plants (bean, melon, cucumber, carrot, tulip, and so on).<br>It is more frequently observed on these hosts in the greenhouse, particularly on cucumbers grown in or out of soil. It seems capable of multiplying in the roots of numerous plant species which help to make its presence in crops permanent. |

# Nematodes

## General information

The phylum Nemata describes approximately 20,000 species which take their nourishment from their congeners, humans, bacteria, fungi, and plants. It is made up of animals which have digestive, nervous, excretory, and reproductive systems. They do not have any respiratory or circulatory system. Their size may vary from a few millimetres to over 8 metres.

**Phytophagous nematodes** are **minuscule worms**, more or less transparent, which are also known as eel worms. They are usually invisible to the naked eye; but can easily be distinguished under a light microscope. As with numerous animals, they have a **fairly complete digestive system** starting with a mouth and ending with an anus. All phytopathogenic nematodes have a hollow **buccal stylet** which allows them to puncture cells in order to absorb their content. In numerous species, we find both male and female nematodes; they basically differ in their genital apparatus. In others there are no males and sexual reproduction does not take place. Reproduction in nematodes results in the formation of eggs; it can be sexual, hermaphrodite, or parthenogenic.

They are commonly **found in the soil** in a free state, feeding on the surface of roots and underground organs. They survive and multiply on susceptible alternative hosts. The eggs and some of the larva stages are capable of surviving for several years in the soil in a dormant state. The cycle of development of nematodes is relatively simple. The **eggs**, whose hatching is sometimes favoured by root exudates, produce **larvae** which grow; each of the 4 larval stages end in metamorphosis. The last larval stage differentiates into an **adult nematode** which can be male or female. If environmental conditions are favourable, the complete cycle takes place within 2–3 weeks. Infectious larva stages and adults present a different parasitic process according to species.

Numerous nematodes have been reported as affecting l.s.v.; for additional information we suggest that you refer to page 203. Only a few species cause significant damage on these plants. Losses are more significant on lettuce, and more restricted on endive. Several **ectoparasitic nematodes** can cause varying degrees of damage. They are content to feed at the expense of the surface root cells without normally penetrating the tissues (*Longidorus africanus*, *Rotylenchus robustus*, and so on). The same does not apply with regard to the **endoparasitic nematodes** which sometimes penetrate tissues very deeply. Once established, they can remain localized and cause the formation of root **nodes** (*Meloidogyne* spp.) or **cysts** (not in l.s.v.) or move around inside the tissues causing **brown lesions** (*Pratylenchus penetrans*). The injection of saliva in cells while they are feeding causes the majority of the damage. In addition to the damage they cause on roots of l.s.v., some of them can interact with other soil-based pathogenic agents, as is the case with various plants.

The development of nematodes in soil is influenced by its humidity, ventilation, and temperature. The presence of a film of water is essential for the larvae or adults to move around in the soil or on the organs they attack using undulating movements. They are spread by tools, agricultural machinery encrusted with particles of contaminated soil (*Pratylenchus* spp., *Meloidogyne* spp, and so on), by drainage and irrigation water, and sometimes after splashing.

**The following fact files will only deal with *Meloidogyne* spp. and *Pratylenchus* spp., as these are the nematodes found most frequently and cause the most damage to l.s.v. As far as other species of nematodes which attack these plants are concerned, we suggest that you consult page 203 for more information, as well as the 'control' section of the fact file *Meloidogyne* spp. for information about controlling them.**

# *Meloidogyne* Goeldi spp.

(Nematoda, Tylenchida, Tylenchoidae, Heteroderidae)

# Root knot nematodes

## Principal characteristics

• **Frequency and extent of damage**

Root knot nematodes are present worldwide and affect over 3,000 plants. Several species of *Meloidogyne* attack l.s.v. and have been reported in numerous countries. Certain species have a rather semi-tropical distribution. ***Meloidogyne arenaria* (Neal) Chitwood** and ***Meloidogyne javanica* (Treub) Chitwood; *Meloidogyne incognita* (Kofoid & White) Chitwood** is present in intermediate areas whereas ***Meloidogyne hapla* Chitwood** attacks more in the north.

Given the polyphagous nature of these nematodes on vegetable crops, they are found in France in the soil of many fields which have formerly been used as market gardens. Their damage can sometimes be serious in fields where management of crop rotation is not appropriate and if the soil is unhealthy.

Four races of M. *incognita* have been identified, two in M. *arenaria*, and one in M. *hapla* and M. *javanica*.

• **Principal symptoms**

Regular root knots of varying sizes on the roots characterize the presence of these nematodes. The nature and extent of the root knots depend on the species and the level of inoculum in the soil. The root knots produced by M. *hapla* are fairly small as it invades the apical meristems and therefore affects a lower proportion of the roots, whereas those caused by M. *arenaria* are the size of a pearl and affect almost all the roots. The two other major species produce large root knots sometimes covering the entire root system.

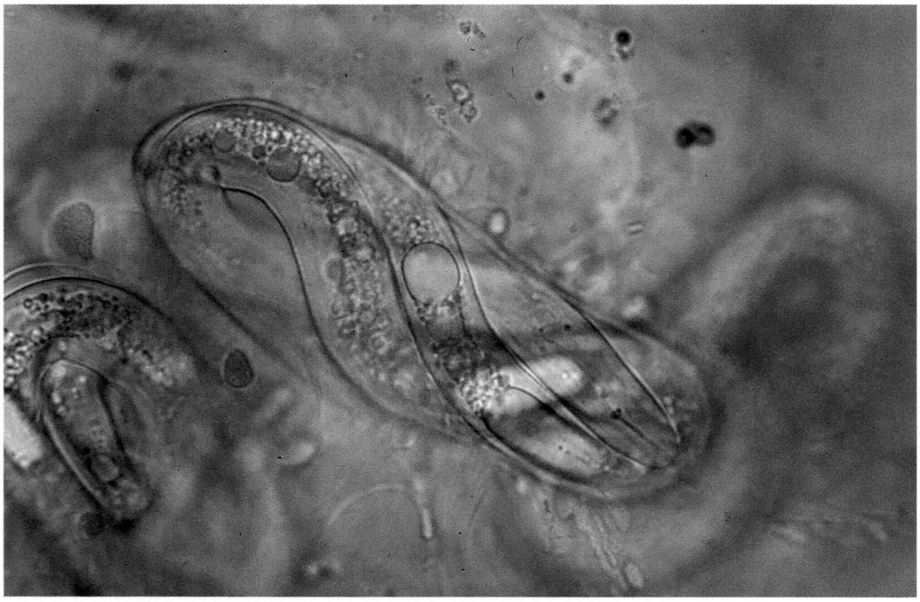

**495** *Meloidogyne* sp. eggs containing a young larva with a clearly visible stylet.

This root damage disturbs the development of plants, which then present reduced growth. In addition, some leaves may be chlorotic and wilting may sometimes occur at the hottest times of the day.

A cross section of the root knots reveals mature females thus confirming the parasitism of these nematodes.

(See **386, 412–416**)

### • Biology, epidemiology

**Survival, sources of inoculum:** These nematodes survive in the soil for over 2 years, in the form of masses of eggs protected by a mucilaginous coating. These white to brown masses may be perceived on the surface of the root knots using a stereoscopic microscope. M. *hapla* can survive in frozen soils.

These nematodes are very polyphagous and attack numerous plants, both cultivated and uncultivated, on which they multiply, guaranteeing their survival.

**Penetration and invasion:** Stage II larvae, attracted by the root exudates, penetrate the roots and migrate through the cortex to the vascular system. As they puncture cells they secrete enzymes and hormones which will cause giant cells to develop. These will be used for their nutrition. Larval development continues while the root swells. Ultimately, a root knot surrounds a massive pear-shaped female (**489**). She produces numerous eggs (on average between 400–500) which are expelled to the outside of the root, surrounded in a gelatinous coating (**416**).

**Dissemination:** Numerous eggs and larvae can be transported passively from diseased plants by run-off, drainage, and irrigation water. Larvae move actively over short distances in damp soil. Dissemination is possible through dust from contaminated soil; they are carried by the wind to nearby plots. Contaminated plants, ploughing tools, and machinery can also carry out this function.

**Conditions favourable to their development:** M. *incognita*, and M. *arenaria* appreciate relatively high temperatures (18–27°C) found in light, sandy soils. M. *javanica* tolerates higher temperatures, whereas M. *hapla* prefers more coolness. Generally speaking, their activity is greatly reduced, or even halted below 5°C or above 38°C.

## Protection*

### • During cultivation

No control method is really effective while crops are growing.

If attacks occur in the nursery, affected seedlings must be eliminated. If not, planting them in the field will help to ensure the dissemination of nematodes and the contamination of healthy soil.

In the open field it is essential for the root systems of plants which have been attacked to be removed from the plot and eliminated, so as to avoid enriching the soil with nematodes. If this measure is not practicable, the roots can be left in the open-air where they can be subjected to the effects of the sun. In the same way, digging over the soil several times during the summer will help to expose the nematodes to heat and kill them.

### • Subsequent crop

To be effective, control of root knot nem-atodes must involve, in a supplementary way, all the methods of control proposed here.

Crop rotation is frequently recommended for delaying the appearance of nematodes, and even limiting their spread. It is not always easy to implement, particularly for certain polyphagous nematodes such as *Meloidogyne* spp. or *Practylenchus* spp. In fact, clearly resistant plants suitable for use in crop rotation cannot always be found. To be effective, crop rotation must result in at least a 4-year absence of susceptible hosts in a plot. In Martinique, short rotation periods with the fodder vegetable crop *Mucuna pruriens* have been able to reduce the populations of M. *incognita* and *Rotylenchus reniformis* in the soil. Keeping the land fallow may be recommended in some countries, but this causes problems with soil erosion.

* In order to give the proposed control methods a 'universal' nature, we have produced a fairly comprehensive list of the methods and fungicides reported in the various grower countries.
**It is clear that these proposals must be modified according to the country concerned and the pesticide legislation in force.**

There are a certain number of nematode 'trapping plants' or 'nematicide' plants (*Tagetes* spp.: *Tagetes erecta*, *Tagetes patula*, and so on) which are not yet extensively used in rotation with l.s.v. Burying certain composts or green manure in the soil just before a planting a salad crop may also reduce damage from nematodes. For example, compost based on coffee pulp reduces the number of root knots and masses of *Meloidogyne incognita* eggs on lettuce. The same is true for *Azadirachta indica*. *Sorghum sudanense*, rye, and oats (hosts which are barely or not at all susceptible to *M. hapla)* are used as cover crops and green manure. Adding chitin to the soil has some effectiveness against *M. hapla*.

It is essential to obtain healthy seedlings. They will be produced preferably on sheets in a disinfected substrate. They can be placed on the soil provided that it is covered with clean plastic mulching free from tears. If you are in any doubt about the quality of the soil in your nursery, it must be disinfected.

Several nematicide products can be used. Their choice will depend on the pesticide legislation in force in your country and the financial means you have for carrying out this disinfection. A list of active materials which can be used throughout the world to control nematodes on crops, as well as some information relating to these, are listed in the table below.

Using these products has several disadvantages:

- many of these products are toxic to humans and the environment;
- they are not very specific, or not at all specific, and disturb biological equilibrium;
- they are costly and sometimes require special equipment.

In very sunny countries solar disinfecting of the soil (solarization) can be used, particularly for cleaning up plots at reduced cost. This technique consists of covering the soil to be disinfected, which has previously been carefully prepared and well moistened, with a polythene film 35–50 μm thick. This will be kept in place for at least 1 month, during a very sunny period of the year. It will increase the temperature of the soil and encourage the activity of microbial antagonism. This helps to reduce the level of inoculum of numerous phytopathogenic micro-organisms in the soil, particularly certain nematodes. The use of nematicides and composts is sometimes combined with solarization in order to increase its effectiveness on the *Meloidogyne* spp. in particular.

Nematodes are sometimes controlled by flooding future plots which are already contaminated for 7–9 months. This flooding can be continuous or broken by periods during which the soil dries out. Under these conditions the soil becomes deprived of oxygen and accumulates

| Active materials | Spectrum of activity | Miscellaneous information |
|---|---|---|
| **Fumigants**<br>Methyl bromide<br>1,3-dichloropropene (1,3-D)<br>Dichloropropene-dichloropropane (D-D)<br>Dazomet<br>Metam-sodium<br>Methyl isothiocyanate | Products which are often fairly versatile in addition to being nematicides (fungicides, insecticides, herbicides). They are generally more effective in well-drained, porous soils. They act directly on the nematodes. | These products are mainly used in the nursery. They are more effective than the non-fumigants. Methyl bromide is the most effective. 1,3-D and D-D have a weak action on weeds. Dazomet and metam-sodium are not used very often and are less efficient than methyl bromide. A number of these are used in the open field. |
| **Non-fumigants**<br>Ethoprophos<br>Fenamifos<br>Thionazin<br>Aldicarbe<br>Carbofuran<br>Oxamyl | Products which are more specific to nematodes and soil-based insects. They are active, either directly on the nematodes present in soil, or via the plant when feeding. They often temporarily inhibit nematode development but do not necessarily kill them. | Their use is less costly. They are mainly used in the open field. Their effectiveness is weaker than the previous products. They are either added at the pre-planting stage, or sprayed during subsequent weeks (applies to oxamyl and fenamifos). |

substances which are toxic to the nematodes, such as organic acids, methane, and so on. This method is only effective if carried out during a hot period of the year. It involves certain risks associated with the possibility of spreading the nematodes at the same time.

Several cycles of digging over the soil, early planting, and planting on raised beds are recommended in order to limit the effects of nematodes. The same is true with regard to the use of large sowing blocks for producing seedlings. The tools used for tilling the soil in contaminated plots must be carefully cleaned before being used in other healthy plots. The same is true for tractor wheels. Careful rinsing of this equipment in water is often sufficient to remove the soil as well as the nematodes which contaminate it.

Weeds must be totally controlled in future plots, as a number of these harbour and multiply nematodes.

It is advisable to control tightly the fertilization of plants as well as their irrigation.

There seem to be some differences in sensitivity to *Meloidogyne* spp. amongst l.s.v. In lettuce, differences in egg production have been noted for M. *incognita* race 1 and M. *javanica*. A monogenic resistance to M. *incognita* has been found in the 'Grand Rapids' lettuce cultivar in Brazil. *Lactuca saligna* and *Lactuca dregeana* proved to be resistant to a population of M. *hapla* in the greenhouse.

A number of micro-organisms which are parasites of root knot nematodes have been used in experiments on l.s.v.: *Paecilomyces marquandii*, *Verticillium chlamydosporium*, *Streptomyces costaricains*, *Bacillus thuringiensis*, and so on. For example, *V. chlamydosporium* infects the second larva stage and the eggs of M. *hapla*.

## *Pratylenchus* spp. Filipjev

(Nematoda, Tylenchida, Tylenchoidae, Pratylenchidae)

# Reddish root rotting of lettuce (Lesion nematodes)

## Principal characteristics

- **Frequency and extent of damage**

The nematodes responsible for root rot cause less damage to l.s.v. than the *Meloidogyne* spp. They are found worldwide and affect over 500 plant species. They have been recorded on l.s.v. in numerous grower countries: USA, Canada, Japan, Europe (Italy, Germany, Great Britain, and so on), but their damage varies from one country to another.

In France, rare attacks have been detected over recent years on l.s.v. involving ***Pratylenchus penetrans* (Cobb) Filipjev & Schuurmans Stekhoven**. It is possible that sometimes, in the past, there has been some confusion in diagnosis; consequently the effects of the *Pratylenchus* spp. may have been rather under estimated.

Other species of *Pratylenchus* attack l.s.v.; particularly ***Pratylenchus crenatus* Loof**.

- **Principal symptoms**

Numerous lesions, varying in size, appear on the roots. Their colour varies from yellowish to reddish-brown. The latter colour is associated with the formation of phenolic compounds in the tissues. Ultimately, fairly significant portions of the cortex decompose. If attacks are severe, a high proportion of the root system may disappear.

When in this situation, plants do not grow vigorously; a variable proportion of leaves may become chlorotic.

Observing the roots under the light microscope often reveals nematodes, with a clearly visible stylet, in the tissues or close to them.

(See **398–402**)

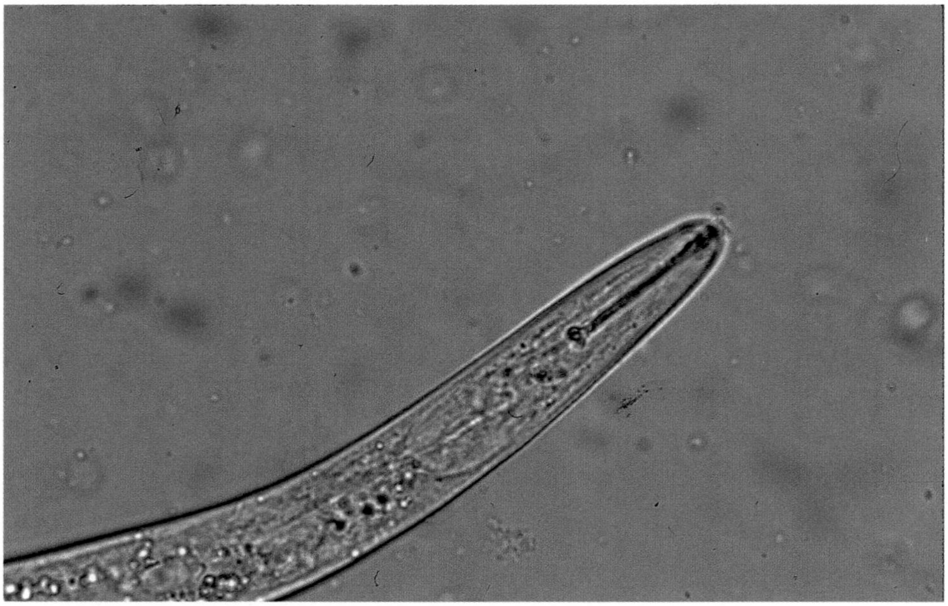

**496** The stylet of this adult nematode is clearly visible. ***Pratylenchus* sp.**

### • Biology, epidemiology

**Survival, sources of inoculum:** These nematodes are capable of surviving the winter in the soil, on and in living or dead roots. A large number of hosts harbour them, ensuring that they can survive and multiply throughout the year. *P. penetrans* has been reported on over 160 different plants.

**Penetration and invasion:** The nematodes (larvae or adults) penetrate the roots by digging holes with their buccal stylet. They progressively invade the cortex, destroying cells and digging cavities as they feed. All stages of these nematodes are mobile inside the tissues of the cortex; this is why they are called **'endomigratory' nematodes**. Ultimately, damaged tissues harbour a large quantity of eggs, larvae, and adults. A significant proportion of these adults go back into the soil.

**Dissemination:** Numerous eggs and/or larvae and/or adults can be passively transported from diseased plants by run-off, drainage, and irrigation water to other plants. Larvae move actively over short distances in damp soil. Dissemination can take place via dust from contaminated soil, carried by the wind to neighbouring plots. Contaminated plants, ploughing tools, and vehicles also fulfil this function.

With *P. penetrans*, the duration of a full cycle (from one adult to another adult) can last for 30–86 days, depending above all on temperature conditions.

**Conditions favourable to their development:** Their thermal requirements fluctuate according to the species of *Pratylenchus*. As far as *P. crenatus* and *P. penetrans* are concerned, the higher the temperature in the soil, the more active they are, within a range of temperatures between 18–30°C. *P. crenatus* prefers rather heavy, silty soil, whereas *P. penetrans* appreciates sandy soils. *P. penetrans* survives more successfully in dry soils than in similar damp soil. It reproduces most successfully at a pH of 5.2–6.4.

In general, growing conditions which are unfavourable to l.s.v. encourage the parasitism of these nematodes.

## Protection*

***Many methods proposed for controlling* the Meloidogyne *spp. can also be used for numerous other nematodes, particularly for* the Pratylenchus *spp.; find out about these in fact file* 36.**

### • During cultivation

No method of control is capable of preventing attacks of *Pratylenchus* spp. on l.s.v.

If damage occurs in the nursery, affected seedlings must be eliminated. If not, planting them out in the field will help to ensure the dissemination of the nematodes and the contamination of healthy soil.

Tools used to work the soil in contaminated plots must be carefully cleaned before being used in other healthy plots. The same is true for tractor wheels. Careful rinsing of all equipment in water will often be sufficient to remove the earth and the nematodes which contaminate it.

Once the crop has finished growing, it is essential for the root systems of l.s.v. to be removed from the plot and eliminated to avoid enriching the soil with nematodes.

### • Subsequent crop

To be effective, the control of migratory endoparasitic nematodes, just like that of root knot nematodes, must involve, in a supplementary way, all the methods of control proposed below.

Crops involved in rotation must be carefully chosen, given the numerous potential hosts of these nematodes. Very often, they only have very limited action. As an example, here are a few cultivated plants likely to harbour these *Pratylenchus* spp.: potato, numerous types of vegetables, sweet potato, sorghum, barley, sunflower, rice, maize, fruit species, and various weeds. Spinach, beet, and the Curcurbitaceae are probably relatively tolerant to *P. penetrans*, whereas the onion appears to be extremely susceptible. Allowing the land to lie fallow during the winter probably reduces the nematode population levels in the soil.

* In order to give the proposed control methods a 'universal' nature, we have produced a fairly comprehensive list of the methods and fungicides reported in the various grower countries.
**It is clear that these proposals must be modified according to the country concerned and the pesticide legislation in force.**

It is essential to produce healthy seedlings. Preferably, they should be grown on sheets in a disinfected substrate. They can be placed on the soil provided that this is covered with clean plastic mulching free from any tears. If you are in any doubt about the quality of soil in your nursery, it must be disinfected. Several nematicide products can be used (see fact file 36).

In countries where there is plenty of sunshine, solar disinfection of soil (solarization) can be envisaged, particularly for cleaning up plots at a reduced cost. This technique is described in detail in the *Meloidogyne* spp. fact file.

The resistance of l.s.v. to *Pratylenchus* spp. does not seem to have been investigated.

# Summary of control methods for leafy salad vegetable bio-aggressors in nurseries and during cultivation

## Some advice to adopt in the nursery

- **Prepare the seedling environment very carefully**

If nurseries follow one after another in the same shelter, you are advised to **'keep the shelter empty'** (absence of any plants for a certain period) between two seedling crops. In the course of this process, the **greenhouse surfaces**, structures, equipment (tools, seedling shelves, pots, crates, and so on), must be **disinfected** in order to destroy any spores and propagules of pathogenic agents.

Several products are traditionally recommended for crops grown under protection:

- bleach (sodium hydrochloride at 48° Chl) at 0.3–0.5% for non-soil systems, 4–7% for structures;
- formalin solution at 2–5% (starting from a commercial solution of 38% formaldehyde);
- a solution of quaternary ammonium (Bioroche at 0.25 L/100 L, Sernet at 0.4 L/100 L, Hortiseptyl POV at 1 L/100 L, Hortinet at 1 L/100 L).

For most of these applications, careful rinsing in clear water will be essential. Spraying of surfaces and structures in the greenhouses must take place at high pressure, soaking them thoroughly. During the disinfecting process precautions must be taken (gloves, masks, and so on) as these products are toxic (by contact and vapour) and corrosive to varying extents.

If the seeds are scattered in the soil or if seedlings grown on sowing blocks are placed on the soil, it must be **disinfected** with a fumigant (methyl bromide*, dazomet, and so on). However, we recommend that by preference you should isolate sowing blocks from the soil using a plastic film. This film must be replaced every year or as soon as it is ripped.

Between two nurseries, careful attention must be paid to the **storage** of new **equipment** intended to be used in the production of seedlings (polystyrene trays, plastic honeycombs, loose or bagged substrates, and so on) or to the storage of the same material disinfected and re-used. In fact if this equipment is not properly stored, in a place which is dirty and/or open to the elements, pollution with soil dust is possible.

You must avoid using the same shelters to produce seedlings for other market garden species, as they will be able to multiply and preserve pathogenic agents which also affect l.s.v. It will also be **advisable to eliminate weeds** all round the nurseries so that pests (direct predators or virus vectors) cannot take refuge there and then spread to the inside of the shelter and to the seedlings.

- **Use quality 'ingredients'**

To obtain good seedlings, you must first of all use **quality seeds** and a **healthy substrate**. Fairly often we encounter growers who, when confronted with a plant health problem in the nursery, rather hastily blame either the quality of the seeds, or the compost they use. Is this attitude, which surfaces from time to time whatever the crop being grown, fair in the case of l.s.v.?

In many countries, the **seed quality** for l.s.v. is good. Generally speaking their varietal purity and germination characteristics are checked before they are marketed. Obviously it is advisable to use seeds which have been checked as being free from LMV. Growers then have seeds which germinate well, and are normally free from any pathogenic agents.

(* Product will be banned from 2005 onwards.)

It is difficult to check whether the **substrate** is completely healthy. Cases of contamination by various fungi have actually been reported in publications. For the time being these situations remain marginal, given the volumes of substrate marketed, and only concern a few sources. You must bear in mind that when compost or sand is mixed with a peat substrate, these are sometimes contaminated. For example, *Pythium* spp. have already been isolated in washed river sand. In this case, it is not the compost which lies at the origin of the contamination, but the other substrate which has been incorporated into it.

In publications the recommendation is sometimes to disinfect substrates. Specialists frequently use a fumigant such as methyl bromide* or steam. As far as peat-based substrates are concerned, drying these out excessively leads to the formation of micro-pouches of air which have an adverse effect on the satisfactory distribution of fumigants. In the case of steam, at the time of carrying out the treatment we recommend a water retention capacity of 1/3 and maintenance of a temperature of 100 °C for 30 minutes. Metam-sodium is not recommended for organic substrates, but it can be used for treating minerals substrates.

You should be aware that **disinfecting the soil** is not always a harmless intervention; sometimes it causes material problems and may involve a number of disadvantages:

- destruction of natural micro-organisms antagonistic to certain pathogenic agents;
- increased receptiveness of disinfected compost to parasites;
- appearance of toxicity phenomena (excess of exchangeable manganese, excess ammonium following more or less complete blockage of nitrification, and so on).

The **quality of irrigation water** is as important as that of the seeds and substrate. You must be very vigilant, because water from the river or storage tank can frequently be contaminated by various fungi such as *Olpidium brassicae*, several species of *Pythium* and *Phytophthora*, *Thielaviopsis basicola*, sometimes by bacteria, or it may be polluted by abandoned herbicide packaging.

- **Provide the best possible conditions for seedlings and protect them carefully**

Once sowing has taken place, plants must be placed in the most favourable **growing conditions** possible for their satisfactory development, and the interventions recommended by your plant specialists must be carried out. Sometimes, lack of time, and various material constraints mean that these are not always carried out in full.

We would just like to remind you of those we feel are essential:

- ensure **optimum irrigation** of plants and good drainage;
- carefully monitor the **moisture** of your substrates as well as their temperature; a number of accidents are associated with plants which have been watered too much and then placed in a cold environment;
- if a **problem** occurs, have this positively identified, so that it can be rectified as quickly and effectively as possible;

(* Product will be banned from 2005 onwards.)

- follow the **instructions** recommended by your plant **specialists**; in particular, **ventilate shelters** properly and don't forget that lettuce mildew, *Bremia lactucae*, is just as serious in the nursery as in the open field, so carry out treatments;
- avoid producing **seedlings** which are too vigorous or too etiolated and do not make them wait too long before planting; under no circumstances must you plant contaminated seedlings;
- **control** the development of **predators** and **weeds** in nurseries; their presence may produce viral contamination;
- before carrying out any treatment, **check** that your **equipment** has been properly rinsed and that you are using the correct product concentration;
- **eliminate** all **plant debris** from the shelter and do not store it close to crops;
- check the health quality of any seedlings you have bought, before planting them in the field. To do this, it is a good idea to wash a few root systems and observe them closely.

In conclusion, you should remember that the quality of the seedlings is a guarantee of the good development of the future crop, so this quality must be unquestionable. Together we have seen that the quality of lettuce seeds and that of horticultural substrates should normally meet this requirement. In fact, the success of raising seedlings will depend almost entirely on putting the recommended prophylactic measures into practice, and on the care taken by the grower at this essential phase in growing l.s.v. Producing young seedlings is not easy and growers should never forget that they are dealing with a particularly susceptible and vulnerable plant material.

## Spectrum of efficacy of principal methods and measures of protection used on crops against bio-aggressors

| Possible measures of protection | *Aerial fungi* | *Soil-based fungi* | *Aerial bacteria* | *Soil-based bacteria* | *Phytoplasma* | *Viruses with 'aerial vectors'* | *Viruses with 'soil-based vectors'* | *Nematodes* |
|---|---|---|---|---|---|---|---|---|
| **Crop rotation** (cereals, green manure, sorghum) | +/– to + | + to ++ | +/– to + | + to ++ | 0 | +/– | ++ | ++ |
| **Avoid the proximity of already affected or susceptible crops** | +/– to ++ | 0 | + | 0 | + | + | 0 | 0 |
| **Levelling or drainage of soil** | + | ++ | + | ++ | 0 | 0 | ++ | + |
| **Soil disinfection** | +/– | +/– to ++ depending on fumigants and fungi | +/– | + | 0 | 0 | +/– to ++ depending on fumigants | ++ |
| **Plastic mulching** | 0 | + on some | 0 | 0 | 0 | 0 | 0 | 0 |
| **Aluminized mulching** | 0 | + on some | 0 | 0 | +/– | +/– | 0 | 0 |
| **Use healthy or tested seeds** | ++ *S. lactucae* only | 0 | 0 | 0 | 0 | ++ LMV only | 0 | 0 |
| **Use resistant varieties** | ++ *B. lactucae* only | 0 | 0 | 0 | 0 | ++ LMV only | 0 | 0 |
| **Check the health quality of seedlings** | + to ++ | ++ | + | 0 | 0 | ++ | + | + |
| **Shelter protected against insects** (nets obstructing openings) | 0 | 0 | 0 | 0 | ++ | ++ | 0 | 0 |
| **Insect net co:ering plants** | 0 | 0 | 0 | 0 | ++ | ++ | 0 | 0 |

0: measure having no effect
+/– : insignificant maesure
+ : fairly important measure
++ : very important measure

| **Possible methods and measures of protection** | ***Aerial fungi*** | ***Soil-based fungi*** | ***Aerial bacteria*** | ***Soil-based bacteria*** | ***Phytoplasma*** | ***Viruses with 'aerial vectors'*** | ***Virus with 'soil-based vectors'*** | ***Nematodes*** |
|---|---|---|---|---|---|---|---|---|
| **Respect planting densities** | + | 0 | + | 0 | 0 | 0 | 0 | 0 |
| **Control fertilization, particularly nitrogenous** | +/– | +/– | + | + some | 0 | 0 | 0 | 0 |
| **Avoid excess water** (use drill, tensiometer) | + | ++ | + | ++ | 0 | 0 | ++ | + |
| **Water mainly in the morning** (so that plants dry out rapidly) | ++ | 0 | ++ | 0 | 0 | 0 | 0 | 0 |
| **Avoid spinkler irrigation** | ++ | 0 | ++ | 0 | 0 | 0 | 0 | 0 |
| **Ventilate shelters and heat them if necessary** (in order to lower humidity or temperature) | ++ | 0 | ++ | 0 | 0 | 0 | 0 | 0 |
| **Eliminate weeds** (crop and surroundings) | +/– some | +/– some | +/– some | 0 | ++ | ++ | 0 | 0 |
| **Eliminate initial diseased plants** | + | 0 | + | 0 | + | ++ | 0 | 0 |
| **Eliminate plant debris** (during and at the end of the crop) | ++ | ++ | ++ | ++ | 0 | 0 | ++ | ++ |
| **Avoid working when plants are wet** | ++ | 0 | ++ | 0 | 0 | 0 | 0 | 0 |
| **Direct chemical protection possible** | +/– to ++ | +/– to ++ | +/– | 0 | 0 | 0 | 0 | 0 |
| **Anti-vector chemical protection possible** | 0 | 0 | 0 | 0 | +/– | 0 to + depending on vector and virus | +/– | 0 |

0: measure having no effect
+/– : insignificant measure
+ : fairly important measure
++ : very important measure

497

498

499

500

501

## Growing l.s.v.: from the nursery to the crop

**497** A few seedlings from a carefully tended nursery.

**498** Seedlings stored under bad conditions – close to crops which may harbour pathogenic agents and/or common pests.

**499–501** Crops of l.s.v. growing under protection.

# Appendix

## Some information about l.s.v. and their resistance to pathogenic agents

A large range of edible plants are grouped together under the description of salad vegetables, even if this category is restricted to leafy salad vegetables. In fact, depending on regions and methods of production (**497**, page 352), various species can be used in salads: lettuce, endive, cress, lamb's lettuce, rocket, dandelion and, diversifying, we also find the young leaves of various types of mustard, purslane, spinach, ice plant, and so on. These plants belong to different botanical *families* (*Asteraceae, Brassicaceae, Chenopodiaceae, Portulaceae, Aizoaceae*) and have different biologies and growing methods. However, as they are all eaten fresh, the consumer requires the same quality from all of them: leaves must appear free from any trace of parasitic attack with no residues of plant health products. In addition, these different types of product supply, to varying degrees, the expanding market of ready-prepared, pre-packaged and refrigerated fruit and vegetables.

Amongst these plants, we will only discuss the most important in this book, namely lettuce, chicory, and endive. These belong to the family of *Asteraceae* (formerly known as the *Compositeae* because of the anatomy of their flower), to the subfamily of *Cichorioideae*, and the tribe of *Lactucae*. This family is one of the largest botanical families; it contains other cultivated edible species, such as sunflower or artichoke, as well as floral species such as chrysanthemum or gerbera.

Lettuce is the top salad consumed. You should be aware that the name lettuce is often incorrectly restricted to one of its forms: butterhead lettuce. In fact, there are several types of lettuce marketed: iceberg, romaine, oak leaf, and so on. Lettuce is grown on every continent; but the type marketed varies according to the region (Tables 28 and 29). The main production regions are North America, particularly California, and Europe, especially Spain, France, and Italy. To meet the consumer's demand, namely to have a healthy product without any pesticide residues, resistance to pathogenic agents is an important criterion for breeding.

Endive is the second most popular salad variety consumed fresh, particularly escarole and frisée. Various forms of chicory are also important. Production areas are much more restricted and essentially located in Europe.

**Table 28: Principal countries producing leafy salad vegetables** (Sources: USDA and Eurostat 1998, personal communication)

| Country | Types | Production in 96/97 (1,000 tonnes) | Country | Types | Production in 96/97 (1,000 tonnes) |
|---|---|---|---|---|---|
| **USA** | Lettuce: | 3,958 | **UK** | Lettuce | 231 |
| | Iceberg | 3,116 | | | |
| | Romaine | 422 | **Germany** | Lettuce | 144 |
| | Leaf and butter | 420 | | | |
| **Spain** | Lettuce (primarily iceberg) | 925 | **Netherlands** | Lettuce | 110 |
| | Endive (escarole and frisée) | 50 | | | |
| **Italy** | Lettuce: | 420 | **Japan (1998)** | Lettuce (primarliy iceberg) | 520 |
| | Endive (escarole and frisée) | 235 | | | |
| | Chicory (Italian type) | 227 | | | |
| **France** | Lettuce: | 480 | **Australia (1999)** | Lettuce (primarliy iceberg) | 99 |
| | Endive (escarole and frisée) | 139 | | | |
| | Witloof chicory | 244 | | | |

502

503

504

505

506

***Lactuca sativa*: a few types of cultivated lettuce**

**502** Butterhead lettuce.

**503** Batavia lettuce.

**504** and **505** Cutting lettuce (oak leaf and lollo).

**506** Stem lettuce.

(See page 365 for iceberg lettuce.)

Table 29: **Production of leafy salad vegetables in the USA**

| Types | Year | USA (1,000 tonnes) | State | (1,000 tonnes) |
|---|---|---|---|---|
| Iceberg | 2003 | 2,944 | California | 2,142 |
| | | | Arizona | 771 |
| | | | Colorado | 24 |
| | | | New Jersey | 7 |
| Romaine | 2003 | 1,215 | California | 995 |
| | | | Arizona | 220 |
| Leaf and butter | 2003 | 527 | California | 453 |
| | | | Arizona | 74 |
| Endive (escarole and Frisée) | 2000 | 37 | California | 17 |
| | | | Florida | 13 |
| | | | New Jersey | 7 |

Source: USDA 2004, California Lettuce Research Board 2004

## Lettuce and similar species

**• Biology of lettuce and various types**

Cultivated lettuce (*Lactuca sativa*) (**503**, page 354) is a herbaceous, annual plant with two very distinct phases:

- the vegetative phase, resulting in the formation of a head varying in firmness, is the stage used for marketing;
- the reproductive phase, in the course of which the central stem lengthens (going to seed), that finally flowers and produces seeds.

During the vegetative phase, the plant firstly forms a rosette of leaves. Next comes the period during which the head forms, in the course of which lettuce types are differentiated (*Table 30*). The young leaves may overlap and be more or less closed in at the top. The plants then correspond to lettuce with heads, with two principal cultivar groups, **butterhead lettuce** with rather soft leaves and **crisphead** lettuce with more crunchy leaves. Other leaves may remain more open and the plant is classified as lettuce without a head or **cutting lettuce**. There are also more local types such as 'Craquante' in the Midi and even stem lettuce, neglected in France, but still consumed in Egypt and Asia. All these types, belonging to the same species, are therefore intercompatible; crosses between them are currently being carried out by breeders in order to transfer resistance genes to pathogenic agents, or to create material with an original morphology. In almost all these types, there are ranges of varieties of different colours, from golden to dark green, some of which are red.

The different types of lettuce are therefore defined by a few morphological characteristics at the head and flower stage (shape of the ligules). The importance of each type varies depending on the country and region (*Table 29*):

- Crisphead lettuce is subdivided into two types of fairly different shapes. Iceberg lettuce has a very firm head along with lower, or wrapper, leaves that are usually separated from the head when harvested.
- Varieties of this type produce very crunchy leaves with broad veins. Where crisphead lettuce is grown (USA, UK, Spain, Austrialia, Japan, and so on) it is almost exclusively the iceberg type. Batavia lettuce is most popular in France. The head is smaller, softer and less crisp than iceberg.

**Table 30: Principal types of lettuce grown and their characteristics**

| | Butterhead | Crisphead | Romaine | Latin | Leaf |
|---|---|---|---|---|---|
| **Head** | Fairly firm | Very firm | Pointed | Firm, small | No head |
| **Leaves**<br>**Veins** | Soft<br>Feathered | Broad, crunchy<br>Parallel | Elongated, crunchy<br>Feathered | Broad, crunchy<br>Feathered | Soft<br>Feathered |
| **Ligules*** | Short serrations | Deep serrations | Short serrations | Short serrations | Short serrations |
| **Sub-types** | | Iceberg<br>Batavia | | Crunchy | Oak leaf<br>Broad crisp leaf<br>Lollo |
| **Importance in US** | Low | Iceberg most<br>Batavia low | Second | Low | Third |
| **Importance in UK** | Second | Iceberg most<br>Batavia low | Third | Low | Fourth |

* Five joined petals

- **Romaine lettuce**, with oblong leaves, a small head and conical in shape, is popular in the Mediterranean area and in the USA.
- **Latin lettuce** with thick leaves often forming a small tight head is only available from regional markets in France and Spain.
- **Leaf, or cutting lettuce** is composed of broad leaf, oak leaf, and lollo types. It is popular in baby lettuce mixtures (mesclum).

Consequently, the majority of salad leaves are lettuces in the botanical sense of the term (*Lactuca sativa* species) whose types have been described previously. There is often confusion between lettuce and endive because some characteristics are shared by these two botanical species: plants consumed at the vegetative stage essentially in salads, used in the category of ready-prepared, pre-packaged, refrigerated fruit and vegetables, shared pathogenic agents (for example several viruses including LMV). In fact, several characteristics clearly differentiate the two species *Lactuca* and *Cichorium*: impossible to successfully carrying out crosses, different floral biology, relative importance of parasites.

After the vegetative phase, the stem lengthens and a branched panicle develops. The plant **goes to seed** with varying degrees of rapidity depending on climatic conditions and genotype. It is encouraged by long days and high temperatures. This is why breeders have created varieties which are resistant to going to seed for summer crops. **The flower**, yellow in colour, or slightly red in certain genotypes, is actually a set of 12–15 florets grouped together in capitula (characteristic of the Asteraceae). The anatomy and development of the flowers strongly favours autogamy. Their small size and rapidity of opening make emasculation tricky. Manual hybridization can only be carried out in selection programmes for combining characteristics of different varieties. The varieties cultivated are inbred line varieties.

Quality **seeds** must be produced protected from aphids vectors of lettuce mosaic virus (LMV) which can be transmitted by seeds. Greenhouses are also effective in protecting flowers from bad weather; in fact, rain can cause seeds to fall and a fungus, *Botrytis cinerea*, to develop. In addition, wind, combined with dry weather, encourages dispersion of seeds because of their anatomy (naked seed with pappus).

- **Species similar to lettuce: the *Lactuca* genus**

*Lactuca* is a vast genus characterized by the presence of 'milk', the latex; this white, sticky liquid, which flows from areas of damage to leaves or stems, has given it its name. Wild types form a rosette with leaves which are often spiny; then, the same year or the following year, the stems elongate and flowering takes place in summer in most regions. As the capitula are wide open, resembling those of mature dandelions, seeds are dispersed by the wind.

The *Lactuca* genus probably originated in the Middle East or Egypt with an area of diversification within the Mediterranean basin. In Europe, according to *Flora Europaea*, over 100 species have been identified amongst which a group of compatibility has been defined. This involves diploid species with 9 pairs of chromosomes comprising, in addition to cultivated lettuce (*L. sativa*), 3 other species with yellow flowers (***L. serriola***, ***L. saligna***, ***L. virosa***). Amongst these species, *L. serriola* is often present along the roadsides and in fallow land (**509**). It can constitute an alternative host for certain lettuce pathogens. These species have been used as sources of genes for improving lettuce. Thus, numerous genes of resistance of current varieties have come from *L. serriola*, a species which is easy to use because it is totally compatible with lettuce. The genes identified in *L. saligna* and *L. virosa* are much more difficult to transfer to the lettuce. In fact, hybrids are tricky to obtain and often sterile or not very fertile. Laboratory techniques have been developed to facilitate the use of the species and thus to broaden the variability available for improving lettuce. Thus, *in vitro* cultures of immature embryos allow hybrids to be obtained (*L. sativa* × *L. saligna*) and facilitate the obtaining of back-crosses between lettuce and hybrids (*L. sativa*/*L. virosa*). When carrying out these crosses or obtaining their descendants, unfavourable characteristics can delay, or even prevent, the transfer of resistance to lettuce. For example numerous hybrids between *L. sativa* and *L. virosa* are necrotic and die at the 4–5 leaf stage.

Other species, such as ***L. perennis***, are specific to certain pedoclimatic systems. This species grows on chalky soils in the Mediterranean area; these are perennial plants with purple flowers, not compatible with lettuce, which are traditionally picked and consumed in salads.

Cultivated lettuce can be distinguished from wild forms by several morphological characteristics, referred to as domestication characteristics:

- formation of a head or at least a marked vegetative stage with a large number of leaves producing a 'cluster' varying in tightness;
- absence of spines under leaves;
- reduction of latex and bitterness;
- tight capitula in the bracts allowing seeds to be retained on the mature plant.

Lettuce has been consumed since ancient times. In the Middle Ages lettuce with heads and romaine lettuce were grown. The number of varieties would increase considerably from the 17th century onwards. Breeding in the US and in Europe is currently very intense, especially in France and the Netherlands. Many new varieties appear each year. Alongside these, amateurs are still growing old varieties from the end of the 19th century.

- **Lettuce selection and resistance to pathogenic agents and pests**

Breeding has greatly modified the characteristics of cultivated lettuce. It has allowed plant material to adapt to the different growing techniques and thus spread production throughout the year. In addition, it is now offering a certain amount of diversification in terms of colours and shapes, particularly with heads more suitable for the modern market. In addition to these presentation and adaptation characteristics, very significant efforts have been made since the 1950s to introduce varieties resistant to pathogenic agents.

**Nature of resistance and progenitors used**

An enormous research study has been devoted to ***Bremia lactucae*** Regel resistance. This, begun in the 1950s, is still continuous given the ability of the 'fungus' to adapt. At the end of the 1960s, research into **LMV** resistance resulted in the creation of varieties tolerant to this potyvirus.

Genetic resistance to insects has also been used in breeding: the *Ra* resistance gene to the root aphid (***Pemphigus bursarius***), the *Nr* resistance gene to an aphid affecting aerial parts (***Nasonova ribisnigri***). Other resistances have been identified in several genotypes of wild species; for example **CMV** resistance in *L. saligna* PI261653, **BWYV** resistance in *L. virosa* PIVT280, and one to LMV-13 in *L. virosa* PIVT1398. In addition to resistance, tolerances or reduced susceptibility have been revealed, often in old European varieties: for example 'Batavia blonde à bord rouge' is tolerant to *B. lactucae*, several icebergs ('Pacific', 'Bay View', 'Spreckels' and so on) are tolerant to **big vein** disease.

507

508

509

510

## *Lactuca serriola*: a weed used as a source of resistance to bio-aggressors

**507** *L. serriola* is a weed which is common worldwide. It is found in crops of l.s.v. or in the immediate environment of plots.

**508** This weed can be infected by several viruses which attack l.s.v. It also harbours some of their vectors, for example aphids; it therefore constitutes a significant source of bio-aggressors.

**509** *L. serriola* at the vegetative stage.

**510** *L. serriola* in flower, in fallow land.

Breeders initially used resistance genes identified in the cultivated species. Thus, the first *Dm* resistance genes to *B. lactucae* come from old European varieties (*Dm2* from Meikoningen, and *Dm3* from Gotte for forcing, in 1950). The *mo1*[1] resistance gene to LMV was found in the Argentinian variety Gallega de Invierno (hence its name, gene *g*, in the original article) and the *mo1*[2] gene in the genotype PI251245 (*L. sativa* found wild in Egypt, classified originally as *L. serriola*). Then breeders looked for new genes in the species *L. serriola*. So, a large number of genes for resistance to *B. lactucae* originated from this species (*Dm 5/8* in 1945, *Dm11* in 1960, *Dm16* in 1980, *R18* in 1970). More recently, other compatible species, *L. saligna* and *L. virosa*, have been used. However, in this case, the work for creating resistant varieties is very long and complicated. A link had to be broken between the *Nr* gene and dwarfism of plants in order to obtain varieties resistant to *N. ribisnigri* of commercial quality. Important work on selection allowed the transfer of *R36* to *B. lactucae* from *L. saligna* and the *Nr* gene identified in L. *virosa* PIVT280 to be transferred to varieties of lettuce. Research into other resistances identified in the species is in progress.

For certain pathogenic agents, no interesting resistance has been found in the genotypes of the *serriola* group of compatibility (**TSWV** resistance for example). In this case, the variability of species referred to as incompatible must be exploited (for example, resistance to several bio-aggressors has been identified in *L. perennis*) or a new variability must be created through mutagenesis or transformation. With this aim in mind, tools for growing *in vitro* and for use in biotechnology have been adapted to lettuce: creating crosses between very different species through fusion of protoplasts, and through the transformation of lettuce by *Agrobacterium tumefaciens*. To date these tools are only used at research level. They have already produced encouraging results: flowering of hybrids (*L. sativa*/*L. perennis*), transgenic plants resistant to TSWV.

**Example of resistance of lettuce to *B. lactucae***

The history of breeding programs in Europe to develop resistance to *B. lactucae* is discussed in the following pages.

Since the 1960s, breeders have produced varieties with resistance genes to *B. lactucae*. These are specific genes known as *Dm* genes, which protect lettuce against certain strains which do not possess the corresponding factors of virulence. Interactions between the fungus and the lettuce are represented in tables showing spectra of fungal virulence. For each resistance gene of lettuce, these tables show the behaviour of each strain of *B. lactucae* described. So, table 31 shows strains of *B. lactucae* identified in Europe between 1960 and 1990, alongside the use by breeders, of new resistance genes.

**Table 31: Spectra of virulence of *Bremia* strains characterized in the Netherlands between 1960 and 1990**

| Resistance strategies | NL strains of *B. lactucae* | Sources of resistance used: number of *Dm* genes | | | | | | | | | | | | | |
|---|---|---|---|---|---|---|---|---|---|---|---|---|---|---|---|
| | | 1 | 2 | 3 | 4 | 5/8 | 6 | 7 | 10 | 11 | 13 | 14 | 15 | 16 | 18 |
| **1960s** *Dm2, Dm3, Dm7* | NL1 | + | + | – | + | – | – | – | + | – | + | + | – | – | – |
| | NL2 | + | + | + | + | + | + | – | + | – | + | + | – | – | – |
| | NL3 | – | – | – | – | + | + | + | + | – | + | + | + | – | – |
| | NL4 | + | + | – | + | + | – | + | + | – | + | + | – | – | – |
| **1970s** *Dm7, Dm11,* avec *Dm2/Dm3* | NL5 | + | – | + | ? | – | – | + | + | – | + | ? | + | – | – |
| | NL6 | + | + | – | + | + | – | – | + | + | + | + | – | – | – |
| | NL7 | + | + | + | + | – | + | + | + | – | + | + | ? | – | – |
| **1980s** *Dm2, Dm3, Dm7, Dm11* avec *Dm6/Dm16* | NL10 | + | + | + | + | + | + | + | + | – | + | + | – | – | – |
| | NL11 | + | – | – | – | + | + | + | + | – | + | + | + | + | – |
| | NL12 | + | – | – | + | + | + | + | + | + | + | + | + | + | – |
| | NL13 | + | – | + | – | + | – | – | + | + | + | + | – | – | – |
| | NL14 | + | + | + | + | + | + | – | + | + | + | – | – | – | – |
| | NL15 | + | + | + | + | + | – | + | + | + | + | – | – | – | – |
| | NL16 | + | + | + | + | + | + | + | + | + | + | ? | – | – | – |

+: Compatible reaction (sporulation)
–: Incompatible reaction (resistance)
*Dm* genes used in selection and overcome by the strain
*Dm* genes used in selection effective against the strain
*Dm* genes effective against all strains identified over the decade

During the 1960s, breeders used the three genes of resistance, *Dm2*, *Dm3*, and *Dm7*, individually or in combination. If you look at table 31, you can see that the *Dm2* gene gives resistance to the pathotypes NL3, but this is overcome by the strains NL1, NL2, and NL4. Likewise, the *Dm3* gene gives the varieties resistance to NL1, NL3, and NL4 but it is overcome by the strain NL2. As far as the gene *Dm7* is concerned, this is effective against NL1 and NL2, but overcome by the pathotypes NL3 and NL4. So, at the end of the 1960s, we had strains of *Bremia* capable of overcoming all of the genes used in commercial varieties. Breeders then identified, in *L. serriola,* a new resistance gene, known as *Dm11*, which protects varieties against 4 strains, NL1–NL4. However, the use of this resistance, in the 1970s, encouraged the appearance and spread of NL6 strains which combat *Dm11*. The 'race' between the breeders and *B. lactucae* continued. Thus, a new gene, known as *Dm16*, was identified in *L. serriola* LSE18; it confers resistance to all strains NL1–NL7. However, NL11 and NL12 strains, characterized in the 1980s, call into question the resistance of *Dm16*.

Each of the *Dm1–Dm16* genes was therefore attacked by at least one of the strains of *B. lactucae* identified in Europe and named NL1–NL15. However, combinations of several of these *Dm* genes nevertheless allowed the mildew to be controlled in the 1980s since no strain attacked all these genes. So, *Dm2* was effective against NL11–NL13, *Dm3* against NL11, and NL12, *Dm6* against NL13 and NL15, *Dm7* against NL13 and NL14, *Dm11* against NL10, and NL11, and finally *Dm16* against NL10 and NL13–NL15. We should point out that the *Dm7* and *Dm11* genes are probably alleles or very closely associated; consequently no fixed variety can possess both genes.

The appearance of the NL16 pathotype changed the situation as it attacks these 6 *Dm* genes used in commercial varieties. So, at the beginning of the 1990s, only one resistance was effective against strains NL1–NL16 identified in Europe; this was the resistance obtained from an *L. serriola* and introduced into the open field variety, Mariska, during the 1970s. This resistance, known as *R18*, was transferred by several seed companies into different varietal types during the 1990s. At the same time several companies had identified other resistances which were effective in combating the strains NL1–NL16. All these varieties were described as 'varieties resistant to *Bremia* races NL1–16' in seed catalogues. It was not possible to distinguish these different resistances since all these varieties are resistant to all the pathotypes NL1–NL16.

At the end of the 1990s, attacks of mildew were reported in several regions of Europe on certain varieties declared as being resistant to NL1–16. In France, a working party brought together the breeders, Seed Variety Control Group, and the INRA, together with a Dutch group, exchanged information about these strains (the strains which are the most threatening for European crops and those allowing a distinction to be made between varieties). The knowledge of the spectra of resistance of several new varieties towards these strains revealed the different resistances used by the breeders: *R18* in numerous varieties, but also *R36* from Ninja or *R38* from Argelès. The 7 strains retained are known as Bl: 17–Bl: 23 (*Table 32*).

Pathotypes Bl: 17, Bl: 18, Bl: 20, and Bl: 22, which appeared in Northern Europe (Sweden, Great Britain, Germany, and Benelux respectively), circumvent the resistance of *R18*, but *R36* from Ninja and *R38* from Argelès are effective in combating these strains, strains Bl: 19 and Bl: 21 appeared in Italy, Bl: 19 on Argelès and Bl: 21 on Ninja. Since the emergence of these strains, strains of closely related spectra have been reported in certain regions of France. None of these, NL1–Bl: 23, attack the 18 resistances which can be used by the breeders (*R17* is not taken into account; this resistance, identified at the INRA in *L. serriola*, has not yet been introduced into commercial material). However, the emergence of a strain which could circumvent these resistances cannot be excluded. Faced with this risk, research laboratories are continuing to look for new resistances.

We should point out that these Bl races often appeared in crops where preventative schedules for plant health protection had not been followed, either through negligence, or through inability in the case of biological cultures. We should stress that effective protection against *B. lactucae* must combine the following: growing resistant varieties, implementing prophylactic measures, and carrying out preventative fungicidal treatments. Among the prophylactic measures, we must remember the importance of eradicating the very first source or sources of mildew by totally destroying any diseased plants.

**Table 32: Spectra of virulence of *Bremia* strains that appeared in Europe during the 1990s**

| NL or Bl strains of *B. lactucae* (1) | Varieties and *Dm* genes or *R Resistance* factors (2) | | | | | | | | | | | | | | | | | | |
|---|---|---|---|---|---|---|---|---|---|---|---|---|---|---|---|---|---|---|---|
| | Lednicky | UCDM2 | Dandie | R4T57 | Valmaine | Sabine | *LSE57/15* | UCDM10 | Capitan | Hilde | Pennlake | UCDM14 | *PIVT1309* | *LSE/18* | LS102 | Colorado | Ninja | Discovery | Argelès |
| | 1 | 2 | 3 | 4 | 5/8 | 6 | 7 | 10 | 11 | 12 | 13 | 14 | 15 | 16 | 17 | 18 | 36 | 37 | 38 |
| NL15 | + | + | + | + | + | – | + | + | + | + | + | – | – | – | – | – | – | – | – |
| **Bl: 16 = NL16** | + | + | + | + | + | + | + | + | + | + | + | – ? | – | + | – | – | – | – | – |
| **Bl: 17** | – | + | + | – | + | – | + | + | – | + | + | + | + | – | – | + | – | + | – |
| **Bl: 18** | + | + | – | + | + | + | + | + | + | + | + | – | – | + | – | + | – | – | – |
| **Bl: 19** | + | + | + | + | + | + | – | + | + | + | + | + | – | – | – | – | – | – | + |
| **Bl: 20** | + | + | + | + | + | + | + | + | + | + | + | + | – | + | – | + | – | – | – |
| **Bl: 21** | + | + | + | + | + | + | + | + | + | + | + | – | + | + | – | – | + | + | – |
| **Bl: 22** | + | + | – | + | + | + | + | + | + | + | + | + | + | – | – | + | – | – | – |
| **Bl: 23** | + | + | + | + | + | + | + | + | + | + | + | – | – | + | – | – | – | – | + |

(1) European strains were called NL because their description was produced by a laboratory in the Netherlands. Now, in order to emphasize the varying origins of strains, the symbol Bl for Bremia of Lettuce has been adopted.
(2) resistance genes are called *Dm* (for Downy mildew) when clearly characterized (1 single gene). Resistance factors, due to a genetic characteristic which is less clearly genetically defined (possibility of groups of genes), are known as *R* (for Resistance).

+: Compatible reaction (sporulation)
–: Incompatible reaction (resistance)
■ *Dm* genes used in selection and overcome by the strain
● *Dm* genes used in selection effective against the strain
■ *Dm* genes effective against all strains identified during the decade

**Example of resistance to lettuce mosaic virus (LMV)**

Another pathogenic agent has been taken into account by the breeders since the 1960s and 1970s: lettuce mosaic virus (LMV). Two recessive genes have been identified in lettuce (*g* in Gallega de Invierno, an Argentinian variety, and *mo* in a wild lettuce from Egypt). They confer tolerance to the common strain of LMV (LMV-0) by reducing multiplication of the virus, leading to a significant reduction, or even absence, of symptoms. In addition, they prevent this strain from being transmitted by the seed. Genetic studies have shown that these genes were alleles (or very closely associated genes), which excludes the possibility of bringing them together in the same fixed variety. Initially, we believed that these two genes were identical and depending on the region, one or the other was used in selection (*g* in Europe and *mo* in the USA). The identification in France, during the 1980s, of strains causing pronounced symptoms on varieties carrying *g*, without attacking the American varieties carrying *mo* (LMV-1 and LMV-9), showed that the two genes were probably different alleles which were named $mo1^1$ and $mo1^2$. Secondly, the Ithaca variety, a crisp type, susceptible to LMV-0, appeared to be resistant to strains of virus isolated in Greece (Gr4 and Gr5) and in the Yemen (Yar); this resistance was called *Mo2*. Strains circumventing all these resistances were then isolated in Europe: LMV-E in Spain in 1984, then LMV-13 in France in 1989. Studying these strains on varieties carrying different resistance genes and analysing their transmissibility by the seed allowed us to establish a table of their biological characteristics (*Table 33*).

Consequently, in the 1990s, almost all varieties grown in the open field in France possessed one of these alleles. Nevertheless, since the end of the 1980s, tolerant varieties have been attacked in various regions by virulent strains transmissible by seed of the LMV-13 strain (France, Tunisia, Brazil). Propagation of the LMV-13 strain can be facilitated by its capacity to be transmitted by the seed even in varieties carrying tolerance to the virus. The only resistance effective against these strains has been identified in a genotype of *L. virosa*. The introduction of this resistance (*Mo3*) in the lettuce is tricky because of unfavourable characteristics transmitted to descendants (necrosis, sterility, and so on); obtaining commercial varieties possessing the *Mo3* gene will therefore be a very long process.

In the USA, LMV is controlled primarily by the use of virtually LMV-free seeds. However, several varieties of iceberg and romaine lettuce carry the $mo1^2$ allele of resistance.

511 

 512

 513

**_Cichorium endivia, C. intybus_: a few cultivated types**

**511** Frisée endive.

**512** Escarole endive.

**513** Chioggia belonging to the Radicchio cultivar groups, also known as Italian chicory.

Table 33: A few characteristics of the different strains of LMV

| Resistance genes | LMV strains | | | | |
|---|---|---|---|---|---|
| | Yar ; Gr4 ; Gr5 | LMV-0 | LMV-1 ; LMV-9 | LMV-E ; GrB | LMV-13 ; Aud |
| *mo1*[1] | + [(1)] | tol [(2)] | + | + | + |
| *mo1*[2] | + | tol | R [(3)] | + | + |
| *Mo2* | R | + | + | + | + |
| Transmission by seed | ? | yes | no | yes | yes |
| **Origin of strains** | | | | | |
| Host plant | Lettuce; *P. echioides*; chicory | Lettuce | Lettuce; chicory | Lettuce; ? | Lettuce; *mo1*[1] |
| Year of detection | ?; 1990; 1983 | 1978 | 1980; 1983 | 1984; ? | 1989; 1993 |
| Country of origin | Yemen ; Greece | France | France | Spain; Greece | France |

[(1)] + : susceptible; [(2)] tol : tolerant; [(3)] R : resistant

Consequently, rigorous prophylactic measures must be applied to restrict the spread of the virus: using seeds which have been checked as virus-free, weeding around edges of plots to destroy wild plants carrying LMV. These measures have been applied effectively in the Salinas Valley in California. The use of varieties described as being resistant to LMV, in fact tolerant to LMV-0, is also a very favourable factor for limiting the virus.

## Principal resistances of lettuce: origin of genes and commercial names

| Pathogens or pests | Original source of resistance | Resistance genes or factors | Resistant commercial varieties | Catalogue descriptions |
|---|---|---|---|---|
| ***Bremia*** | *L. sativa* | *Dm2, Dm3* | Under shelter and in the open field | Resistance to *Bremia* races: NL1, 3–6 , 11–13 or Bl: 1, 3–6, 11–13, 18, 22 |
| | *L. serriola* | *Dm5/8, Dm11, Dm16, R18, R38* | Under shelter and in the open field | Resistance to *Bremia:* races (list of strains, between NL1 or Bl: 1 and Bl: 22, depends on genes of the variety) |
| | *L. saligna* | *R36* | A few varieties | Resistance to *Bremia*: races Bl: 1–20, 22 |
| | *L. virosa* | | No (research stage) | |
| **LMV** | | | | |
| LMV-0 | *L. sativa* variety Gallega de Invierno *Variety* | *mo1*[1] | Open field in Europe | Tolerance to LMV (LMV II) - or: tolerance LMVO: tolerance LMV – 0 or: tolerance VML – or resistance: lettuce mosaic virus (LMV) |
| LMV-0, LMV-1 and LMV-9 | Wild *L. sativa*: PI 251245 | *mo1*[2] | American varieties, iceberg | |
| All strains identified | *L. virosa* | *Mo3* | No (research stage) | |
| **Aphids** | | | | |
| Pemphigus (root aphid) | | *Ra* | A few varieties in the open field | Tolerant to the root aphid |
| *Nasonovia ribisnigri* | *L. virosa* PIVT280 | *Nr* | A few varieties in the open field | Tolerant to aphids<br>(*) *N. ribisnigri*<br>Tolerance to red aphids (*N. ribisnigri*)<br>Resistance to the aphid (*N. ribisnigri*) |

## Genetic links between genes of resistance to various pathogenic agents in lettuce

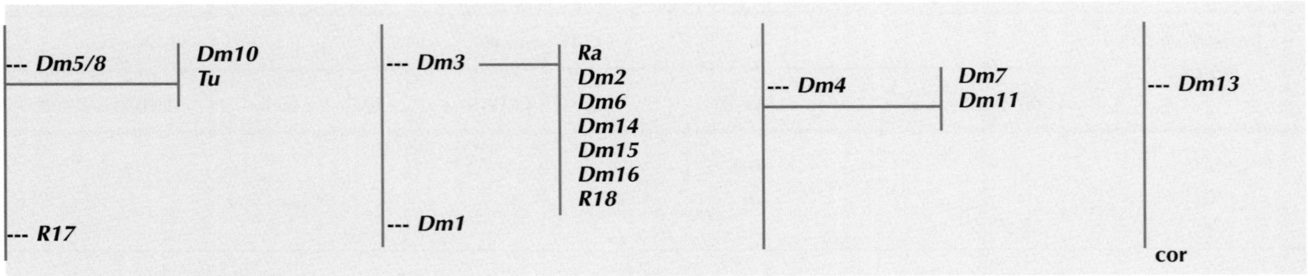


Resistance to *Bremia* (*Dm* genes or *R* factors) have been located on 4 linkage groups. Several of these genes are even alleles to the same locus (*Dm7* and *Dm11*) or very closely associated genes (Group *Dm2–R18*); in this case, it is not possible to combine them in one variety (homozygous genotype).

Other resistance genes have been located on these same linkage groups: *Tu* (resistance to TuMV) is probably closely associated with the allele to sensitivity for *Dm10* and *Ra* (resistance to root aphid) is closely associated with the 6 genes *Dm2–R18*. In addition, resistance to *Rhizomonas suberifaciens* (*cor* gene for 'corky root') is probably in the same group as *Dm13*.

## Endive and chicory

Cultivated endive and chicory belong to two very close, intercompatible botanical species (diploid species with 9 pairs of chromosomes): *Cichorium endivia* L. and C. *intybus* L. (**511–513**). Like lettuce, these plants have a vegetative phase, the leaves which are consumed, and this is very distinct from the reproductive phase. The capitula are blue and visited by insects. Endive and chicory seem to have been consumed in the Mediterranean area since olden times, initially no doubt gathered in the wild. A clear distinction between the use of the two species only appears at a late date (end of the 16th century).

**C. *endivia***, endive, includes the plants with narrow leaves which are marketed under the name of **frisée**, and plants with broad leaves known as **escarole**. They are grown mainly in the Mediterranean (Italy, France, Greece, and Spain). However, some crops are grown in the USA. The development of the market for ready-prepared, pre-packaged, refrigerated fruit and vegetables has led to many changes in the consumption of l.s.v. This species is physiologically very close to lettuce: annual plants, a rosette stage, then going to seed in the spring-summer. Self-fertile flowers are grouped together in capitula of 16–19 violet to mauve florets, with a rate of cross-fertilization which can be high, following pollination by insects. Selection essentially concerns criteria of presentation (serration of leaves, density of head), adaptation to summer crops (resistance to going to seed) and whitening of the head. The varieties grown are imposed lines.

**C. *intybus***, 'chicory', includes some very different types. One type, known as witloof in Flemish, is obtained by forcing the root while keeping the plant out of the light. The tuberized root of this chicory is pulled up in autumn and placed in the earth or in a container to allow the bud to grow and produce the head which is eaten. Production is high in France, Belgium, and the Netherlands. The second is an endive produced without forcing and includes several types, with **Sugar loaf** and the **Radicchio** cultivar groups (**Trevise**, **Verone**, **Chioggia**) also known as Italian chicory. The main producer of unforced chicory is Italy. The market for ready-prepared, pre-packaged, refrigerated fruit and vegetables has led to a rise in popularity of these types, particularly chioggia. The species C. *intybus* is a biennial or perennial and requires vernalization in order to encourage growth of the floral stem; flowers are often self-incompatible. An important research and breeding study obtained F1 hybrids of witloof, which are very homogenous and was able to develop a method of forcing this variety indoors in containers supplied with nutrient solution (hydroponic forcing). Selection was directed towards characteristics of quality (shape of head, tolerance to internal browning, reduction in bitterness). F1 hybrid varieties of chioggia, also very homogenous, have been obtained. There are varieties of green chicory and red varieties. Recently, a cross between a variety of red Verona and a witloof chicory has succeeded in creating red forced chicory.

Few genetic studies concerning resistance to pathogenic agents have been carried out in relation to all types of chicory. A dominant gene for resistance to *Alternaria* has been identified in C. *intybus*; resistance to turnip mosaic virus (TuMV) has also been identified in this species. Currently, no commercial variety has proved to be resistant to a pathogenic agent.

# Additional material supplied by Dr Edward J. Ryder,
## USDA Agricultural Research Station, Salinas CA

514

515

516

517

518

519

**514** Varying stages of lettuce dieback on romaine plants. One plant appears healthy, the others are mildly yellowed and stunted or severely stunted or dead.
**Lettuce dieback (TBSV, LNSV)** (see also page 153)

**515** Romaine plants with lettuce dieback. The plants are either dead or all lower leaves are dead with only leaves of the shoot remaining green.
**Lettuce dieback (TBSV, LNSV)** (see also page 153)

**516** An iceberg lettuce (left) and a field of the iceberg lettuce variety 'Salinas' in the Salinas Valley, CA.

**517** Heads of iceberg lettuce are being packed in cardboard cartons holding 24 heads each. These will be loaded on a truck and taken to a vacuum cooling facility.

**518** Wedge-shaped dark brown to black lesion on one side of the root of a romaine plant.
***Phoma exigua*** (see also page 181)

**519** Healthy plant on left. Plant on right shows one-sided growth. Phoma lesion is below the side with the leaves of reduced size.
***P. exigua*** (see also page 181)

# Glossary

**Acervulus:** asexual fruiting body in the form of a wide open receptacle producing short conidiophores and conidia; it characterizes the Melanconiales.

**Aecidiospore:** binuclear spore produced in the aecidium (aecidiolum).

**Aecidiolum (aecium, aecidium):** fruiting body formed by the rust fungi in which binuclear spores are produced.

**Aggressiveness:** quantitative component of the pathogenic power of a micro-organism.

**Alternative host:** another plant host of a given micro-organism that sometimes allows it to complete a full cycle.

**Anamorph:** asexual form of a fungus, also known as imperfect stage, frequently resulting in the formation of conidia.

**Anastomosis:** fusion between mycelium hypha belonging to the same thallus or complementary thalli.

**Antheridia:** cryptogamic male structure responsible for the formation of gametes.

**Anthocyanosed:** describes a plant organ, for example, which has taken on an abnormally purplish hue.

**Anthracnose:** disease caused by fungus whose anamorph form of reproduction is an acervulus (Melanconiales).

**Antibody:** specific protein produced by an animal in response to an antigen.

**Antigen:** foreign molecule frequently of protein nature inducing the formation of antibodies when injected in an animal.

**Apothecium:** structure in the shape of a disc or trumpet on which ascospores form in the Ascomycetes.

**Appressorium:** extremity of a hypha or germination allowing a fungus to fix on its host and penetrate it.

**Ascocarp:** sexual fructification of the Ascomy-cetes.

**Ascomycetes:** group of fungi producing their sexual spores, ascospores, via asci.

**Ascospore:** spore resulting from sexual reproduction in the Ascomycetes, forming inside an ascus.

**Ascus:** cell in a 'bag' form in which generally 8 ascospores are formed, and which characterizes the Ascomycetes.

**Avirulent:** describes, for example, a strain of a micro-organism incapable of infecting a given cultivar.

**Basidium:** cell on which basidiospores differentiate.

**Basidomycetes:** group of fungi producing their sexual spores, the basidiospores, on basidia.

**Basidiospore:** spore resulting from sexual reproduction in the Basidiomycetes, forming on a basidium.

**Binuclear:** containing two nuclei.

**Capsid:** protein envelope of viruses which contains their nucleic acid.

**Canker:** necrotic lesion, more or less localized.

**Chlamydospore:** spore resulting from asexual multiplication, with a thick wall guaranteeing its protection and survival under unfavourable conditions.

**Chlorosis:** describes, for example, a plant organ which has taken on an abnormally yellow hue.

**Cleistothecium:** fully closed ascocarp, sometimes sculpted with conidiophores, bursting open at maturity.

**Conidiophore:** the specialized hypha on which one or several conidia form.

**Conidia:** spore resulting from asexual multiplication and formed at the extremity of a conidiophore.

**Contamination:** first stage in the development of a disease during which the pathogenic agent penetrates the host using its own resources.

**Cortex:** parenchyma tissue located between the epidermis and the phloem of the stem and roots.

**Cotyledon:** embryo leaf located in the seed, and contributing, thanks to its reserves, to initiation of the seedling.

**Cultivar:** cultivated variety.

**Cuticle:** waxy waterproof layer covering the epidermal cells of leaves, stem, and fruits.

**Damping-off:** destruction and rapid disappearance of seedlings frequently associated with damage affecting the crown and/or roots.

**Dissemination:** final stage of a disease during which the inoculum will be liberated, disseminated over varying distances, and will be used to contaminate healthy plants.

**Dominant:** describes an allele which is expressed phenotypically when present in a plant.

**Enation:** The outgrowth of a leaf forming on certain portions of veins.

**Endemic:** describes a disease whose frequency is low and relatively constant in time, sometimes with localized distribution.

**Epidemiology:** study relating to the environment of initiation, development, and dispersion of the disease.

**Epidermis:** layer of cells covering plants sometimes surmounted by a cuticle.

**Endoconidia:** conidia formed inside a mycelium hyphae.

**Enzyme:** protein substance catalyzing a specific biochemical reaction.

**Epiphyte:** describes an organism living on the surface of a plant: epiphyte bacteria.

**Fasciation:** malformation (increase in size, hypertrophy, flattening, fusion of several organs) affecting a short, floral organ.

**Flagellum:** filiform, elongated structure ensuring the mobility of bacteria and zoospores of fungi.

**Fruiting body (ies):** produce spores; or spores produced by a fungus.

**Fumigant:** pesticide acting in a gaseous form inhibiting growth or killing various micro-organisms or predators, particularly those attacking in soil.

**Fungicide:** a substance which kills fungi or inhibits their growth or the germination of their spores.

**Gene:** hereditary unit located on a chromosome, a plasmid or a cytoplasmic organelle, coding for a protein.

**Haploid:** describes a cell or an organism which only has one single full set of chromosomes.

**Haustorium:** extension of the mycelium inside the cells, whose role is to withdraw the nutrient substances indispensable to fungi, while preserving their host.

**Hermaphrodite:** individual with both male and female sexual organs.

**Heterothallic:** describes a fungus whose male and female gametes are formed on different thalli.

**Homothallic:** describes a fungus which can form male and female gametes on the same mycelium.

**Hyalin:** describes, for example, a structure without colour, transparent.

**Hybrid:** descendant of two parents of different genotypes.

**Hydathode:** specialist structure of the leaf epidermis where water is secreted or exuded.

**Hypha:** corresponds to an isolated filament of fungus mycelium.

**Immune:** describes a plant which, confronted with a given pathogenic agent, does not show any contamination.

**Incubation:** period separating contamination from manifestation of initial symptoms of a disease.

**Infection:** process in accordance with which a micro-organism penetrates and multiplies in a plant.

**Inoculum:** parts, structures of a micro-organism capable of infecting a plant.

**Isolate:** pure culture of a micro-organism obtained without special cloning.

**Larva:** juvenile form of certain animals preceding the adult stage.

**Latent:** describes, for example, a pathogenic micro-organism present on a plant but remaining invisible and/or inactive.

**Latency:** period separating contamination from the differentiation of the initial fruiting bodies.

**Microsclerotium:** sclerotium of very small dimensions.

**Monogenic resistance:** resistance determined by one single gene.

**Mutation:** hereditary genetic modification occurring in a cell.

**Mycelium:** filament or mass of filaments constituting the basic structure of fungi.

**Nematicide:** substance causing the death of nematodes.

**Oogone:** structure containing one or several female gametes in the Oomycetes.

**Oospore:** spore resulting from sexual reproduction between an antheridium and an oogone in the Oomycetes (*Pythium*, *Phytophthora*, and so on).

**Ostiole:** the circular orifice allowing pycnidia and perithecia to liberate their spores.

**Parasite:** living to the detriment of another living organism.

**Parthogenetic:** describes a type of reproduction which does not involve a sexual phenomenon.

**Pathotype:** strain of a micro-organism with one or several pathogenic genes.

**Peritheces:** globular or rather flattened receptacle containing asci, fairly frequently opened by an ostiole.

**Phialid:** terminal cell of a conidiophore or conidiophore with one or several terminal openings across which conidia are issued in a basipetal way.

**Phloem:** vascular tissue consisting of tubes full of holes, companion cells, and parenchyma, transporting and storing the elaborated sap.

**Physiological race:** see pathotype.

**Primary inoculum:** inoculum at the origin of an epidemic.

**Procaryote:** Unicellular organism without any nucleus.

**Propagule:** elementary structure of an organism capable of being disseminated and of reproducing a disease.

**Protoplast:** plant cell without any wall.

**Pycnidia:** fruiting body, often spherical, containing conidiophores and conidia, frequently opened by an ostiole.

**Recessive:** describes an allele which does not express itself phenotypically when it is accompanied by a second, dominant, form of the same gene.

**Rhizosphere:** micro-environment of roots in soil.

**Saprophyte:** living at the expense of organic matter, decomposed to varying degrees.

**Sclerotia:** compact mass of bunched mycelium, often brown to black in colour, adapted to survival under unfavourable conditions.

**Secondary inoculum:** inoculum guaranteeing the progress of the disease.

**Selection pressure:** constraint exercised by a plant population, a pesticide, etc, against a population of a micro-organism causing a change in the genetic composition of certain individuals from this population.

**Serum:** part of the blood containing antibodies.

**Sorus:** compact mass of spores located under the epidermis of leaves of plants affected by rust.

**Specialist form (special form, sp.):** the same species of an organism can have several forms which are differentiated by their host specificity; this is, for example, the case with species *Fusarium oxysporum* in which we are aware of numerous specialized forms: *lycopersici*, *melonis*, *lactucum* and so on.

**Sporangia:** 'fungal structure' producing asexual spores, often zoospores.

**Sporangiophore (sporocystospore):** the structure carrying the sporangia in 'fungi'.

**Spore:** unit of fungal reproduction consisting of one or several cells.

**Sporodochia:** fairly dense group of conidiophores sometimes mixed with an acervulus.

**Stomata:** opening of the epidermis surrounded by guard cells allowing diffusion of gases.

**Strain:** pure culture of a 'selected' micro-organism from an isolate, sometimes having one or several special biological characteristics.

**Stroma:** condensed mycelium in which various fungal fruiting bodies can form.

**Telomorph:** sexual form of a fungus, also known as perfect stage, resulting, after nuclear fusion, in the formation of ascospores, basidiospores, and so on.

**Thallus:** a unit of mycelium filaments in a fungal colony.

**Tolerance:** characterizes plants undergoing the attacks of a pathogenic agent without their yield potential being altered.

**Toxin:** toxic substance of biological origin.

**Uredospore:** spore produced by rust inside the sorus.

**Virion:** mature virus.

**Viroid:** the smallest known infectious agent, consisting of a single nucleic acid.

**Virulence:** qualitative component of pathogenic power.

**Viruliferous:** describes an insect or nematode carrying a virus and therefore capable of transmitting it.

**Xylem:** vascular tissue providing transport for raw sap.

**Zoospores:** fungus spore carrying one to two flagella and capable of moving in water.

# References


AGRIOS G.N. (1988) *Plant Pathology* 3rd Edition, Academic Press, p. 703.

ALFORD D. (1994) *Atlas en couleur. Ravageurs des végétaux d'ornement: arbres, arbustes, fleurs (Colour atlas. Pests attacking ornamental plants: trees, bushes, flowers)*. IRNA Editions, Versailles, p. 464.

BENNETT W.F. (1993) *Nutrient Deficiencies and Toxicities in crop plants*. APS Press, p. 203.

DAVIS R.M., SUBBARAO K., RAID R.N., KURTZ E.A. (1997) *Compendium of lettuce Diseases*. APS Press, p. 79.

EVANS K., TRUDGILL D.L., WEBSTER J.M. (1993) *Plant Parasitic Nematodes in Temperate Agriculture*. J. M. Webster CAB International, p. 656.

FARR D.F., BILLS G.F., CHAMURIS G.P., ROSSMAN A.Y. (1989) *Fungi on plants and plant products in the United States*, APS PRESS, The American Phytopathological Society, St. Paul, Minnesota USA, p. 1,251.

FLETCHER J.T. (1984) *Diseases of greenhouse plants*. Longman Inc, New York, p. 351.

HAWKSWORTH D.L., KIRK P.M., SUTTON B.C., PEGLER D.N. (1996) *Dictionary of the fungi*. CAB International, p. 616.

MESSIAEN C.M., BLANCARD D., ROXEL F., LAFON R. (1991) *Les maladies des plantes maraîchères (Diseases of market garden plants)*, 3rd Edition, INRA Editions, p. 564.

REGENMORTEL M.H.V. VAN, FAUQUET C.M., BISHOP D.H.L., CARSTENS E.B., ESTES M.K., LEMON S.M., MANILOFF J., MAYO M.A., MC GEOCH D.J., PRINGLE C.R., WICKNER R.B. (2000) *Virus taxonomy: classification and nomenclature of viruses*. Seventh report of the International Committee on Taxonomy of Viruses. Academic Press. San Diego, USA, p. 1,162.

RICHARD C., BOIVIN G. (1994) *Maladies et ravageurs des cultures légumières au Canada, Un traité pratique illustré (Diseases and pests affecting vegetable crops in Canada. An illustrated practical treatise)*. La Société Canadienne de Phytopathologie, Société entomologie du Canada, (*Canadian Phytopathology Society, Canadian entomological society*) p. 590.

ROORDA VAN EYSINGA J.P.N.L., SMILDE K. W. (1981) *Nutritional disorders in glasshouse tomatoes, cucumber and lettuce*. Centre for Agriculture Publishing and Documentation Wageningen, Netherlands, p. 130.

SEMAL J., *et al.* (1989) Traité de Pathologie végétale (*Treatise on plant pathology*). Les Presses Agronomiques de Gembloux, A.S.B.L., p. 621.

SHEPHERD J.A., BARKER K.R. (1990) *Plant parasite nematodes in subtropical and tropical agriculture*. CAB International, p. 648.

SHERF A.F., MAC NAB A.A. (1986) *Vegetable Diseases and their control*. John Wiley & Sons, New York, p. 728

THICOIPE J.P. (1997) *Les laitues (Lettuce)*. CTIFL/SERAIL, p. 281.

# Index

## Micro-organisms cited

## • Bacteria



## • Phytoplasma



## * Viruses

## Predators and parasitic plants

## Photos of damage caused by predators and parasitic plants

### • Nematodes



### • Insects



### • Animals



### • Parasitic plants



## Photos of symptoms caused by non-parasitic diseases